T0139918

The Rise of Artificial Intelligence and Big Data
in Pandemic Society

Kazuhiko Shibuya

The Rise of Artificial Intelligence and Big Data in Pandemic Society

Crises, Risk and Sacrifice in a New World Order

 Springer

Kazuhiko Shibuya
Tokyo Metropolitan University
Tokyo, Japan

ISBN 978-981-19-0952-8   ISBN 978-981-19-0950-4   (eBook)
https://doi.org/10.1007/978-981-19-0950-4

This Springer imprint is published by the registered company Springer Nature Singapore Pte Ltd.
The registered company address is: 152 Beach Road, #21-01/04 Gateway East, Singapore 189721,
Singapore

Preface

At the brink of the COVID-19 pandemic, this book was conceptualized by the author. At the time before the outbreak (February 20, 2020), a book on digital and computational social sciences, which was edited by the author, had just been published (Shibuya, 2020). In succession, at the time of confronting with such global pandemic (March 11, 2020), next, the author started writing a new book on social phenomena of such pandemic by standing on own philosophical and sociological foundations.

Namely the main goal of this book serves some of the fundamental questions that the author has been contemplating. *"Can the humanity coexist beyond the social evil inevitably engendered by the pursuit of the highest good (Summum bonum) among individuals?"*, *"how to cooperate with each nation beyond the decoupling of the world order?"*, and *"how to minimize entire sacrifices at confronting with any tragedy, if possible, how to avoid such situations in advance?"*.

In the past, Voltaire said "History is only the register of crimes and misfortunes". Human history could be certainly traced as somehow endlessly survival process from lethal crisis, and it means that human history had been inscribing into the record by survivors in each time. However, every time the humankind overcame each crisis such as infectious disease, disasters, and others, our ancestors had made a lot of the progress since the birth of our civilizations. For example, during the plague epidemic in Europe in the seventeenth century, Newton's discovery of universal gravitation and calculus had been established.

So, what lessons did humanity learn from the COVID-19 pandemic? However, humanity had not even been able to solve such situation. Here, the pandemic of COVID-19 was claimed as an incomparable emerging disease in recent decades. The most important was that the ordinary risk management and quarantine regime was unable to prevent the COVID-19 pandemic, and indeed it posed the challenge of whether it can adequately and effectively save people when another infectious disease (such as Ebola hemorrhagic fever) that is more lethal than COVID-19 spreads in the future. Further, aftermath of the COVID-19 revealed the inside of the humanity facing such extraordinary and emergency situations. We observed many issues such as blames against the others with infected symptoms, repeatable claims about epidemic policies among the citizens, mindless exclusiveness between multicultural hetero-geneity, and further global tensions among the nations after the pandemic due to triggering the balance of powers between China and the USA. Many of the citizens literally missed "*social distance*" to sustain the coexistence among the others.

If there was one fact that COVID-19 exposed, it might have been the reality of the revenge of the poor. Many of the poor and minorities became victims of COVID-19. Normally, it is the same as always, the nameless who are not cared about priceless life by the government that are impoverished, but when countless poor people suffer and die, it is clear that not only do we lose the labor force that supports the bottom of the social economy, but also the social economy stops functioning at its very core. The wealthy had no choice but to splurge in search of investment opportunities. Citizens, whose survival continued to be threatened, feared unemployment, sought stable employment opportunities, and intensified their criticism of the government. Such a situation led to growing unrest in the society as a whole, and mutual distrust led people to riot and engaged in prejudiced fake news. The hard work of medical doctors and specialists was more than gratuitous self-sacrifice. As domestic order and governance diverged from the hopes of citizens, the nation and the world as a whole stood on the brink of collapse. In other words, when the people who have always been responsible for sacrifices are no longer there, no longer able to do so, or cannot be replaced, the entire world turns into chaos. In fact, in the Corona disaster, what was seen in many parts of the world was the fact of who is responsible for the daily sacrifices and the absurdity of the reality by the causality.

Regarding such fundamental matters, this book especially devotes to conducting a study on the human nature confronting with the crisis. And it is mainly edited to synthesize and discuss rich idea from sociology, risk studies, and digital social sciences, because the author convinces that it requires to reconsider the common depth of both the COVID-19 pandemic and international order in the age of After Corona. China has been accused by global countries, where the COVID-19 pandemic originated. And international tension on economics and military aspects between China and the USA has already been invoking before the pandemic. It is historically considered as a derivation of "*The Thucydides Trap*" between them. And then, China and its historical culture seems to hold the keys for understanding how to coordinate with the Western sides. This book intended to explore the possibility of conflict or cooperation with the emerging Asian values (i.e., Eastness) in the current situation where the values of Western society (i.e., Westness) are said to be disappearing (i.e., Westnessless).

During the COVID-19 pandemic, the dichotomy of either health care or economic policy was also prevalent in the world. Whichever one was chosen, not only many of those who were infected but also those who were impoverished were killed, regardless of which one they prioritized. Why is it too difficult for the humanity to cooperate together to confront with the crisis? Can be a selfish act of escaping an infection at the sacrifices of the others justified? Why don't the great powers join hands for global cooperation? Moreover, despite the many sacrifices undertaken by countries and citizens in the wake of the COVID-19 pandemic crisis, which has become a new historical fact, the essence of the problem remains unresolved.

"*Quid pro Quo*". Indeed, the history of humanity was not just a sort of visible "glories", rather a more realistic history built on countless negative facets of the consequences such as "victims", "loss", "costs", "burden", "exploitation", "tolerance", "patience", "failures", and others. And then these concepts can be paraphrased as the "*sacrifices*".

In welfare economics, especially, many of such economic conditions can be stabilized as a *Pareto efficient* situation. However, principally there are often the extreme cases that cannot rely on either Pareto or fairness principle in the crises and risky situations such as pandemic, disaster, other natural hazards, and human-made ones (e.g., conflicts, warfare). Further, whether the government leaders accurately decide what to do (not to do) or not, sacrifices would be inevitably undertaken by someone.

The author calls it "*distribution of sacrifice*", which can be modeled to envision a part of social phenomena (see Chapters "A Worldview Seen from Sacrifices" and "Crises, Risks and Sacrifices"). Sacrifices make especially salient in each crisis, emergency situations, and uncountable events such as warfare, pandemics, environmental disasters, international conflicts, political revolutions, and other social events. Namely such sacrifices almost imply not only the total necessary cost to achieve and solve something but also each *life* of individual. In other words, the author frequently wonders how to equalize the total value of both sacrifices and final consequences whether these are achievements or not. Perhaps, both of them cannot be evaluated as equal values, because those are not always insured an equivalent exchange between total sacrifices and final results. If so, who and what such *difference* (absolute value: results minus sacrifices) would be undertaken? The author considers that such inevitable difference should be called as real sacrifices.

Hence, what did Japan and other countries sacrifice and what did they gain against the corona? It is difficult to compare the future that people were supposed to obtain in the face of a crisis with the future that was lost. But if we compare them with the actuality that remains in people's hands, it clearly sacrifices something valuable. As a result, the actuality we got is often not the future we desired. It would be hard to accept such consequences, nevertheless we have to face the actuality.

With this background in mind, first, the author intends to unveil facts of the COVID-19 pandemic in the context of own social scientific investigation (see Chapters "A Crisis of COVID-19 and Its Sacrifices" and "Formalizing Models on COVID-19 Pandemic"). Especially, the author serves own views, which stands on the "*distribution of sacrifice*". In politics, especially, "*distribution of sacrifice*" is always included to politician's job (rather duty), and "*making decisions that involve sacrifice*" is definitely a reason for a politician's existence. They are to make the necessary decisions in the event that they have to make a decision that involves the sacrifice at the crisis rather than "*making a decision that doesn't involve sacrifice*". However, politicians who often deceive the *crisis* do not tell the truth about latent *sacrifice* that necessarily implicates. When talking about the *sacrifice* is frequently inevitable, there is no acknowledgment of their responsibility. In the first place, they can be criticized for the imposition of sacrifice, but they will not be prosecuted for such reasons on sacrifices. And there is no such thing as adequate compensation for any sacrifice. They often said that "*it should be equally undertaken by everyone*" and further they claim against another *scapegoat* (sacrificial sheep) that can be the target of its attack. In addition, they exaggerate for the sake of the national security, national integration, and other agitations.

When the COVID-19 pandemic reaches the end, it is good to remember what responsibility politicians have taken and what compensation and sacrifices have been made. There was no one to take responsibility for it. Because their job is to "*make decisions that involve sacrifice*", they just fulfilled their responsibilities in the face of the COVID-19 pandemic regardless of their accurateness to achieve the goal of the policies. However, if the citizens examine it to include the unseen sacrifices, how many sacrifices have been made in the world as a whole? Even if the approval rating did not drop, does this mean that every citizen did not recognize the fact that so many people died as *sacrifice*, or they accepted the government's response?

Secondly, the author discusses the historical significance of the humanity due to differences in cultural values (see Chapters from "AI Driven Scoring System and 'Reward or Punish' Based on a Theory of Han Fei" to "A Living Way in the Digitized World"). For example, it seemed that the recent outbreak of the COVID-19 has highlighted not only the quality of domestic medical services and public health policies, but also the differences and adaptations between Western and Eastern cultural values in the age of artificial intelligence. As of March 2020, the WHO's global infection statistics exhibited that the EU countries, the UK and the USA, made up the top ten with the exception of China. It was severe fact that freedom of mobility, the emphasis on private rights, and the supremacy of personal values, which are the cornerstones of Western values directly related to democracy and individualism, were malfunctioned in a state of emergency. The collapse of medical institutions had led to economic recession, exhaustion, and confusion not only in one country but all over the world. On the other hand, countries that follow East Asian values were getting over the infection situation under control even in China which was the origin of the outbreak, and the situation had not become serious in Asian countries including Taiwan, South Korea, and Japan yet. It was possible that the value system that prioritizes relationship, social values, and cooperation throughout the society, which is called collectivism or totalitarianism, may be working effectively in the spread and recovery of infectious diseases.

And thirdly, the author examines the possibilities for conflict or cooperation in the international politics in the "After Corona" (see Chapters from "A New World in Motion" to "For Strategies in the Age of After Corona"). Once, at the time of the Cold War, an international political scholar Morgenthau articulated China as a country that would emerge in the future, regardless China was only a Communist bloc state. Now, as he expected, China has been further increasing its global presence with the advent of the AI and big data technologies. It has made no secret of its movement, which can be taken as gaining global hegemony, not only in the socioeconomic sphere but also in the exercise of cultural influence and sharp power over the world. There is a persistent view that the "*One Belt One Road (OBOR)*" initiative will also erode European traditional values. As for digital cryptocurrencies, there are other concerns that the Chinese government is leading the way and that the dollar-based international currency market will be losing meaning. The world now appears to be entering an era of qualitative phase transition from a clearly Western nations led regime. Actually, as of August 2020, Esper who is Secretary of Defense of the USA announced *"The Pentagon Is Prepared for China"*[1] and *"PLA (China's People's Liberation Army) modernization is a trend the world must study and prepare for—much like the U.S. and the West studied and addressed the Soviet armed forces in the 20th century"*. As Morgenthau predicted, the time has come for Western society to once again learn deeply about China, their values, cultural and historical background as well as advanced technological and military trends.

With all these points in mind, will the Western nations literally fade away as it is the status quo, and will Western values go on self-destructing? And, what kind of prominence is China about to achieve in the new hegemony? Further, what is the course that Japan should take in the face of such global trends?

Tokyo, Japan   Kazuhiko Shibuya
2022

---

[1] https://www.wsj.com/articles/the-pentagon-is-prepared-for-china-11598308940.

Acknowledgments

A part of book contents was expanded from an invited presentation of the author ("*A Revaluation for The Morgenthau's Prospects in the Digital Society*") at international workshop ("*Transdisciplinary Approaches to Good Governance*") held by the Center for Southeast Asian Studies (CSEAS), Kyoto University, Japan, on February 7, 2020. And, it also included own lecture notes on risk models, epidemics, disasters, political, and international issues of "*Risk Management*" in Tokyo Metropolitan University, Japan.

Additionally, this book included a part of the research results which were officially supported by JSPS KAKENHI Grant (Number 26590105: Grant-in-Aid for Challenging Exploratory Research: *An Exploring Study on Networked Market Disruption and Resilience*) and ISM Cooperative Researches (Number 19-0008: *A Study on the Research Evaluation of Science & Technology and the Rationality of Decision Making*, and Number 25-0014: *A Study on Social Representation and its Networking Dynamics*, and Number 28-0017: *A Study on Mathematics for International Cooperation*).

About This Book

This book broadly categorized the issues as follows, and it elaborates them in each chapter.

Introduction and Principles of Sacrifices

In the first part, with the Preface to this book in mind, the author defines risk, crisis, and sacrifice and outlines the methodologies pertaining to these matters. And the author's own perspective on the *"distribution of sacrifice"* principle articulates in line with the COVID-19 pandemic and "After Corona" era.

1. Introduction: Chapter "A Worldview Seen from Sacrifices"
2. A principle of "distribution of sacrifices" and relative methodologies: Chapter "Crises, Risks and Sacrifices"

COVID-19 Crisis and Its Sacrifices

The second part of the book scrutinizes the COVID-19 pandemic as a social phenomenon. In addition to examining statistical data and mathematical models, the nature of the crisis and sacrifice will be clarified from the perspective of "distribution of sacrifice" proposed in this book.

1. A comprehensive study on COVID-19 crisis and its sacrifices: Chapter "A Crisis of COVID-19 and Its Sacrifices"
2. Simulations of the spatial model on epidemic processes: Chapter "Formalizing Models on COVID-19 Pandemic"

An AI-Enhanced Society After the COVID-19 Crisis

In the third part, the author discusses the qualitative changes brought about by cutting-edge technologies, focusing on sociocultural and economic aspects, such as AI and big data technologies, which are thought to have been promoted by the Corona pandemic. And it brought to light issues such as people's behavioral principles, social morality, stagnant economic development, socioeconomic disparity, and human resource development. In particular, the author deepens own technological vision and consideration of what China's promotion of AI and big data will bring, contrasting it with Western culture.

1. Sociocultural aspects in COVID-19 pandemic in terms of the digital transformation: Chapters "AI Driven Scoring System and 'Reward or Punish' Based on a Theory of Han Fei"–"Synchronizing Everything to the Digitized World"
2. Economic aspects in COVID-19 pandemic in terms of digital transformation: Chapter "A Living Way in the Digitized World"

International Affairs Against Crisis of the After Corona

The author gave an invited talk on system dynamics and simulation studies on the global politics at Kyoto University on February 7, 2020. At that time, it had not become the COVID-19 pandemic yet. However, international tensions between China and the USA were already high. In light of such a global situation, the author discussed a potential for the worse of the relation between China and the USA, and cyber-warfare, and particularly lectured on the risks and crises of military conflict, balance of power, intervention in democratic politics, and manipulation of public opinion.

Here, this part expanded an above lecture and further explores the international issues of the After Corona world with an eye to the aspects of these crises. Accordingly, given that it views as an international issue, the basis of the decoupling world represented by the two great powers between China and the USA will engender the difference in preferences and values over the system of national governance. And it will fluctuate the structure of the balance of power, in which mutual strategies dictate future in the world of the After Corona.

1. A new world in motion caused by the aftermath of COVID-19: Chapter "A New World in Motion"
2. Political aspects and cyber-warfare: Chapter "Digitized Shifts of Regime and Hegemony"
3. Simulations on the balance of power among the stakeholders: Chapter "On Balance of Power"
4. Global trends and international cooperation in the After Corona: Chapter "For Strategies in the Age of After Corona"

Conclusion

This part summarizes all contents of this book.

1. Conclusion: Chapter "Conclusion"

Contents

AI and Our Society

Introduction

A Worldview Seen from Sacrifices

1 Posing the Questions

What we have to face is actually what we have to resolve now. An invisible problem means a serious problem for us. Many of the problems that have become visible suggests already too late. Philosophy and social scientific disciplines on risk and crisis shall explore the possibility of the survival of the humanity by anticipating such situations and revealing its true nature. Namely, it is for this reason that in this book, the author contemplates negative facets of the possibility and the actuality as well as the imaginary hopes. Then, there is a structural or changing crisis going on at every level. Recent COVID-19 pandemic and the international affairs demonstrated serious potentials to reach the point where humanity's future is at brink of existential survival.

The main goal of this book serves some of the serious questions that the author has been contemplating. Namely, central questions can be enumerated below.

- Can the humanity coexist beyond the "social evil" inevitably engendered by the pursuit of the highest good (*Summum bonum*) among individuals?
- How to minimize entire sacrifices at confronting with any crises and emergency situations, if possible, how to avoid such situations in advance?
- How to cooperate with each nation for pursing the common goal beyond the decoupling situation of the world order?

As one of the survivors of the COVID-19 pandemic, the author conceptualized and articulated such matters in particular. First, this epidemic tragedy revealed the inside of the humanity facing such extraordinary and emergency situations. We observed many issues such as blames for the others with infected symptoms, repeatable claims about physical distancing among the citizens, mindless exclusiveness between multicultural heterogeneity, and further global tensions among the nations after the pandemic due to triggering the balance of powers between China and the USA. Many of us literally missed "social distance" to sustain the coexistence among the others.

© The Author(s), under exclusive license to Springer Nature Singapore Pte Ltd. 2022
K. Shibuya, *The Rise of Artificial Intelligence and Big Data in Pandemic Society*,
https://doi.org/10.1007/978-981-19-0950-4_1

2   Unveiling the Reality by a Perspective of "Sacrifices"

2.1   Invisible But Important

Before starting own discussion on *sacrifice* corresponding with each *crisis*, let me give you a simple but serious example. According to the Bible, Noah *selected* or *prioritized* a coupling pair of each species before coming the big-flood. Reversely saying, it means that Noah totally *sacrificed* the residuals, that included those innocent animals and species as well as guilty peoples (and, at that time, it can deduce a fact that Noah's ancient ark perhaps brought not only ordinary plants and animals but also much *mutable virus of the diseases*).

In actual, politics frequently impose sacrifices on the individual without mercy in order to bless the public interest with mercy. At emergency, similarly, such situations like the pandemic of the COVID-19 enforces the humanity to prioritize those who select firstly. If you are either a medical doctor or government leader, how should clients be prioritized? Like Noah's ark, there are always the limited spaces and resources for the survivors. For example, medical doctors did not exist enough, and any vaccines and panacea have not been produced yet, and there were not sufficient beds and hospitals for innumerable patients. Thus, it will be inevitably required the value standard to include or exclude someone for pursing the survival of the humanity. When facing at such situation, it is necessity rather than guilty, because these situations usually can be regarded as a pattern of "*Plank of Carneades*".

Next, similarly the author poses some questions as bellow.

[*National supports for drug company and patients*].

As a thought experiment, given that a new drug cures more than 10 million patients suffering from emerging infections, however, the development of this drug is enormously too expensive than a drug company expected. Because the company has invested a lot of expenditure, they desire to sell it at a higher price. It costs more than $500 per patient. The government leader intends to make it publicly funded to prevent the spread of infectious diseases, but it negatively affects other health care budgets comes out. How is that policy fair and justifiable?

[*Public health and disaster*].

The WHO, UNDRR (United Nations Office for Disaster Risk Reduction), and other organizations have been working on the global response to infectious diseases, medical issues, and disasters. Is it a fair burden to allocate more money and experts than the risk of damages to each country?

For actual example, in early 2021, vaccination began in many countries. However, there was controversy in the medical field regarding the order of treatment, planning of vaccination, and distribution of vaccines between developed and developing countries. Not only do developed countries have the upper hand in buying up vaccines, but the EU has set export controls.[1] Such situation had been described as a *"catastrophic moral failure"*[2] (WHO Director-General, 2021). Four countries, the US, UK, Israel, and China, began vaccinating by January 2021. These four countries represented less than 6% of the world's population, yet the US, UK, and Israel administer more than 45% of the world's vaccine doses. The WHO's call for international cooperation had been ignored. These[3] exposed the fact that their democracy and fairness they advocate, though backed by their own capital and scientific and technological capabilities, was a self-justification to secure national interests based on their self-centeredness. How should we think about the problems underlying humanity in crises like this?

[*Business and Taxation*].

The IT companies known as GAFA continues to increase its revenue in the Corona pandemic. This is because AI and Big Data related services are doing well, coupled with the opportunity to make more progress in digital transformation. The problem here is the imbalance and inequity in the tax system. Should companies be given the right to decide which country to pay their taxes in? How should we think about proper, fair and equitable tax burden and collection for global corporations? And to what extent should corporations and wealthy individuals like GAFA, which is an oligopoly of the world's wealth, cooperate relieving the burden of nations and citizens economically impoverished by the Corona pandemic?

[*Budget for Peacekeeping*].

The cost of establishing, maintaining, and managing international cooperative relationships, as promoted by the United Nations and other organizations, is often questioned in terms of the proportion and amount of costs to be shared. The reason for this is that the cost effectiveness of each country's investment is not necessarily proportional to the amount of investment. Moreover, it is common sense that it is inherently forbidden to have influence over the United Nations agencies such as the WHO according to the amount of investment.

Someone is responsible for a lot of costs and sacrifices every day for peacekeeping forces and other measures against conflicts, terrorism, and disasters that occur around the world. So how should we think about the fair burden of deploying US troops to NATO? How much of the sacrifice is fair to bear? What kind of military balance should be struck in the world as a whole? How should we consider the development of military equipment (AI-driven robots, drones, missiles, etc.) among allies?

Of course, it's not limited to above examples. The same is true of the COVID-19 pandemic. For other examples, there were a number of medical workers who

---

[1] https://ec.europa.eu/commission/presscorner/detail/en/ip_21_307.

[2] https://www.forbes.com/sites/tommybeer/2021/01/29/who-vaccine-hoarding-would-be-a-catastrophic-moral-failure-that-keeps-pandemic-burning/?sh=3669df3e15ac.

[3] What it will take to vaccinate the world against COVID-19, https://www.nature.com/articles/d41586-021-00727-3.

died on duty.[4] Even if their sacrifices saved enormous citizens and the country, it is impossible to reward their sacrifices. If they sabotaged or went on strike, even more citizens would have been alternatively died. The consideration for those sacrifices is essentially never going to be balanced no matter how much you offer them. Even if you assume a lifetime income of each medical doctor and pay the difference, medical doctor can not receive it when he dies. Is it worth risking and sacrificing own lives to save them? In any case, the country has also been able to revive its economy at the sacrifices of its medical doctors and victims.

The most important is that the current risk management and quarantine regime is unable to prevent the COVID-19 pandemic, and indeed it poses the challenge of whether it can adequately and effectively save people when another infectious disease (such as Ebola hemorrhagic fever) that is more lethal than COVID-19 spreads in global.

2.2   The Distribution of Sacrifices

"*Quid pro Quo*". This phrase directly expresses both the weight of the responsibilities and the seriousness of the nature that nothing can be decided without any sacrifice. In welfare economics, especially, many of such economic conditions can be stabilized as a *Pareto efficient* or optimum situation. Since economics can basically handle with only an economic value eliminated from the humanity and social objects, and it usually does not to consider either a life or the dignity of each individual even if welfare economics.

However, the author intensively articulates such matters in detail through this book, and its contemplation based on philosophy, ethics, sociology, and policy studies must include both viewpoints of the humanity and quantification in the crisis and unavoidable conditions. Because, principally there are often the extreme cases that cannot rely on either *Pareto* or *fairness principle* (Rawls, 1971, 2001) in the crises and risky situations such as pandemic, disaster, other natural hazards, and human-made ones (e.g., conflicts, warfare). Further, whether the government leaders accurately decide what to do or not, sacrifices would be inevitably undertaken by someone. Above cases such as Noah's ark and the COVID-19 pandemic are good examples for the author's intention. And other cases also indicates the nature of sacrifices, and then those events must be delved into more depth.

The author calls it "*distribution of sacrifices*", which can be modeled to envision a part of social phenomena. The "*sacrifice*" almost means to be observed as any negative synonyms such as "burden","exploitation","duty", "responsibility", "obligation", "loss", "costs", "efforts", "budget", "time", "victims" and other negatives in various contexts. Namely, a "*sacrifice*" can be found anywhere and anytime, especially, when anyone confronts with somewhat "crisis" and "risk". Frequently missing

---

[4] https://www.sciencemag.org/news/2021/02/unprotected-african-health-workers-die-rich-countries-buy-covid-19-vaccines.

such important facts, each *crisis* must be recognized corresponding with *sacrifices*. And it can be arranged with the other problematic concerns such as *decision making, prioritization, value, preferences* and somewhat *social standard* whether it handles in each individual level or collective group level. Some points to fairness and procedural justice (Rawls, 1971, 2001) are quite needed. But no matter what decisions are made, there will always be sacrifices.

In sociological studies, as mention later chapter, even though technical terms such as "crisis" and "risk" are repeatedly picked up (Habermas, 1976; Luhman, 1984, 2002), they have not been crucially investigated and theorized "sacrifice" yet. On the contrary, but, economics, international studies as well as strategic studies of military based on game theory have frequently observed such negative meaning of terms (e.g., cost, loss, demerit, victims, etc.) by comparing with positive terms (e.g., merit, profit). In addition, medical sciences and operations researches also face the inevitable situations to choose and prioritize between several alternatives (i.e., triage in the emergency), and it means to abort one of them at least. Namely, as ethical studies noted, what *"trolley problem"* (ethical confrontation between two choices in one of the thought experiments) and other issues has been discussed may be nearly similar to the author's considerations (Ackeren & Archer, 2020; Nozick, 1974). In computer science and AI studies, it may sometime manage such problems as "resource allocation" and "optimizing resources".

In such serious situations, it is often ignored whether or not the consideration is properly done, but the *"distribution of sacrifices"* becomes particularly salient when *"both individuals and the society face a crisis of the existence if it is not properly and promptly executed"*. Of course, not only in the emergencies and crises, but also it is naturally present in everyday life (Habermas, 1976; Roeser et al., 2012). In times of emergency, the significance of the decision based on distribution of sacrifices will be keenly recognized. However, as it is empirically supported by psychology that humans are vulnerable to such risky decisions, either individual or collective decision making is always a challenging matter (Kahneman & Tversky, 2000; Vose, 2008).

As also discussed in detail at later chapters, the COVID-19 pandemic brought to light the nature of human existence (Shah et al., 2020), and then the author contemplated own perspective on such *"distribution of sacrifice"*. It indeed weights on negative facets rather than *"distribution of capital"* in economics and *"distribution of fairness and justice"* in law (Rawls, 1971, 2001). And further, the author discusses such matters interlinking with the cultural differences between Eastness (i.e., countries based in Asian culture) and Westness (i.e., western cultural countries such as the USA and Europe) (Root, 2020). In the age of After Corona, because it will be a decoupling situation between Eastness and Westness in the world, and the clash between the two great powers such as China and the USA will divide the world in the new international politics. Indeed, this is also nothing less than a configuration in which the two countries and their alliances impose *sacrifices* of various things on each other. Without investigating such facts, no one can save the economic damage, reparations and human suffering caused by the pandemic and the various forms of endurance and impoverishment of daily life.

Thus, the universal problem of the humanity can be indeed summarized this issue of "*distribution of sacrifices*". The perspective is effectively true of environmental issues, food shortages, financial risks, and the current issues of infectious disease control and international politics in global. This is a problematic structure that we face regardless of whether we are in a democracy or a tyranny. In brief, it implies a structure of the power based on game theory that imposes certain *sacrifices* on each other. Politics through democracy, in particular, tends to be determined to impose *sacrifices* on those who are absent. If such alternatives are impossible, it would inevitably choose a situation "*everyone must undertake the burden equally*", but that would only lead to increased grumbling among the citizens. International diplomacy is similarly about imposing *sacrifices* on each other so that participation in international conferences is quite important and not only our own country is not bearing unnecessary *sacrifices*. The most terrible is that "*those who are absent at the place for decision making*", and they are often relatively poor people in developing countries or people of the future generation who have not been born yet. Conversely, we also have been forced to make such sacrifices by previous generations (e.g., post-war reconstruction, compensation, pension obligations for them). Namely, "profit sharing" and "equalization of burden" are prioritized and minimized only for the convenience of the people who are currently living, and the sacrifice will be inevitably imposed on those who are invisible and absent here. Furthermore, active generations often forget that globalism can exist today at the sacrifices of the earth's environment and other living existences. Such situation suggests why unsolved global issues are a far-reaching cause to chain further negative consequences.

3 Life and Money

3.1 Equal and Indiscriminate Measurement

There is a discourse that "*life*" and "*money*" are commonly one of the most equal and indiscriminate measurement in the world. The former can be said to be equal among all individuals situated by the fact that "embodied physical existence" is constrained in the "space" and "time" (Heidegger, 2008). On the contrary, the latter is because there is no other factor separating them than the fact that anyone who owns "money" (which recently includes cryptocurrency, electronic currency and points) has the necessary requirements for the establishment of a sale or purchase. In short, it is said to represent a state of freedom from all forms of "*stereotypes*", "*prejudice*" and "*discrimination*", and then such status can be recognized as the non-discrimination.

But here, is it discrimination or prejudice to impose further "*constraints*" on such primordial beings, or to select them based on "*someone's preferred values*"? Those constraints are based on the perceptions, namely it is called as "*preferences*" by which each value for particular objects. It prioritizes "*specific value order*" based on their attributes and various types of information based on specific criteria. It has also different meanings depending on whether the object is *"each individual"* or the *"entire society"*.

Each society relies on the specific cultural background that has different value standard. Western societies (i.e., Westness), which represented by individualism and capitalism, individual rights to life and money (i.e., private rights) are given the highest priority. In other words, everyone who living in the society principally "prefers" such value. And then, it is a culture that allows individuals to be as "freedom" as possible in how they live and what they spend their money on (further, how they earn and invest their money at the same time).

On the other hand, East Asian societies (i.e., Eastness) usually characterized by collectivism and cooperative values, and the rights of each individual to handle "life" and "money" (i.e., private rights) are basically approval by the government, but they are often subject to certain social restrictions. These can be "social systems" or "laws", or it defines as a kind of "value system" (whether explicit or latent) as some kind of "common sense". It can be said that this is a culture that dislikes the fact that when individual private rights are allowed without limit, the "freedom" can become a "state of irresponsibility", and consequently it becomes difficult to take control of the society as a whole (e.g., *anomie*).

As noted earlier, the author said that "life" and "money" are commonly one of the indiscriminate measurements. However, there are many situations in the medical field where patients must be prioritized for treatment according to the severity by medical *triage* (prioritizing patients to be saved and deciding who to sacrifice). Similarly, even though the "possession" and "use" of money itself is "nondiscriminatory", but "amount of possession", "means of earning" and its underlying attributes of educational background and origin of each individual are certainly constrained through the social living and culture, and the rights and obligations of each individual has been regulated with them.

3.2 Same Democracy But Different Meaning

Democracy in the digitized age must be reexamined again in various context (Hyde, 2020; Weaver & Prowse, 2020). Prehistorically, in western cultural background, the ancient Greece philosopher Aristotle's book *Politics* (Πολιτικά) seemed to have envisioned a democracy and a state system based on ancient *polis (πόλις)*. Additionally, because the ordinary citizens took for granted that slavery and own private property rights were based on communal living, they probably considered the rights, duties and status of ordinary men (it was not all people) to participate in the political process as necessary attributes of living as the human beings. In other words, they handled both their common goods as well as slaves to be own disposable property. Democracy in ancient Greece was clearly based on such common sense.

In contrast, for example, modern American democracy is quite different, especially, after Lincoln's Emancipation Proclamation their slave-based social structure ceased. However, the broad division of the workers can be continually observed in this era and in every country. In other words, those with a university degree or higher who can engage in intellectual property and scientific research, or politicians and business administrators who participate in managements are in a different position from those who work including immigrants and low-wage workers who employed as a simple labor. The place occupied by the slaves was only replaced by the latter's workers. The similar can be said in many of the developed countries (e.g., Japan, the EU). Indeed, to the extent that both are granted basic human rights and suffrage, the democracy of the modern society has progressed from the democracy of ancient Greece. And current capitalist economy might realize the ideals of private property and statehood as an extension of that envisaged since the days of ancient Greece. As economic development through globalism progressed, the international framework developed in line with such traditional frames.

Indeed, the COVID-19 pandemic seemed to expose its true nature. Namely, those who are belonged in establishment position of the higher status has benefited the most from globalism. On the contrary, the COVID-19 pandemic spread along the economic network of globalism, it was the poor working peoples that suffered the most. With that precondition, the workers of the Rust Belt were replaced by the lower wage labor force in China. This is because even individual dignity has been also replaced by the numerical measurement as the cost. Was it truly unfair (Rawls, 1971)? Rather, was it just the necessary sacrifice?

3.3   Inequality as a Result? Or An Institutional Flaw?

The sacrifice is apt to be concentrated on relatively weaker side. Weaker countries, people of low social status, people without the right to vote, and neglected beings. Of course, the side in the stronger position often undertakes a reasonable burden or fulfills their obligations, but even then, it's not enough or not functioning well enough as a system. And consequently, all those sacrifices may be for naught.

Then, there is also a need to reconsider inequality and institutional fairness. Let's recall here the COVID-19 pandemic. What separated those who were able to receive treatment in NY State from those who were not (Gonzalez-Reiche et al.,2020)? Those with enough money were able to escape from New York State to save their own lives. People with no money were forced to stay in New York State, and without medical insurance, they had to work outside, risking their lives, in order to make ends meet. On the other hands, there have been other cases in which medical doctors and medical professionals with much money and prestigious status have taken the initiative in dealing with large number of the patients, risking their own lives, and as a consequence, some of them have been infected or died.

Could these tragedy events have been directly infected by the factors of "discrimination" or "prejudice"? Certainly, there may have been some possibility in the process of getting infected. What is important here, however, is that in extreme conditions such as the COVID-19 pandemic, the factors that should be the most distant from discrimination, such as "life" and "money" are also given some sort of "priority"

(Barrett et al., 2021), namely it is the crucial factors pertaining to the "value of living existence" of each individual with own dignity (Shibuya, 2022). Shouldn't those who are placed at the bottom of the priority list, those who, by some measure, leak out of medical care and public support, be called "*sacrifices*" in such crisis obviously? Rather, they were left unattended before the pandemic, and such problem did not be resolved as a latent crisis.

On the contrary, the infection and mortality rates of the COVID-19 were significantly lower in Japan and many other Asian countries. It assumes that this is due to some critical factors directly related to "life" and "money" such as the presence and number of poor people, immigrants and illegal workers, the generous medical and long-term care insurance coverage for them, and the poor housing and working conditions. Many of the Asian countries such as Japan and other Asian countries relatively still have lower cases. And some of them are close to a single ethnicity, and then they are less likely to be linked to persistent racism and labor discrimination than others. Further, some of the countries have strong medical and scientific skills, and then it can serve more responsive to public health and initial treatment. In the USA and Europe, in contrast, there seemed to be relatively many poor people, immigrants, refugees, and illegal workers, and discrimination against them was still evident before the pandemic, which caused more serious problems in these situations. In this concern, there was also a discourse that China's political system of governance was superior to Western democracies. It is probably, as noted above, that differences in the nature of what is prioritized as the most valuable in the society rooted in historical tradition and culture may have contributed to this pandemic.

Namely, a simple factor "the amount of capital" exclusively determined who sacrificed, and such society majorly occupied by those who perceived it as "fairness" and "justice". Regarding those views, the goal of "*distribution of sacrifices*" is firstly to offer another recognition for such situations, and explore the opportunity to minimize and optimize such situations in advance, and the extinction of such situations if possible. But such solutions would be hardly achieved by the humanity.

4 For Living with the Others in the World

4.1 Recognizing Each Dignity

Generally, the problem on the human existence firstly begins with a reawakening of the differences between self and the others, and both of them must coexist with their mutual corporeality as physical existence in the same environment. Social distancing strategy to prevent infection of the COVID-19 would recognize the citizens such simple and important facts.

But, every infection always comes via "the others". As Spinoza (2008) said, "*Homines ex natura hostes sunt*". The others are clearly no other than an "*enemy*" who intends to invade the self, who will become also someone's enemy. Such anxieties and conflicts among them can shape various forms of discrimination, prejudice, and emotional expression in the city as well. And *"Leviathan"* described by Hobbes (2009), it said *"Bellum omnium contra omnes"*. This phrases crucially indicate that the malicious conditions can be always engendered by interacting with the others.

4.2 Emergent Evils Derived from the Highest Good

In the world, in the studies of philosophy and ethics, the divine is the being who embodies the highest good (*Summum Bonum*), and that the goodness is "All living existence desires to be ongoing alive". All values are hierarchically and descending ordered from the highest good. Therefore, "suicide" and "killing others" are not present in this value order. They are inherently contradictory and self-contradictory and are not acceptable behavior (Heidegger, 2008; Shibuya, 2020). This daily world that anyone lives in is made up of only those who "continue to be alive". For if "those who desire to die and commit suicide", they will be absent from our everyday world. When people make the will to "live" and actually make that choice on a daily basis, they have no choice but to realize that they are with others and face all the problems that arise in such relationship. The question is whether the various problems emerges when the self and the others encounter can be avoided by coordinating and adjusting each other's goodness at the interface between the self and the others. This is a problem that arises as a result of an unspecified number of human beings constituting a society. In the daily life when we are constantly confronted with such situation, an event called "evil" emerges.

In other words, all problems as *social evils* arise as one of the *emergent properties*. All actions performed by each individual are essentially own "goodness" if they lead to the highest good (living as long as they are performed by each individual alone). However, these issues arise only when human beings form a society and an unspecified number of peoples coexists.

For example, individual A's damming of the river upstream is essential to his survival, but for individual B, who lives downstream, the act of damming itself becomes a major threat to B's living. In other words, in society, "evil" is something that is born in the course of living a social life, and is often referred to as an *"emergent property"*. Therefore, in the eyes of an individual, all "evil" always comes via the others. Because of the existence of the others, things and actions that were nothing on their own can become "evil". Hence, it cannot be fully predicted in advance. This means the world handled by complexity science, which cannot be perfectly controlled. But it can explain the formation of a certain structural order that is the essence and nature of the world through the repetition of countless interactions between individual elements as control parameters and factors.

Namely, the essence of the humanity is definitely rooted in evidence that socially creates "evils" in the surrounding environment and society through its own actions, and they create numerous *problems* with the others in order to do its own goodness (i.e., to be alive). And further they had historically repeated actions to solve or improve these problems. And it will continue to do in what could be felt an eternity.

Therefore, politics, which defines as a stage for solving these problems together, serves always a place to be serious conflict of interests among the stakeholders, and it tends to be talked about only gains and profits. However, it tramples on the others who have fallen short by various acts that seek to maximize their own viability through the "*distribution of sacrifice*", that is, the imposition of sacrifice. Such decision making in politics is a kind of the necessary evils in the world. Even if decision making conducted by majority system in democracy is applied, minority and silent citizens will be often ignored (Mill, 1989). When examined from the perspective of the "distribution of sacrifice", the structure of the problem becomes more real, because in the end, someone inside or outside of the country has to make a sacrifice, and the natural environment and future generations are not often in the scope of consideration. Even if a vaccine or a panacea will be produced, we are faced with a situation in which the human existence itself is evaluated and prioritized over its priority.[5] In addition, many citizens are oppressed by corporations and the national government because of their economic interests and political intentions. Rather than "distribution of capital" or "distribution of justice", all of the world's emerging problems should be firstly scrutinized from the perspective of the "distribution of sacrifice", and then it should be scientifically examined to unveil whether it is possible to minimize or optimize sacrifice and whether it is possible to solve it in the first place. If it is possible to do so, the decisions must be made quickly and effective actions must be taken. Otherwise, if it is certain that the problem cannot be solved, the humanity will be required to be inevitably faced such tough conditions.

In short, politics always stays one of the *necessary evils*. The varieties and methods of solving problems and decision-making in the political context are not the essence of the problem, but the fact that the others simultaneously coexist becomes one of the social evils and then it indeed means an essence of all of the problems. If and only if a society where only the self exists can be possible, such problems never exist. As long as the self and the others simultaneously coexist in the same environment, politics is inherently incapable of solving such emergent problem. One's own "freedom" to coexist in the world determined by the others is just trivial, regardless of whether it is democratic or tyranny. Even a peaceful daily life of each individual can be easily threatened by the invisible others and collapses at few moments. We were inevitably reminded by the COVID-19 pandemic that it was the "actuality" that can end at any time, even if the "world" of yesterday had continued until a few hours ago.

---

[5] The unequal scramble for coronavirus vaccines ? by the numbers: https://www.nature.com/art icles/d41586-020-02450-x.

5   Condemnation and Atonement

By independent research think-tank [e.g., IPPR[6] (The Institute for Public Policy Research)], some of the reports accused that the negligence and complacency of China and the WHO[7] had caused people around the world to be infected and killed or injured by COVID-19 virus, which has led to the collapse of the global economy and the spread of *suspicion* about international cooperation. Hence, the pandemic caused by their initial policy entailed proving that they were guilty of a penal offense. As consequentially, from individual relationships to all over the international society, the invisible fear of the others suffered many of them.

At the time of the COVID-19 pandemic, many of the global citizens might feel that China government leaders and WHO leaders commonly prioritized only their own survival, and as a result, they entirely sacrificed both the world and its global citizens. At least, it doesn't deny a fact that they prioritized and over-weighted the temporary economic loss and faces of the leaders rather than priceless lives of the citizens around the world. The irreversible result of saving faces of such few leaders, not utilitarianism and not just nationalism, was that the negative aspects of both the individual dignity of Westness (i.e., saving only top leaders) and despotism of Eastness and collectivism (i.e., political ways in China government along with WHO) emerged simultaneously. In other words, the superiority of either has no meaning in a real crisis, only human nature lastly remains to be visible.

Unfortunately, many of those sacrifices will not be compensated at all. To be sure, it was undeniable that the bilateral serious tension between the USA and China before the pandemic was the primary cause. This pandemic, however, will lead to a situation that exacerbates such an aggravated situation and place an inevitable new burden on future generations.

6   Triggering Global Tensions to Rise Conflicts

If the great conflict exemplifies warfare, and the dead-end of "no-solution is a solution" in the international politics may also induces warfare. Then it would be hypocritical to speak of peace alone, since all problems and social evils are possibly to arise through daily living of the humankind. The reason why peace is so difficult to establish and keep is that people do not seek the idea of peace, but rather the minimization of sacrifice as a practice, that is, a solid economic foundation and a monotonous, repetitive life based on it, and these are not fulfilled yet. As long as this is not fulfilled, living as the highest good cannot be established, and eventually it will further develop into a conflict among oneself, the others, many groups, or nations. There are various expressions such as competition and self-responsibility,

---

[6] https://www.ippr.org/coronavirus-response.

[7] Why did the world's pandemic warning system fail when COVID hit?, https://www.nature.com/articles/d41586-021-00162-4.

but the root of the all matters is based on the political and economic breakdown over profits and sacrifices. Self-responsibility can be seen as having already abandoned the political solution. Because it is tantamount to saying "*the winner is the truth*", and it will abandon coordination and mediation, which are supposed to be political functions. If it gains an insight into the problems and conflicts that arise in every human being, the root of such matters is the conflict for the realization of the highest good of individuals.

Despite the fact that the realization of the goodness that each individual seeks for own survival itself is always truth and goodness, the more all of the citizens organize the entire society, the more they engender social evils, and the more they have these factors of conflict. And such conflict always inherent in the goodness based on their living. Given that imposing sacrifices with weapons is evil, and imposing sacrifices through discussion and majority voting becomes also inherently evil. Since the goodness that means each individual stays to be alive, brings the fact that the root of the all problems is unavoidable, and it is impossible to completely eliminate smaller problems or even more major wars. Trying to establish a law and come up with a minimum standard or a solution will not lead to a fundamental solution, since we have to choose to live in the society. Moreover, if politics loses or abandons its essential functions, war (or a state of war called competition) may be inevitable at the worst cases (known as the failure of politics).

7 Repeating Historical Events as Déjà Vu

To be sure, it should clearly reconsider the latent depth of both the COVID-19 pandemic and international order in the age of the After Corona. Because China and its historical culture seems to hold the keys for understanding how to coordinate with the Western sides. These backgrounds are investigated to explore the possibility of conflict or cooperation with the emerging Asian values (Eastness) in the current situation where the values of Western society (Westness) are said to be disappearing (Westnessless).

Some previously argued that the cultural differences between the Westness and the Eastness, and the conflicts between them made the situation difficult when we viewed as international issues. The overtly confrontational structure between the USA and China has been worsening not only in politics and economics, but also in science and technology and military affairs. While signs of the EU's internal decline are also visible, we are also beginning to see a power shift in the balance of power due to the rise of emerging economies. The crises over the world's political and economic system will inevitably bring about innumerable sacrifices through the process of transition.

On the other hand, the *crises* over the global environment must be considered in the same way as the major factors such as infectious diseases, greenhouse gas emissions, and countless other factors (Inoki, 2014). We need to know exactly how this affects the environment and what the linkages coexist. The COVID-19 pandemic also reminded us that the globalism in network of economic transactions cannot be independent of the environment and biosphere on the earth. Economic networking for transactions around the world, digitized interconnecting networks of information, linkages to generate and maintain energy, and linkages in the material cycle are all part of a system that needs to reveal both each risk as well as the overall crisis. And it must improve understanding of those interactions. A global environmental management system based on the concept of sustainable development has evolved gradually in recent decades. There are several levels of globalization of these crises such as governmental, private, non-governmental, and multinational. To be sure, the COVID-19 pandemic also made it clear that more efforts are needed to ensure that there is consensus on the severity and importance of each crisis.

For pursing those purposes, it must recall a fact that there are both positive and negative aspects of each probability. And in essence, risk management is about taking the most scrutiny of the negative aspects rather than the positive aspects of profit, fairness and justice, and taking measures to deal with future problems and such issues arise. In fact, both economics and politics should always focus on the future rather than the status quo, but the humanity is too irrational to solve any problems separating with the past and the status quo [i.e., *bounded rationality* (Gintis, 2014)], and they often tend to fall into the risk aversion of avoiding the actuality (Kahneman & Tversky, 2000; Shibuya, 2006). The leadership's risk aversion can be observed anytime, but they must be required to be a quick decision that correctly recognizes what the true cost is. However, the tendency to get stuck in a mental condition as *group think* (Herek et al., 1987; Janis, 1982) and various psychological biases has been frequently identified as a problem (Gibson, 2012a, 2012b). Whether it is an economic emphasis, a health care policy emphasis, or a major decision involving military conflicts, the essence of the decision is that it requires a higher-order judgment of what to sacrifice and who to sacrifice. How many politicians are capable of being prepared and aware of such a fact? Despite the fact that both the COVID-19 pandemic and the rise of China had been predicted and warned about by the scholars' advice, many of the politicians are reluctant to make a move unless the problem is confronted and the damage is actually done. It's very similar to waking up from a dream and knowing the reality. Such result is indeed too late.

When we look over the history of the humanity, these actions and inevitable causes and effects, similar to the many fractals repeated by many people, are nothing more than déjà vu. In other words, it can express "*solving the present problems by the seeds that becomes the future problems*", namely it means that nobody clearly settles problems yet. And such situation looks the Möbius Circle as a history of the humanity is continuously repeated. What lies beyond our gaze, is there a way out of the recurring causes and effects of history? The clue to solve is not yet in the hands of the humanity.

The world would have been decent a long time ago if someone could believe in the God or religion and be saved. The reason why the world has not become a decent one is simply because the humankind has not been able to solve the fundamental of the problems. Therefore, the God is not inherently necessary. As philosopher Nietzsche said, it is not that "*the God is dead*". The only thing that is needed to solve the problem is the practical action by the humanity. The virus of the COVID-19 has not been dwindling away to nothing on the earth yet, and then the humankind shall be prepared to conquer future attacks by the emerging risks.

8 As a Contemporary Review of the Social Theories

A sociologist Habermas belongs to the Frankfurt School, who has been advocating for the public sphere and the crisis. However, despite his experience with German post-war reparations and reconstruction, he merely analyzed the various sacrifices that respond to the crisis and its nature, the quantification of uncertainty around these issues, and the pathways in the social system including the difference of interests. As a result, it was not clear who was ultimately bearing the sacrifices. On the other hand, "risk" was also explored by sociologist Luhman (2002) from the perspective of social systems theory, but again, no specific analysis and solutions were provided.

But, as a sociology in general, an argument in terms of "sacrifice" in each crisis has not been clearly examined yet. Criticism is necessary to capture consequences as sacrifice and relate various factors to each crisis in those dynamics. A historical event means, in other words, a one-time event in the human history. But, even though these are the same crisis, these are clearly not the same sacrifices what details. Namely, the scrutiny of the actual nature of the sacrifice, conversely, critically highlights the inherent identity and historical uniqueness of each crisis.

In addition, Rawls (1971, 2001) had deliberated own theory of justice in political philosophy, and his renown concept "distribution of fairness" has a significant relation to a concept of the sacrifice. But he also had not specialized his thought on sacrifice nevertheless he weighed on consideration on both justice and fairness. Certainly, he could notice negative facets on political process and evaluating its fairness in any levels, and his books may be read that he laid weight on rather concepts such as equality, justice and fairness than as sacrifices. Probably, his deliberations on justice and fairness were based by examining negative sacrifices, because the actual cases what he discussed can be understood as a pattern of the sacrifices. Particularly, the arguments on child bearing by woman and social welfare policy could be dealt with rather distribution of sacrifice than distribution of fairness.

With these points in mind, some of the issues that the author has been conducted were related to crises and sacrifices such as the Fukushima nuclear accidents (Shibuya, 2017) and the issue of the identity of the human in the AI society (Shibuya, 2020). The AI society and the After Corona have rapidly changed our living world. Confronting the consequences of the myriad crises that have threatened the human existence in the form of sacrifice, it realizes that it is not only a matter of the present generation, but also the need to think about future generations in the same way. In the solving process sacrifice in each crisis, what problems should we undertake and what should we leave behind for the future (IPBES, 2020)?

Hence, the author indeed convinced that, in relation to recognizing both *crisis* and *sacrifice*, an understanding based on the nature of the humanity is now quite required.

In this book, with this framework in mind, the author would like to detail *sacrifices* in the *crises* in political, economic, socio-cultural, and international affairs from own perspective pertaining to the pandemic crisis of COVID-19. And the author further contemplates both crises and sacrifices that integrates as a meta-frame in an auxiliary line of Rawls's considerations.

References

Ackeren, M., & Archer, A. (2020). *Sacrifice and Moral Philosophy*. Routledge.

Barrett, A. E., Michael, C., & Padavic, I. (2021). Calculated ageism: Generational sacrifice as a response to the COVID-19 pandemic. *The Journals of Gerontology: Series B, 76*(4), e201–e205.

Gibson, D. R. (2012a). Decisions at the brink. *Nature, 487*, 27–29.

Gibson, D. R. (2012b). *Talk at the brink: Deliberation and decision during the cuban missile crisis.* Princeton University Press.

Gintis, H. (2014). *The bounds of reason: Game theory and the unification of the behavioral sciences (Revised Ed.).* Princeton University Press.

Gonzalez-Reiche, A. S., et al. (2020). *Introductions and early spread of SARS-CoV-2 in the New York City area*, Science, eabc 1917. https://doi.org/10.1126/science.abc1917

Habermas, J. (1976). *Legitimation crisis*. Heinemann.

Heidegger, M. (2008). *Sein und Zeit (English translated edition)*. Harper Perennial Modern Classics.

Herek, G. M., Janis, I., & Ruth, P. (1987). Decision making during international crises. *Journal of Conflict Resolution, 31*(2), 203–226.

Hobbes, T. (2009). *Leviathan*. Oxford University Press.

Hyde, S. D. (2020). Democracy's backsliding in the international environment. *Science, 369*(6508), 1192–1196.

Inoki, T. (2014). *A history of economics,*. Chuokhoron-shinsha (in Japanese).

IPBES. (2020). *IPBES Workshop on Biodiversity and Pandemics*. https://ipbes.net/sites/def ault/files/2020-12/IPBES%20Workshop%20on%20Biodiversity%20and%20Pandemics%20R eport_0.pdf

Janis, I. (1982). *Groupthink: Psychological studies of policy decisions and fiascoes* (2nd ed.). Houghton Mifflin Company.

Kahneman, D., & Tversky, A. (2000). *Choices, values, and frame*. Cambridge University Press.

Luhman, N. (2002). *Risk: A sociological theory*. Aldine Transaction.

Luhman, N. (1984). *Social systems*. Stanford University Press.

Mill, J. S. (1989). *On liberty*. Cambridge University Press.

Nozick, R. (1974). *Anarchy state and Utopia*. Basic Books.

Rawls, J. (1971). *A theory of justice*. Harvard University Press.

Rawls, J. (2001). *Justice as fairness a restatement*. Harvard University Press.

Roeser, S., Hillerbrand, R., Sandin, P., & Peterson, M. (Eds.) (2012). *Handbook of risk theory: Epistemology, decision theory, ethics, and social implications of risk*. Springer

Root, H. L. (2020). *Network origins of the global economy. East Vs. West in a complex systems perspective*. Cambridge University Press.

Shah, S. K., et al. (2020). Ethics of controlled human infection to address COVID-19. *Science, 368*(6493), 832–834. https://doi.org/10.1126/science.abc1076

Shibuya, K. (2006). Actualities of social representation: simulation on diffusion processes of SARS representation. In C. van Dijkum, J. Blasius, & C. Durand (Eds.), *Recent developments and applications in social research methodology*. Barbara Budrich-verlag.

Shibuya, K. (2017). *An exploring study on networked market disruption and resilience*. KAKENHI Report (in Japanese).

Shibuya, K. (2022). An "Artificial" concept as the opposites of human dignity. In T. Sikka (Ed.), *Science and technology studies and health Praxis: Genetic science and new digital technologies.* Bristol University Press.

Shibuya, K. (2020). *Digital transformation of identity in the age of the artificial intelligence.* Springer.

Spinoza, B. (2008). *Ethica, ordine geometrico demonstrata,* BiblioLife (English translated edition).

Vose, D. (2008). *Risk analysis: A quantitative guide* (3rd ed.). Wiley.

Weaver, V. M., & Prowse, G. (2020). Racial authoritarianism in U.S. democracy. *Science, 369*(6508), 1176–1178.

Crises, Risks and Sacrifices

1 On Crisis, Risk and Sacrifice

1.1 How to Think About Crisis, Risk and Sacrifice

As crises and risks are real matters in the world, the only way to deal with them must be through realism. In considering crises, risks, and sacrifices, the first thing to do is to abandon a feeling of omnipotence to change or adjust the world and phenomena according to one's own subjectivity. Based on only scientific manners, it is desirable to deal with problems quickly, in the right order, at the right time, and if possible, before it occurs. In sequential processes, all unintelligible factors such as political reasons, opinions from the establishment class, and emotions must be eliminated. In addition, agreements and discussions should not be made after a crisis has occurred, rather before it occurs.

Actually, nevertheless we have limited time, effort and funds, the humanity has been endlessly surrounding with various risks and crises (Table 1). These patterns can be categorized as natural-origin, human-made, or, of course, mixed one.

For example, according to CDC (Center for Disease Control, USA), they estimated the number of deaths from seasonal influenza in the world is constantly to be 291,000 to 646,000 per a year, although the number varies depending on the type of virus being spread. In the USA alone, more than 30,000 people die annually. On the other hands, larger scale natural disasters and human-made hazards have globally potentials to corner to the crisis of the humanity. Because our global society has been endlessly threatened by various disasters (UNISDR, 2018), more than 160 million citizens have been globally harming by natural disasters every year. And this data contains more than 100,000 deaths per year. Natural disasters are almost interrelated to numerous factors such as climate, demography, environment, and anthropogenic events (Shibuya, 2020a, 2020b).

As Table 1 shows, each risk and crisis in each item should be paid attention. But the researches and practical solutions on risk and crisis related to medical and epidemics have to conduct more comprehensive efforts assembled by heterogeneous experts

Table 1 A part of risks in various situations

Risks	Items
Biological	Bio-terrorism, biotechnological hazard, genome editing, genetically modified foods (Shibuya, 2013), etc.
Chemical	Pollution of air, water, and soil. Harmful side effects of a medicine or drug. Damage risks by Daily necessary products. Nuclear substance and exposure by radioactivity, etc.
Economics	Stagnation, inflation, deflation, market disruption, systemic risk, financial risks, etc.
Human Made Disasters	Plant accidents, traffic accidents, conflicts, severe cases caused by criminals, polluting environment caused by the human (Shibuya, 2017a, 2020b), etc.
ICT	Information security, privacy leaking, cracking, social engineering, severe damage against infrastructure and database, hacktivism, etc.
Medical and Public Health	Epidemic diseases, pandemic, emerging virus, etc
Military	International tension, conflict, warfare, hard power, soft power, sharp power, cyber-warfare, terrorism (Fierke, 2012), armament, AI driven military weapons, etc.
Natural Disasters	Storms, floods, quakes, cracks in the ground, tsunami, volcano explosions, landslide, drought, locust damage, blizzard, extreme weather, crustal movement, etc.
Social	Social stratification and class, poverty, prejudice, refugees, immigrants, panopticon driven by the AI, threatening human rights, etc.

and professionals. CDC[1] of the USA actually intends to deal with *all-hazards* in any risk and crisis (CDC, 2013), and then they laid weight on intelligence, security, risk communication and logistics at larger part of the disease controls.

In risk researches on those issues, *risk management* in such emergency situations can be defined as "*a cycle of forecasting, planning, improvement and verification to address risks before they occur*". It defines a process managed by PDCA (plan-do-check-analysis). It is principally necessary to *estimate* the probability of risk occurrence and *analyze* the risk, and to *decide* whether the level of risk is *acceptable* or not, or whether the risk should be *avoided* or financially *insured* based on possible factors.

Nevertheless, risk management involves deciding what or who to sacrifice and it absolutely how to acceptance such sacrifices as negative outcome in the present and future. How should we prepare for natural disasters that can be estimated to be frequent? And simultaneously, should extreme rare cases, which estimated to occur once in a thousand years be well prepared? Similarly, in the case of infectious diseases, the response to a seasonal influenza with normal toxicity is naturally different from the response to an emerging infectious disease with a higher lethality rate and infection possibility. Even if it considers many issues in advance, such as the

---

[1] https://www.cdc.gov/.

scale of possible damage and economic loss as well as the potential loss of human resources and national wealth, it cannot always be fully prepared if such a risky event cannot be avoided (United Nations, 2016).

1.2 Definitions and Differences on Crisis, Risk and Sacrifice

Each meaning and definition on crises, risks, and sacrifices are definitively different in various fields (Bostrom, 2013; Bostrom & Ćirković, 2018; Fanelli, 2018; Roeser et al., 2012; Shibuya, 2020a). For further discussion, let me confirm these basic terms in this book.

A "*crisis*" can firstly be defined as a tremendous scale and negative event that has been currently occurring. Next, a "*risk*" literally means a negative event that has not yet occurred, but it has potential to occur in the future. Finally, "*sacrifice*" denotes a negative consequence that will be brought about past, present or in the future caused by daily life events, risky or crisis events. In such respect, both risk and sacrifice include prospectively future event, but in this case, risk can become the cause of sacrifice, not the result.

Besides, according to the *Oxford Advanced Learner's Dictionary*, these terms are defined as follows:

- *crisis*: a time of great danger, difficulty or doubt when problems must be solved or important decisions must be made.
- *risk*: the possibility of something bad happening at some time in the future; a situation that could be dangerous or have a bad result.
- *sacrifice*: the fact of giving up something important or valuable to you in order to get or do something that seems more important; something that you give up in this way.

Further, there is strictly a subtle difference of meaning between sacrifice and risk in technical terms of specific fields. A concept "*risk*" usually denotes a probability to occur a negativity in general cases, but financial risk can be sometime meant including both positive and negative chances. On the contrary, *sacrifice* can be understood as only a negative status or outcome to be occurred in general.

In this book, sacrifice can be defined as, in essence, all that is necessary to achieve the "goal" of accepting, avoiding, correcting, or resolving such problems when "risk" is predicted or when a "crisis" occurs.

By the way, it should be noted that the difference between *sacrifice* and *victim* is semantically considerable (Jacki & Pamela, 2020; Joseph & Jergenson, 2020). In Japanese, the word "*sacrifice*" is used to describe people who suffer damage when a crisis or risk occurs in many situations. However, the English word "*sacrifice*" means a nuance as above, and essentially refers to "something or someone" that is offered as sacrifice in order to achieve a goal. Therefore, in the cases of disasters and crimes, it should call them "victim" rather than "sacrifice". However, there are many cases where it is impossible to make a clear distinction between them, and the discussion in this book takes these semantic meanings into consideration.

1.3 Definitions on the Principles of Distribution of Sacrifices

In the first chapter of Rawls' main book, *A Theory of Justice* (1971), there is the following sentence. "*Justice is the first virtue of social institutions, as truth of systems of thought. ... For this reason justice denies that the loss of freedom for some is made right by a greater good shared by others. It does not allow that the sacrifices imposed on a few are outweighed by the larger sum of advantages enjoyed by many. Therefore, in a just society the liberties of equal citizenship are taken as settled*" (underline appended). In other words, we can understand that Rawls, in terms of *fairness* and *justice*, was also aware of the "*sacrifice*" that the author is now trying to discuss further in this book.

First, the differences between Rawls and the author can be summarized as follows. Rawls desired to save "sacrifice" from "justice" and "fairness". On the other hand, the author, especially in the "crisis", considers "sacrifice" as "something" (or "someone") that is necessary and inevitable in order to solve problems, and tries to consider whether it is possible to minimize and optimize it as much as possible. Indeed, Rawls also sought Maximin strategy and average principle without imposing sacrifice on any particular minority. However, he did not directly face sacrifice in the name of fairness and justice, but only thought of equal distribution of profits and opportunities, in the end, forgetting the fact that we must sacrifice something to save something else. As a result, in the case of the crisis, averaging principle and fairness principle cannot save anyone, not to mention the failure to achieve the goal itself.

Here, let's look back at Rawls' "*Fairness Principle*" (1971, 2001). The main idea can be summarized as follows.

- *First Principle*: "Each person is to have an equal right to the most extensive total system of equal basic liberties compatible with a similar system of liberty for all".
- *Second Principle*: "Social and economic inequalities are to be arranged so that they are both:

    (a)   to the greatest benefit of the least advantaged, consistent with the just savings principle, and
    (b)   attached to offices and positions open to all under conditions of fair equality of opportunity".

However, throughout his life, Rawls had been subjected to criticism of his Fairness Principle. The reason for this was the ambiguity of definition, problem formulation, and concrete solutions. What constitutes "fairness" depends not only on the situation but also on the subjectivity of the person making the judgment. Even if a numerical value or indicator is set, it is meaningless if there is no means to improve or eliminate the causes. Therefore, it is not enough to ask philosophically whether a situation is fair or not, but it should comprehensively consider all the negative aspects as "sacrifices", understand such nature, and explore for its solutions. It is important to realize the means of solving the "sacrifice" problem and its result in accordance with the "goal" to be achieved. It may be necessary to ask whether such process is fair or not, but it is not the essence.

Next, the principles of distribution of *"sacrifices"* proposed by the author can be hypothetically defined as follows. These are not ironies. These are indeed the reality of our world.

- *First principle*: sacrifice is not usually negotiable. Nor is there any guarantee that the reward is always sufficient. Rather, it often forces voluntary and gratuitous.
- *Second principle*: If the person undertaking the sacrifice is undecided, non-existent or inadequate, the problem will never be solved. Moreover, more often than not, the sacrifices required are progressively worse.
- *Third principle*: If possible, the sacrifices per person should be minimized and evenly distributed (thus, it often observes that we all shall accept equally). However, there is no guarantee that such situation can be achieved.
- *Fourth principle*: In decision-making situations, people who are not present are often sacrificed. Sacrifices are often imposed on those who are in a weaker position.
- *Fifth Principle*: On a system, overall performance is determined by the degree to which those who bear sacrifices are robust, tough, durable, and flexible. Conversely, the performance of a system is no better than its vulnerable elements.

In a nutshell, all social problems in the world (i.e., social evils engendered by an unspecified number of citizens as the existential beings) can be summed up as the problem of "distribution of sacrifices" in line with the above principles.

For example, let me show you some situations in the daily life.

Life course: childbirth, parenting, investing in yourself or investing in your children? If you can't have it both ways, you have to give up something (Guillaume & Pochic, 2009).

Work: How much of the extra burden should people carry at home and workplace?

Earthquake Disaster: In the case of the Fukushima nuclear power plant accidents, those who made money from building the nuclear power plants, those who suffered damage of the severe nuclear power plants accidents, and those who bear the brunt of the bill are all separately different. In the economics, such situation is called the externality. The Pigou tax, the Coase theorem, and other studies are well known, and the economic cost of the destruction and resilience of the environment and nature caused by the responsibility of corporations and other entities must be considered.

Pandemic: as discuss later in detail, the gap between rich and poor was exposed in the COVID-19, because the national and global system for sustainable economics as well as prevention and health care are vulnerable.

War: The state of war is one of the worst crises, referred to as the Supreme Emergency, and there is no fairness or justice involved in the sacrifice. As Grotius explained in his book *"The Law of War and Peace"* (*"De jure belli ac pacis"*, English translation by Neff, 2012), since "war is an instrument for the achievement of practical ends carried out under the control of law", the consequential sacrifices are also inevitable. And, there are two moralities in war such as *jus ad bellum* (e.g., the justice of war, right to war) and *jus in bello* (e.g., justice in war, law in war) (Walzer, 2006). However, from a view of the distribution of sacrifices, such a broad distinction is meaningless. In an indiscriminate and all-out war, all sacrifices would be made regardless of whether it is a government leader who directly causes the war, a soldier or an ordinary civilian (including the future generations). For example, to end the Second World War, the politicians in Japan sacrificed also "us" who weren't even born yet. There is no adequate compensation for us. Nor did they get an agreement from us. After the war, even if it was true that economic development had been achieved at the cost of all sacrifices, it will not be balanced with the lost national wealth and the dead.

After all, you will realize above principle on the meaning of "distribution of sacrifice". When you are especially participating to discuss something among the multiple stakeholders, it takes a lot of time to decide who will be responsible for the "sacrifice". The reason why democratic debates take so long is because the person who undertakes the sacrifice is not decided. Indeed, it is better to start the discussion by asking, "Who will be sacrificed?" Even if the situation is ambiguous, the focus of the discussion will be set and intensive discussion will be possible. And when the person who makes the sacrifice has been predetermined, the discussion can proceed smoothly. In short, "sacrifice" is the essence of all social problems.

As a framework for existing discussions, public economics, for example, usually discusses the problem of equitable burden of public goods, and price adjustment mechanism of supply and demand in the economic standpoint. They broadly distinguish between the *benefit principle* (imposing more burden on entities with higher utility from public goods) and the *ability principle* (imposing more burden on entities with higher ability to bear). However, from the perspective of "distribution of sacrifice", in the former, when future generations are expected to have the highest economic utility, there is a concern that future burdens will be imposed on them without their consent, while in the latter, there is no guarantee that the economic utility of the entities with the highest ability to bear will also be high.

Namely, it is not a problem of will or freedom to do or not to do. It is a situation of having to do. It is literally sacrifice because there is no or insufficient compensation for such sacrifice. If there are no or insufficient sacrificers, the problem will not be solved, the situation will become worse, and the sacrifice required will increase. Rawls' "fairness principle" tends to deal with the question of whether the burden or role is fairness, justice in line with the public good. In times of crisis, especially, if we do not take practical measures and solutions that confront "sacrifice", it will only call more severe crises, and it is certainly sacrifice because whether it is fair or fulfills the public good is meaningless for the real solutions.

From the point of view of the author, who has been engaged in research dealing with computation, algorithms, and optimization problems using computers, regardless of the diversity of expressions used to mean "sacrifice", all social problems are a kind of effective utilization, optimization, coordination and allocation of "sacrifices" such as costs, funds, time, labor, human resources, and natural resources. If it can be solved, it should be solved quickly and reliably. If it cannot be solved, there is no choice but to take the next best solution or other measures.

From a practical point of view, the issue is not the benefit for undertaking sacrifice (and in many cases, not even sufficient compensation), but rather whether peoples to bear sacrifices are prepared and has the spirit of self-sacrifice. As long as this is not fulfilled, social problems as social evils will not be solved, but rather will continue to worsen.

1.4 Sociological Categorization on Crisis, Risk and Sacrifice

1.4.1 Categorization

Standing on the sociological perspectives, there are crucially needs to lay weight on both social *structural* and *changing* aspects to examine each risk and crisis. Sociology has been characterized to investigate the human as social-being, their relationships, organized associations or society. Naturally, everything has own specific pattern as a structure and potential to be changing its order during the time passing. Namely, each risk and crisis can be obviously observed somewhat specific problem in terms of sociological interests, and further each sacrifice ought to be recognized through such effortful examinations (Table 2).

Sociologically thinking, each element in both structure and changing has own inherent risks and crises. Namely, "sacrifice" is not only the result of one of the above subdivisions, but also the result related to multiple subdivisions. It will be perceived as a complicated consequence caused by the structural or changing dynamics. For example, in the ontological case, even if an individual forms a solid identity, but there is no escaping way from the risk of own existential anxiety (Heidegger, 2008). Moreover, transformation occurs throughout each lifespan, because of aging, disease, and "*Sein zum Ende*". In addition, the "middle age crisis" (Erikson, 1950, 1959, 1982), identity disorder such as dementia, and psychosis may become also issues.

The frame of the crisis pertaining to political, economic and social aspects are almost noted by Habermas (1976). The same is true for the transformation of social institutions and identities through the technological innovations (Shibuya, 2017b, 2020a, 2021a). In addition, the ideological background of Husserl's "*the Crisis of European Studies and Transcendental Phenomenology*" was one of the historical perspectives at his living age. A word "crisis" might be used to describe his world-view or zeitgeist. It can regard a same phenomenon as the contemporary notion of "*Westnessless*", the structural stability of the western civilization was gradually and steadily losing and the phenomenological anxiety that certainly leads to auto-decay of the structure due to historical changing process to shift to China's domination in the world may be expressed in a straightforward manner.

Hence, as it can recognize, there are either structural or changing risks or crises at every level. The AI and big data invoked an identity crisis (Shibuya, 2020a). Moreover, it should be tackling various crises such as the nuclear accident crisis at

Table 2 Exemplifying a part of the structural and changing patterns on risks and crises in terms of Sociology

	Social structure	Social change
Ontological	Human as physical and living existence, and identity	Development, disease, aging, death, and identity lost
Phenomenological	Perception of the world, perception of risk, and personal experience	Reshaping cognition of the world, representing the reality, and personal attitude for the society
Scientific	Evidence based knowledge Factual understanding	Paradigm shift in scientific revolution
Historic	A period of the civilization, and cultural pattern	A period of cultural changing before a new order
Institutional	Enacting laws, governance, and legitimacy	Distrust in the administration and bureaucracy, regime change, and redesign of the legal system. Legitimation Crisis
National	Establishing a new regime, and sustaining regime of the nation	Self-decaying a regime of the nation, and revolutionary uprising
Interactive	Keeping balance of power among the stakeholders	Collapsing balance of power among the stakeholders
Socio-economic	Living styles of the citizens Global financial markets Trades and supply chain networking	Changing daily life of the citizens Financial systemic risks in global Rewiring trades and supply chains
Technological	Traditional and conservative ways	Innovative process, and creating something new
International	Cooperating with each nation Cosmopolitanism Fragile equilibrium among permanent members in the security council of the United Nations	Salience of severe conflicts among the nations and races Identity politics Decoupling permanent members in the security council of the United Nations
Systemic	Self-organizing system as a maintaining autopoietic process, and complex system as a structural pattern	Self-organizing system as an apoptosis process, and complex system as a variant pattern

Fukushima (Shibuya, 2020b), the increasing magnitude of natural disasters, and the upheaval of the international affairs. These sequential events have reached the point where humanity's survival is at stake. The ravages of further cataclysmic change, with many casualties, are spreading to the socio-economic and daily life.

In particular, the COVID-19 pandemic is a problem that indeed deals with a similarly a number of the sacrifices (Shah et al., 2020). In this case, in global, the number of casualties and economic damage can easily be recalled. The total damage cannot be quantifiable, and the global change will be proceeded in conjunction with the AI and big data revolution.

The fact that China occupies a position of the epicenter of international tensions lets us feel an ironical meaning in relation to the COVID-19 pandemic (Thorp et al., 2020). And while the older hegemony of the USA is declining in its relative international presence at domestic and international affairs. The confrontation with China and the composition of prejudice and discrimination is once again manifesting itself domestically. China, on the other hand, has not presented to the international community a democratic-like global principle and trustful leadership that brings together a pluralistic set of the multiple values yet.

1.4.2 The Nature of Sacrifices

The author is not merely referring to the verification "sacrifice" as something of the "failure". For example, the author does not regard it as a similar pattern of the examinations for the "Space shuttle launch accident and its failure" or the "Fukushima nuclear power plant accidents" (Shibuya, 2017a, 2018, 2020b, 2021b). All possibilities that humanity could choose to take, and all sacrifices based on such outcomes, should be investigated as much as possible. The "risk" deals only with future possibilities, while the "crisis" mainly concerns only the current problems. However, "sacrifice" can be distilled from the causes and effects as the matters of all past, present and future. Of course, human perception and verifiable time are too limited, and then it can't verify everything.

However, it is still quite important to hold an attitude that seeks to consider all sacrifices in the past, present, and future. These are indeed a start to sociological or metaphysical contemplation (Brännmark, 2019; Zack, 2009). What are the respective consequences of success, failure, or doing nothing? To consider in advance whether there will be problems, consequences and possible sacrifices, and to make decisions based on these aspects. This is because it gives us the opportunity to consider what to do or not to do in advance. Human decision-making had been experimentally well observed as a psychological mechanism that avoids negative consequences, but the emotion of "regret" clearly is not unrelated to all of the "sacrificed" results. The consequences by the people's repeating choices will shape own life course, and sometimes drastically change our communal daily lives and the structure of the world. In these structures and changes, it can envision specific "sacrifice". The cause and effect of such historical events shape both social structure and social change, and conversely, through these social structure and changes, it can discover and observe a "sacrifice" in the world.

Especially, political decision making is about distributing sacrifices. Legislators and government leaders are certainly elected to enact necessary legislation in compliance with citizens' demands in the democratic society. But it is not the prerogative of the executives to impose the necessary sacrifices. Some say, "if you put sacrifice on the table, we can't do anything about it, and we can't decide". Or some assert "Well, what should we have done then?" But they are wrong problem setting. Because no matter how it decides, there will always be a sacrifice. Of course, it does not mean that any decision is simply acceptable. It is for this reason that politics is most problematic in terms of the sacrifice.

The pandemic of COVID-19 was no different. Politicians talked about their policies of health care and economic achievements as if they had accomplished something great. Rather, all achievement was accomplished by the medical doctors who risked their lives onsite to treat the patients, and further there were the medical doctors who died on duty. Their sacrifices were strictly not the achievements of the government leaders. The daily lives of ordinary citizens were also greatly restricted, but they felt also sacrifices that imposed by the government. As a result, it is true that infection was prevented. On the other hand, the progress of ICT and other technologies has reshaped our daily life and working environment during the pandemic, but such social changes that resulted from such progress also led to other sacrifices.

The most important is that there is no adequate compensation for the sacrifice. It does not mean that accountability is necessary to compensate for those who bear sacrifices. Ultimately, there is no sufficient economic compensation for them at all, and nobody can offer the equivalent if they asked to do. For example, workers are paid a wage, but there is no guarantee to sufficiently pay for selling off the priceless time of one's life. Namely, there is no human history without these sacrifices, but their sacrifices are never taught in history class. Then, many of the peoples now may understand that it is not decisions but sacrifices that are the most important in the human history. Their efforts should be acclaimed as an incomparable achievement rather than anything.

1.4.3   A Role of Sociology

Sociology is essentially a discipline that examines all social phenomena from a perspective of sacrifice, as the author thinks. Given that, there is no sociological research that does not consider sacrifices in its essence. The revealing of social problems suggests that sacrifices are also unveiled.

However, until now, it seems that sociology had mainly discussed "*sacrifice*" in Durkheim's sociology of religion (Durkheim, 2005; Jones, 1981), and "*sacrifice*" from anthropological perspective (Huberts & Mauss, 1981; Mizruchi, 1998). These sacrifices mean, literally, "*scapegoat*" as a religious offering, which is not the purpose this book focus on. And there is no specific entry on "*sacrifice*" in "the Dictionary of Sociology" published by Oxford University Press (Scott & Marshall, 2014), which suggests that sociology has been little emphasis on "crisis and sacrifice".

This clearly indicates that sociology has overlooked "*sacrifice*" in crises and risk situations because of its narrow view of it. In the first place, the sociology of crime cannot ignore the victims, and the sociology of conflict and warfare deals with the damage of common wealth, the states and civic victims. The sociology of the family cannot fail to articulate the contradictions in family relationships. The sociology of regions and cities also manage the problems caused by the human relationships. The same is true for environmental sociology and research pertaining to science and technology (Bord et al., 1997). Political and economic sociology cannot afford not to clarify the interests of human society and the contradictions associated with them. And sociology of international relations handles various "sacrifices" on the multilateral balance of power and its many problems through its conflicts.

Thus, these studies always involve some kind of "sacrifices", or it probably expresses such term as other synonyms such as "problems", "concerns" and "results".

The "distribution of sacrifices" clearly encompasses *fairness principle* as Rawls argued (Rawls, 1971, 2001), but also makes the nature of the social problem more explicit. The difference between them particularly become clear in discussions that directly relate to the *human dignity* (Ackeren & Archer, 2018, 2020). An example is the issue of women's childbirth and childcare. These are not "fair" issues, no matter how they are handled, and it is impossible to make them fair. If men will not ask women to such maternal responsibility to give birth and raise children, the entire human history will not continue. Moreover, it is essentially impossible to appeal for better treatment or compensation for them. Economic assessments have indeed led to the substitution of childbirth and childcare (e.g., nursery schools, babysitting, etc.). However, as this makes clear, if such actions are 60,000 yen a month service, nobody say that the value of the mother's daily activity is less than such value. Rather, it is the child who is the most victimized, nevertheless the mother's love and living conditions have priceless value, they are undermined by "per month" economic value. The system that converges into a single economic value standard is quite wrong, and it misleads women as replaceable labor forces. Some people said that the existence of children is seen as long-term economic cost and burden itself. The issue of disregard for the human dignity made explicitly the nature of sacrifices in the economic standard.

Mere enumeration of facts or description of a social phenomenon is not a study. Having such an attitude based on the perspective from the "sacrifices" leads to a research standard to exhaustively perceive and examine all possibilities and complicated linkages with time-sequential dynamics. It involves examining and contemplating social phenomena from multiple perspectives. Therefore, it is clearly different from any studies that only talk about the results of social surveys or literature that only talk about their own reflections. A variety of analytical tools will be used to scrutinize, and a number of relevant findings will be comprehensively and analytically verified. The final conclusion will become quite different in terms of both quality and quantity.

2 Measuring and Quantifying Methodologies

2.1 As Social Scientific Studies

It can say that sociological research is, after all, an academic field that examines risks, crises and sacrifices in each context in detail. The diversity of such studies can be also tackled in broad categories of methodologies: theoretical studies, narrative content analysis, interview, social surveys, social media analysis, data quantification, statistical analysis, meta-analysis, systematic reviews, mathematical modeling and computer simulations, for example.

In practice, it is necessary to know the nature of social phenomena through many research methods and indicators (Bach et al., 2020). It will make sure about them (Saltelli et al., 2020). A variety of indices and measurement methods have been developed in other research fields (Haug et al., 2020), and it will be possible to apply sociological examinations.

2.2   Risk Assessment

Risk assessment and its ways of measurement are diversified in various fields. In financial investment, there is also portfolio theory, which is an investment strategy through risk diversification. It is to maintain a balance between profit and loss so that the profits ultimately outweigh the losses. The risk assessment in such models also takes into account statistically variance, standard deviation and expected rate of return to calculate the risk as uncertainty (Hayashi & Lombardi, 2019).

And further, in subjective estimation of psychological experiments, risk perception and public understanding of risk phenomena have been repeatedly examined by mathematical models (Fischhoff et al., 1999). As it is empirically supported by experimental psychology that people are vulnerable to such risky decisions, either individual or collective decision making is always a challenging matter (Kahneman & Tversky, 2000).

According to Sober (2008), data itself can be inferred based on inductive facts and evidence, but there are some different schools in the manners for inferences. Namely, these are Bayesian, likelihood-oriented, frequency-oriented, and other schools. Each of them has become a major trend in the field of statistics, and it is not possible to say that any one of them is easily superior to the other. However, in general, Bayesian school is characterized by the modeling of subjective probabilities of observers and their update its probabilities, which makes it easier to incorporate the perspectives of events and risk assessors. In addition, the perspective of likelihood-oriented and frequency-oriented schools are discerned as inductive data accumulation (e.g., the law of large numbers) and model selection criteria (Akaike Information Criterion (AIC) is a representative of frequency-oriented school).

2.3   Pareto Efficient in Economic Models

As mentioned before, in welfare economics in particular, many of such economic conditions can be stabilized as a *Pareto efficient* or optimum situations. Since economics can basically handle with only an economic value eliminated from the humanity and social objects, and it usually does not to consider either a life or the dignity of each individual even if welfare economics. Their theorems and assumptions for formalization can usually deal with only something quantifiable and comparable to evaluate among the existing phenomena or objects.

Namely, the human dignity itself should not be handled by any economic reasons, because the whole life of each individual cannot divide two or more apart. Even if advanced medical and biological technologies will be possible (e.g., clones), an identity of each individual must be kept in accord with ethical compliance (Shibuya, 2020a). Naturally, economists cannot consider which better is more rational to save an individual or die all, and then they cannot hardly recommend an alternative among the solutions for the government leaders if there are any solutions which includes such the humanity. It may be effective if inter-exchanging indivisible goods in economics (e.g., kidney) can be arranged in line with the disciplinary border of economics (Roth, 2008).

But as mentioned above, sociologists and ethics scholars as well as government leaders should be aware about such severe conditions with various ways to evaluate and ponder for the pursing the goal. If necessary, they must use various measurements and scientific quantification, but all of them shall be understood by intensive and serious decisions.

2.4 Cost–Benefit, Expected Utility and QOL

Next problem is how to qualify and measure them. In economics and its applied fields, such risky situation can be defined as a situation of higher uncertainty. Sacrifice is often converted to something measurable and calculated by relation between benefit and cost.

As first example, economic policy making has formalized any risk situation decided by cost–benefit analysis and its *cost performance* in each model. Simply, when a consequence as a benefit performs over total input as the costs, such situation can be said to be successfully. Of course, we also need to consider the *opportunity cost*. Those analyzing process from formalization to interpretation of the models can be diversified in each problem. Values which can be defined and measured by differences of the model would be depended on the purposes. Particularly, such values clearly vary economic income, financial based evaluation (e.g., discount cash flow (DCF), standard deviation, Gini coefficient), cost shifting and other assessments for real estates. Of course, the severe case requires to compare with cost performance among the clients who hold priceless value as the humanity to be survived, but emergency situations such as medical triage and estimation of survival rate ought to be handled by prioritized judgment by medical doctors for the clients.

Secondly, "expected utility" introduces to measure any future risks and its probabilistic conditions. Because an original contribution was served by both von Neumann and Morgenstern, who also originated *game theory*, it is often called as von Neumann-Morgenstern utility function.

For example, let's assume that there is a risk condition that definitely costs 500,000 at each time. It notates this situation as La = {x; 1}. On the other hand, there is also an uncertain risky condition, assuming that there is a cost of 300,000 with a probability of 0.8 and a cost of 1,000,000 with a probability of 0.2. It notates this situation as Lb = {300,000, 1,000,000; 0.8, 0.2}.

Since the probability of the two risky situations occurring is 0.8 and 0.2, then, expected utility values can be formulated by below equation.

EU (Lb) = 0.8 U (300,000) + 0.2 U (1,000,000) = 440,000.

It can decide which rational is EU(La) = U(x) or EU(Lb). In this condition, EU(Lb) will be more acceptable because it is more reasonable to compare EU(La) to U(x) and choose the one with the least cost.

Additionally, "Cost-loss model" had been proposed by applying above expected utility model in economics (Murphy, 1969). For example, when the weather forecast announces a 50% chance of precipitation, this model considers the cost of bringing an umbrella and the risk (=loss) of not bringing an umbrella. Think of the former cost as the cost of buying an umbrella (e.g., 900) to avoid the risk of getting wet in the rain. The latter, on the other hand, assumes that if it doesn't bring an umbrella and get wet in the rain, it will have to pay for the cleaning of suit (e.g., 2000). In this case, it would be 0.45 (=45%), which means that since a probability of the weather forecast is higher than a calculated result of this model, it is reasonable to bring an umbrella.

Otherwise, financial modeling is necessary to forecast future profits based on discount rates, risk assessment and costing. However, meanwhile traditional economics manage with the demand and supply predictions as well as expected utility hypothesis, behavior economics that relies on psychological empirical evidences tackles nudges as sensible clues and our mental accounting on risks and sacrifices (Camerer, 2003). In particular, *Prospect theory* (Kahneman & Tversky, 2000) explains and predicts that decisions under uncertainty of the human are the tendency to specifically choose risk aversion than the gains. Ordinary peoples usually prioritize the psychologically felt costs and sacrifices over the real costs and sacrifices.

And thirdly, the assessment of QOL (quality of life) is now considered an important factor for daily living. WHO[2] defined quality of life as an individual's perception of his or her own life situation in relation to goals, expectations, standards and interests within the culture and values in which he or she lives (WHO/QOL Basic Questionnaire). It consists of six domains: *physical, psychological, level of independence, social relationships, living environment* and *spirituality/religion/belief*. It is an evaluation method in which the responses are tabulated. In other words, the more these measures are well satisfied, the higher the benefits of receiving medical and welfare services (Kawachi et al., 2008). Conversely, the more people who do not satisfy, the more economic, medical, and social sacrifices they may be forced to stay in miserable situation.

---

[2] https://www.who.int/mental_health/publications/whoqol/en/.

2.5 Balance of Power and Allocation by Contributions

Sacrifice has a strong tendency to be concentrated on relatively weaker side. Weaker countries, people of low social status, people without the right to vote, and neglected beings. Of course, the side in the stronger position often undertakes a reasonable burden or fulfills their obligations, but even then, it's not enough or not functioning well enough as a system. And consequently, all those sacrifices may be for naught.

Here, such situations told an important fact: there should shed light on balance of power among the peoples, and contributing degree among the peoples.

First, balance of power, namely, has frequently been focused on human relationships, international affairs and any organized institutions. There is a Shapley-Shubik index to be considered the balance of power among the stakeholders (Shapley & Shubik, 1954). This matter will mention later chapter in detail.

Secondly, contributing degree can be defined as quantity or quality of its something valuable. But those measuring ways and envy-free system for allocation are often difficult to be determined among the stakeholders (Thaler et al., 2018). In that condition, Shapley value (it is not same above technical term) can apply determining each profit by contributing for total cooperation among the stakeholders (Roth, 1988). His idea intended to be for a fair distribution to the players in game theory. It is a rational way of distributing the total reward value to the players if everyone is to cooperate together.

Otherwise, computer science has also been studying various fair allocation problems such as an envy-free cake-cutting problem (Aziz & Mackenzie, 2020) and DAA (Deferred acceptance algorithms: Roth, 2008). They found an algorithm to be fair resource allocation among the participants, and it ought to be designed for wider area.

2.6 Equality and Relative Deprivation

When sacrifice invokes a further issue, both fairness, equity, and equality should be often cared about at the same time (Dutt & Wilber, 2010; Dworkin, 2005). In lexical concerns of the Oxford Advanced Learner's Dictionary, fairness is defined as "*the quality of treating people equally or in a way that is reasonable*", equity is "*a situation in which everyone is treated equally*", and equality is "*the fact of being equal in rights, status, advantages, etc.*". There are subtle differences of nuances among them, and each status that should be achieved by somewhat value standard may be always difficult. It can be seen as the difference between *equality* as "opportunity" and *equity* as "outcome", and what is sacrificed will naturally differ. The former, by opening up a wide range of opportunities (e.g., fair and competitive races), takes it for granted that there will be disparities between the few winners and the many losers. The latter, on the other hand, suggests equalizing the results according to some value standard, and trying to make the most of the majority as much as possible. Therefore, the axis of evaluation inherent in a society or a nation becomes clear depending on which of such semantic differences it is based on.

Generally, in social psychology, one of the mental biases on fairness was named "*just-world hypothesis*". As long as each individual evaluates facts through his subjective view, even if the facts are of the same quantity, unfairness still remains (Luhan et al., 2019; Niesiobędzka & Kołodziej, 2020). There is a difference between results as facts and its subjective evaluation. Hence, the result of sharing equal sacrifice is likely to leave dissatisfaction. Or, even if no one seems to be disgruntled, the sacrifice often is forwarded to someone who is absent in the decision-making process.

On the contrary, Rawls's book "*Justice as Fairness*" (2001) clearly intended to discuss not only about profit sharing but also about the negative profit sharing as sacrifices (e.g., cost, demerits, burden). It must be read as assuming both benefit and sacrificial allocation in the society. His consideration over principle of fairness is clearly linked to a concept of the highest good in ancient Greece. The moral system represented by the Five Virtues in the East Asia must also be examined at the sacrifice between cultural differences (as discuss it at later chapters). When discussing the benefits that a person with a disability should receive, the person with a disability has already assumed the inherent sacrifice that is a given. The ordinary citizens should take on a different burden as sacrifice in order to *equalize* the difference between them. As he told, maximin principle, which means a strategy that focuses on the worst-case gains from all alternatives and chooses a strategy of the greatest worst-case gains can be also carefully investigated in each case. Then, if such response itself can be called *fair*, how do we define *justice as fairness?* And the question is whether it can be evaluated and achieved in actual society (Brandstedt & Brännmark, 2020; Rawls, 1971).

Secondly, psychologist Adams formalized an *equity* that is simply defined by a rate between outcome and input. Equity exists when "the ratio of rewards to one's inputs is equal to that of others". That is to say, the equity can be mentally determined by social comparison with the others. When person(X) feels inequity when X compares own efforts as input and rewards as outcome with others' inputs and outcome, X is motivated to act in a way that eliminates the inequity and makes it equally. But such simple assumption has critical deficits to manage nonlinear phenomena and psychophysical dispositions. In this point, Anderson (1996) had expanded Adams' model and integrated models on equity and fairness by his theory of information integration.

And thirdly, in medical related studies, equality has specific importance for everyone, and it can be defined as *equality of access, equal treatment for equal need* and *equality of health* (Culyer & Wagstaff, 1993). Equality of access means that everyone can be cared by equally medical services. Equal treatment for equal need is that everyone can be treated by equally regardless their economic level and other social standard. And equality of health intends to reach equalization of health conditions among everyone. In India, for urgently promoting the national health governance not limited to the COVID-19, they aim to percolate own policy for the digital health from the megacities to rural communities (Agrawal, 2020).

Further, as mentioned above, it can be divided broad categories of *equality*: equality of either opportunity or outcome. Even if the principle of fairness is pursued, social conflicts will arise over the disparities that arise from outcomes. The reason why conflicts over social resources are constant is that social consensus over fairness and equality is not always established. This is because the disparity between high and low socioeconomic status and achievement becomes obvious among the members. In particular, the debate over relative deprivation (RD) (Merton, 1968) has been found to be positioned by social comparative processes. Attempting to maintain these status perceptions and their own self-esteem among the members calls further complicated issues. The issue is complicated by socio-economic disadvantage, in the absence of opportunities to acquire social resources and their adequate distribution. Social resentment among the citizens will rise. Not just such negative perceptions and emotions, but also such situations may call the problems such as immigration and racial conflicts. The development of globalism, especially across the national borders, is partly for these reasons.

Social stratification and socio-economic disparities sometimes lead to differences in the spatial distribution of the actuality. When the existence of economic disparities and the lack of social institutions in a particular region has always been empirically known, there will engender not to mention the disparities in slums and ethnic districts, and the differences in social relationships and economic status of certain areas (Josephson et al., 2021).

2.7 Preference and Priority Order

There can be a situation where it can't save anyone unless it firstly sets priorities. In economics and operations research, individual value systems can be defined as *preference*, and they intend to clarify the preference order of any individual. The same is also true when considering them as an organization, and their *bounded rationality* in light of the structure in decision making process (Gintis, 2014). By examining these patterns, it can quantify what it should be prioritized, but conversely, what it sacrificed. It will be also clear what should be proceeded.

Preference-based computation can be often arranged in computational social choice studies (Arrow, 1963; Brandt et al., 2016; Rothe, 2016; Sen, 1970). For example, AHP (Analytical Hierarchy Process) is characterized with applied mathematical procedures. It can be defined by pairwise comparisons between several alternatives among all combinations. Calculation procedures of canonical AHP in detail can refer to Saaty (2013).

2.8 Priority for Economic Utility or QOL

Risk and crisis studies must consider another facet in depth. Furthermore, there are other difficult problems between economics and medical sciences in such emergency situations (Verweij, 2009). For example, these are economic utility and QOL and subjective well-being of each individual (Diener & Suh, 1997; Cenci & Cawthorne, 2020). Such related studies have been developing in welfare economics (Sen, 1970) and examining social capital in health economics (Kawachi et al., 2008). They often handle the values and preferences among the citizens as each individual or the entire society, and such preference-based order of the value can possibly engender many paradoxes and complicated results (Arrow, 1963; Brandt et al., 2016).

Especially, in health economics, it must be paid attention to both dignity of the humanity and economic utility (Andrews & Withey, 2012). The demand is for human health, and necessary services of the medical cares can be extremely accessible and usable for all people. For each individual, the occurrence of illness or injury, the effects of treatment, and the amount of medical expenses are unpredictable and often contain higher uncertainty as latent risk factors. The value of human health is hard measured by money, and health care services are highly depended on client-specific and the price for them is hard to determine in advance.

Here, similarly, let me give you some problematic examples for further discussions.

(1)   A cancer patient X is admitted to the hospital. Should medical doctors discharge X and let X live out the rest of X's life? Alternatively, rather, should they keep X in the hospital and wait for a miraculous recovery, as there is little chance of his recovery if they continue to treat X?

(2)   A cancer patient X is admitted to the hospital. Special drugs have been developed for certain cancers and have a significant effect for X. But it's a very expensive medical expense and such expense will increase a burden of the national budget of medical welfare. Further, in rare cases, it can kill X caused by intense side effects. Should medical doctors discharge X and let X live out the rest of X's life, or rather should they use this special medicine for X?

Concisely, both situations pose same suggestion. Which better is the patient's postoperative quality of life a priority or survival above all else?

2.9 Resilience

The concept of *resilience* specifies the extent to which it can recover from a certain load. In particular, it was used as a concept to verify whether the load on the environment exceeds a certain threshold (Allen et al., 2016). Further, in addition to the problems of economic recovery as well as medical and clinical psychological problems, it has become a concept used in examining the recovery of damaged market or patients (Shibuya, 2017a, 2018).

The sacrifice in these cases is inevitably whether or not it remains within a resilient range. Within a certain threshold range, it is possible for the situation to eventually improve, but if not, the system has the fear of its collapse, life threatening and severe illness. Economically, *circuit breaker* or other ways to prevent systemic risk in the stock market and other markets (Helbing, 2013). On the other hand, in medical and psychological terms, psychological care for clients will be provided through counseling.

2.10 Systematizing Analysis on Crisis, Risk, and Sacrifice

The measurement of risk is diverse. For example, a risk assessment in ICT security is generally based on "*multiplying the probability of an incident by severity of such incident and its results*" (e.g., risk = probability × severity). In contrast, UNDRR[3] (The United Nations Office for Disaster Risk Reduction) defines a risk = hazard × exposure × vulnerability. In this point, the former laid weight on that observation and recognition of risk and crises have a fairly subjective aspect, whereas the latter is to focus on physical damages caused by natural catastrophes. In many cases, the severity of a disaster is the result of a chain of causal factors leading up to the occurrence of the disaster, and depends on the ability to recognize such a "crisis" as soon as possible.

More elaborately, for example, according to WWF (World Wide Fund for Nature),[4] their risk measurement tool has a similar formalization as follows.

The Supply Risk Analysis is a framework used to:

1. *Evaluate risks and potential impacts associated with commodities sourced or financed by companies*
2. *Identify where the risks are greatest*
3. *Highlight where companies can focus their efforts to best mitigate the supply risks*
4. *It is structured into four macro risk themes.*

Supply Security & Governance:

1. *Environmental*
2. *Social*
3. *Economic & Financial.*

To deduce risk, more than 50 indicators are weighted across 30 criteria within these four themes. For the commodity and geography of sourcing, risk is evaluated at the indicator level based on probability of occurrence and severity of threat.

Consequentially, as Table 3 shows, they constitute a matrix that assesses a risk according to the probability and severity of the event's occurrence. For example, when

---

[3] https://www.preventionweb.net/risk/disaster-risk.

[4] https://supplyrisk.org/our-analysis.

Table 3 WWF's risk analysis

Probability of occurrence (Common ↔ Unlikely)				
Severity of impact (Severe ↔ insignificant)	25	24	22	19	15
	23	21	18	14	10
	20	17	13	9	6
	16	12	8	5	3
	11	7	4	2	1

risk A can be measured by an observation data such as a probability of occurrence is 22 and severity is 8, whereas risk B can be understood as 25 and 23 respectively. And then, risk B is assessed more severe than risk A.

Here, it is important to keep in mind that we should not judge the risk or crisis of an event based solely on the probability of the risk occurring. One of the most common discourses in Japan during the COVID-19 pandemic was that COVID-19 has a lower mortality rate than normal influenza, and therefore the domestic economy should be actively promoted. But the risk of COVID-19 was not only mortality. As will be discussed in more detail later in another chapter, the accumulation of severe cases imposed a heavy burden and other sacrifices on medical personnel. In addition to sequelaes of COVID-19, other spillover risks must also be taken into account, and the risks and sacrifices as the total cost of the entire crisis must be carefully weighed. For example, if the mortality rate of an infectious disease and the risk of an accident occurring in a factory are the same at 0.01%, it cannot be said that the two are not a problem because both of them have the same level of risk and danger. Matters to be considered are quite widely, and there are many considerable factors such as the severity of the risk if it occurs, the cost of recovery if it occurs, the availability and cost of recovery and restoration measures, and the availability of alternative measures to prevent the occurrence of the risk. It should not be judged merely on the basis of the immediate probability of occurrence.

In fact, for example, in February 2021, Nature[5] published its recommendations for reducing carbon dioxide emissions, titled "*Eight priorities for calculating the social cost of carbon*". It calls for eight steps: (1) Reverse Trump's changes, (2) Seek broad input, (3) Update 'damage functions', (4) Reappraise climate risks, (5) Address equity, (6) Review discount rates, (7) Update socioeconomic pathways, and (8) Clarify limitations. Although many factors are involved in global environmental issues, we should not be confined to arbitrary probability assessments or superficial risks. In particular, it is essential to maintain a scientific attitude. Furthermore, the estimation of discount rates is an important factor when considering the impact not only on the current generation but also on future generations (Table 4).

Here, Table 4 compares with the COVID-19 and other examples of crises. First, it goes without saying that the accumulation of data on infectious diseases and deaths is important. For preventing the COVID-19, by measuring real-time data

---

[5] https://www.nature.com/articles/d41586-021-00441-0.

with contact history and geolocation of each individual as necessary, each national authority announced such information everyday. The differences of each national policy provoked differences of priority on the number of new positives as infection, the unknown rate of contact history among new positives, the increase rate in positives, the number of critically ill patients, the number of hospitalized patients, the

Table 4 Exemplifying crisis, risk and sacrifices in each event

Crisis	Risks (for example)	Sacrifices (for example)	Severity
COVID-19 (2019 ~ ?)	It can't be called a risk because it has already occurred But as this pandemic becomes new causes, there will be potentials to break down the global economics and international cooperation in progress	Massive death toll Increased health care costs due to a large number of infected patients Economic recession in global, and extensive damage to the economy and industry of each country including GDP Massive unemployment The load on the insurance industry The future cost of economic recovery Widening educational disparities Medical doctors and other medical professionals died on duty Deaths of elderly and low-income people, immigrant refugees, illegal undocumented workers, and homeless people Major constraints and costs to daily life Rising international tensions	Very high
Locust Damage (2019–2020, ~ ?)	*Although the incidence itself was not high, there has been an increasing trend in recent years* Food shortage More people are starving to death Rising food prices Confusion in the futures market over foods	A lot of people starved to death Collapse of agricultural land Further deterioration of the living and natural environment Unemployment among dairy and agricultural workers Investment failures of market participants, such as futures trading	High

(continued)

Table 4 (continued)

Crisis	Risks (for example)	Sacrifices (for example)	Severity
Severe Accidents of Fukushima Nuclear power plants (2011, ~ ?)	*Although the science has repeatedly pointed out the risks of tsunamis and earthquakes, the government and power companies have not responded proactively enough* Larger Earthquakes Nuclear melt-down Tsunami Exposures by Nuclear substance and radiation Severe damages against Citizens' Health	OECD/NEA INES (international nuclear event scale) ranked level 7 (the worst) Explosion and Melt down of Fukushima Nuclear Power Plants Polluted water matters Polluted venting air matters Polluted nuclear substances Longer term of evacuation and migration of the citizens from Fukushima Area Physical damage and mental distress among the citizens A larger amount of Compensations for victims A larger amount of cost for Environment restoration A longer term to purify around the power plants	Very high

positive rate of PCR tests (Polymerase Chain Reaction test), and other measurable index. Such indexes ought to be guided for safety and health of the citizens. But risk communication on viral knowledge and prevention for infection has not been occasionally misunderstood among the naive citizens in various countries, and then such fake-news and misinformation confused them (Escolà-Gascón, 2021).

Secondly, at 2020, the World Food Programme (WFP) was awarded the Nobel Peace Prize. In the midst of the COVID-19 pandemic, the award committee clearly stated that there are still now hundreds of millions of people in even greater need, and that food is a source of conflict and vice versa. For example, from the end of 2019 to 2020, large areas of Africa and the Middle East continues to be heavily infested by a kind of the grasshoppers (Desert locust[6]: *Schistocerca gregaria*). In addition to the fact that there was a serious shortage of labor in the agricultural sector during the COVID-19 pandemic, measures should be taken keeping in mind that there will be a serious global food shortage. In line with risk management and business continuity planning (BCP), it is necessary to build and maintain a supply chain that can respond quickly to many risks and crises.

---

[6] Why locusts congregate in billion-strong swarms — and how to stop them.
   https://www.nature.com/articles/d41586-020-02453-8

In this concern, how much raw materials and foods are imported or exported from other countries through trade and other means? In the case of Japan, its food self-sufficiency rate is only 37% (statistics at 2015 FY), and it is mostly dependent on imports. In the case of the COVID-19 pandemic, it was forced to restrict travel and imports from abroad, but many countries were too late to do so. It can be rather said that the pandemic was brought about by the fact that countries had built a network of interdependent economies through globalism. On the contrary, standing on risk management and BCP, it should be paid much attention to a fact that the amount of enormous damage suffered by each country was embedded in the line of such supply chain of the global trades.

And thirdly, the Fukushima case, which was one of the worst accident cases of the nuclear power plants was caused by the Tohoku Earthquake at 11th, March, 2011 (NRC, 2011; Science Council of Japan, 2011; OECD/NEA, 2016; Shibuya, 2017a, 2018, 2020a, 2020b, 2021b). It was characterized tremendous damages by both natural and human-made disasters. Indeed, when the author studied this Fukushima problem, the author found that radioactive substances would take more than 10,000 years to reach its physical half-life, the problems of disposing of huge amounts of contaminated polluted water and radioactive substances, and an inevitable fact that the required processing after the nuclear accidents would take more than 45 years to complete. These are nothing less than *sacrifices* made by the egoism and greed of the last generation (and before generation), and the generation that created those problems would not be responsible for the sacrifices such as succeeding damages and recovery costs. In other words, the essence of the problem is that different persons undertakes either benefit or damage at different time, and it means the future generations must undertake only larger burden, efforts and costs as sacrifices.

Thus, in above all cases, many variables on the problems that cause the larger scale damages exist first, and these variables combine to generate further events and lastly become a *crisis*. On top of that, a number of potential and future risks should be concerned, and many sacrifices are expected to be made as a result. In other words, occurring serious negative events becomes the main cause from which many more *causes and results* are further linked resulting in various and enormous sacrifices. Of course, there may be occasions when the situation should be solved in the mathematical form of a polynomial or a simultaneous equation rather than a monomial and dealt with concurrently. Nevertheless, if the ruling government leaders don't make decisions quickly and effectively by clarifying the priorities to achieve the goal and discerning the appropriate sacrifices to be made, it will impossible to save anyone in the end. It is occasionally necessary to follow the utilitarian proverb "*The greatest happiness of the greatest number*" rather than the immediate sacrifice.

Therefore, the causal understanding of those involved in health care and public health must be rational. In particular, politicians have to make correct and quick decisions based on scientific evidence.

Significantly, the author is eagerly to deny "*solving the present problems by the seeds that becomes the future problems*". It intends to solve a problem that has already occurred with a risk factor that will become another problem in the future.

A familiar example is the case of a country that has issued another bond to repay its debt, but such policy has not reduced its outstanding debt at all. Such conditions clearly suggest that policy strategies and national governance have failed to handle own risks and its negative consequences, and further they permitted to grow from simple risk to larger crisis (Keech & Munger, 2015; Linnerooth-Bayer & Amendola, 2000).

Additionally, there is another problem of *under-determination*. It is a problematic condition for decision-making how should be made by the government leaders and citizens when the scientific evidence that cannot be determined even by the experts concerned. Such a situation could make the *crisis* more serious, and then this pandemic should be also crucially examined in detail in the future.

2.11 System Modeling and Improving Bottlenecks

As Clade-X project[7] and its epidemic simulation had shown, frequently, for the purpose of pursing to visualize above assessment, system modeling and computational engineering can be applied in the studies on such risky conditions and crises (Meadows et al., 2004). A perspective on system dynamics and complexity science also can contribute to understanding innumerable interactions among complicated and multiple factors in the world. Those complicated systems usually contain various *bottlenecks*, and it means negative constraints, which must be firstly detected and removed. For such purposes, it should try to improve such bottle-necks as constraints solving in operations researches by sacrificing other factors.

In the COVID-19 case, the bottlenecks (i.e., constraints) were various variables related to each national health care such as the number of hospitals, beds, medicines, testing tools, medical doctors, nurses and budgets. Moreover, the failure to meet demand for any one of these items led to even greater pressure against medical capacities. These constraints satisfaction problems appeared to have affected a country's performance in the COVID-19. In the case of the USA, initial bottlenecks were improved by sending military hospital ships and military medics, setting up field hospital facilities, and procuring other necessary resources. Furthermore, a presidential order forced companies to mass produce medical tools. These responses were probably because the USA had the principal staff with such a systemic understanding to settle bottle-necks.

Namely, system dynamics in computational modeling can be recognized those hidden negative factors. It can depict a system model based on feed-back loop and cybernetic systems. Causal pathways and correlational interactions on each crisis ought to be well recognized and examined in those contexts, because those visualized figures and pathways can envision us clear sacrifices.

---

[7] https://www.centerforhealthsecurity.org/our-work/events/2018_clade_x_exercise/.

3 Rule of Conduct of the Science and Technology at the Crisis

Political decision making must be also accountable for future critical investigations.

For example, after the Tohoku Disaster, natural scientists in Japan had been accused by the citizens for their optimism in nuclear energy (Shibuya, 2020a, 2021b). Similarly, the COVID-19 pandemic ought to be also examined by serious investigation. For all that, it is still prematurely to conclude that science and technology itself are not innocence, but it is too naturally that they only will be obligated to solve above emerged and serious matters.

Hence, it is necessary of the ethical rule of conduct to be shared among the members of the government as well scientists. Obviously, the role of scientists and academicians in the crisis and risk management (Pidgeon et al., 2003) had been also focused in the COVID-19 pandemic. In sociology of science, Merton (1968) already proposed his conceptual term CUDOS, namely *Communality, Universality, Disinterestedness* and *Organized Skepticism.*

In this concern, firstly, WHO and China government was criticized their lack of initial effective risk management for the global pandemic before the crisis. Not to mention, scientific committees on public health should be recalled the significance of both Disinterestedness and Organized Skepticism in prospective discussions. The rule of conduct among scientists, anti-hazard operations and risk management need to be critically reviewed whether national crisis or not. Eventually, scientists must be always devoting to strictly scientific governance in their community. These social roles shall be undertaken by professional academicians and researchers in terms of their majors standing on the CUDOS. Similarly, many scientists and scientific communities should solve and tackle problems through international cooperation.

Secondly, in many actual cases, why was it too late to manage against the epidemic progress so much? This is because, globally, elderly political leaders have proven to be a potential bottleneck in identifying the signs of infectious diseases and implementing accurate policies such as ordering preventive measures. Suggestions by CDC of the USA were ignored by the president, and tyranny through unscientific and naïve judgment was allowed to prevail. Such human errors by government leaders should be remedied after they are adequately addressed by evidence-based scientific scenario analysis as well as advice from human experts.

Certainly, as denoted a problem on *under determination* before, in the case of the COVID-19 pandemic, the Chinese government's information disclosure and initial response as well as the WHO's response shall be crucially investigated at the end of the pandemic.[8] Initially, their response was repeatedly to say "no problem", but their initial responses had resulted in more than 10 million people being infected and more than 500,000 dead worldwide (as of June, 2020). If the decision was made solely on the basis of scientific judgment, a serious warning could had been issued to the world at an earlier stage.

---

[8] Scientists call for pandemic investigations to focus on wildlife trade: https://www.nature.com/articles/d41586-020-02052-7.

Above all, the author would like to summarize above issues as following phrases. *No social systems exist without human matters, and no solutions will be properly conducted without the efforts by both professionalism (academician, governors and practitioners) and civic engagements.*

References

Ackeren, M., & Archer, A. (2018). Self-sacrifice and moral philosophy. *International Journal of Philosophical Studies, 26*(3), 301–307.

Ackeren, M., & Archer, A. (2020). *Sacrifice and moral philosophy*. Routledge.

Agrawal, A. (2020). Bridging digital health divides. *Science, 369*(6507), 1050–1052. https://doi.org/10.1126/science.abc9295

Allen, C. R., et al. (2016). Quantifying spatial resilience. *Journal of Applied Ecology, 53*, 625–635.

Anderson, N. H. (1996). *A functional theory of cognition*. LEA.

Anderson, J. E. (2011). The gravity model. *Annual Review of Economics, 3*(1), 133–160.

Andrews, F. M., & Withey, S. B. (Eds.) (2012). *Social indicators of Well-being: Americans' perceptions of life quality*. Plenum Press.

Arrow, K. J. (1963). *Social choice and individual values*. Yale University Press.

Aziz, H., & Mackenzie, S. (2020). A bounded and envy-free cake cutting algorithm. *Communications of the ACM, 63*(4), 119–126.

Bach, K., et al. (2020). Ten considerations for effectively managing the COVID-19 transition. *Nature Human Behaviour, 4*, 677–687.

Bord, R. D., Fisher, A., & O'Connor, R. E. (1997). *Is accurate understanding of global warming necessary to promote willingness to sacrifice?* https://scholars.unh.edu/cgi/viewcontent.cgi?article=1342&context=risk

Bostrom, N., & Ćirković, M. M. (Eds.) (2018). *Global catastrophic risks*. Oxford University Press.

Bostrom, N. (2013). Existential risk prevention as global priority. *Global Policy, 4*(1), 15–31.

Brader, T, Valentino, N. A., & Suhay, E. (2008). What triggers public opposition to immigration? Anxiety, group cues, and immigration threat. *American Journal of Political Science, 52*(4), 959–978.

Brandstedt, E., & Brännmark, J. (2020). Rawlsian constructivism: A practical guide to reflective equilibrium. *The Journal of Ethics*. https://doi.org/10.1007/s10892-020-09333-3

Brandt, F., et al. (2016). *Handbook of computational social choice*. Cambridge University Press.

Brännmark, J. (2019). The independence of medical ethics. *Medicine, Health Care and Philosophy, 22*, 5–15.

Buzby, A. (2016). Locking the borders: Exclusion in the theory and practice of immigration in America. *International Migration Review, 52*(1), 273–298. https://doi.org/10.1111/imre.12291

Camerer, C. F. (2003). *Behavioral game theory: Experiments in strategic interaction*. Princeton University Press.

CDC. (2013). *All-hazards preparedness guide*. https://www.cdc.gov/cpr/documents/AHPG_FINAL_March_2013.pdf

Cenci, A., & Cawthorne, D. (2020). Refining value sensitive design: A (capability-based) procedural ethics approach to technological design for well-being. *Science and Engineering Ethics*. https://doi.org/10.1007/s11948-020-00223-3

Culyer, A., & Wagstaff, A. (1993). Equity and equality in health and health care. *Journal of Health Economics, 12*(4), 431–457.

Diener, E., & Suh, E. (1997). Measuring quality of life: Economic, social, and subjective indicators. *Social Indicators Research, 40*, 189–216.

Durkheim, E. (2005). *The elementary forms of religious life (Oxford World's Classics)*. Oxford University Press.

Dutt, A. K. & Wilber, C. K. (2010). Fairness, distribution, and equality. In A. K. Dutt, & C. K. Wilber (Eds.), *Economics and ethics* (pp. 175–201). Palgrave Macmillan.

Dworkin, R. (2005). *Sovereign virtue: The theory and practice of equality.* Harvard University Press.

Erikson, H. (1959). *Identity and the life cycle.* International Universities Press.

Erikson, H. (1950). *Childhood and society.* Norton.

Erikson, H. (1982). *The life cycle completed.* Norton.

Escolà-Gascón, A. (2021). New techniques to measure lie detection using COVID-19 fake news and the Multivariable Multiaxial Suggestibility Inventory-2 (MMSI-2). *Computers in Human Behavior Reports, 3,* 100049. https://doi.org/10.1016/j.chbr.2020.100049

Fanelli, D. (2018). Is science really facing a reproducibility crisis, and do we need it to? *PNAS, 115*(11), 2628–2631.

Fierke, K. M. (2012). *Political self-sacrifice: Agency, body and emotion in international relations.* Cambridge University Press.

Fischhoff, B., et al. (1999). *Acceptable risk.* Cambridge University Press.

Gintis, H. (2014). *The bounds of reason: Game theory and the unification of the behavioral sciences* (Revised Ed.). Princeton University Press.

Guillaume, C., & Pochic, S. (2009). What would you sacrifice? Access to top management and the work–life balance. *Gender, Work & Organization, 16*(1), 14–36.

Habermas, J. (1976). *Legitimation crisis.* Heinemann.

Habermas, J. (1987). *Eine Art Schadensabwicklung.* Nachdruck.

Haug, N. et al. (2020). Ranking the effectiveness of worldwide COVID-19 government interventions. *Nature Human Behaviour, (*4), 1303–1312. https://doi.org/10.1038/s41562-020-01009-0

Hayashi, T., & Lombardi, M. (2019). Fair social decision under uncertainty and belief disagreements. *Economic Theory, 67,* 775–816.

Heidegger, M. (2008). *Sein und Zeit* (English translated edition). Harper Perennial Modern Classics.

Helbing, D. (2013). Globally networked risks and how to respond. *Nature, 497,* 51–59.

Hubert, H., & Mauss, M. (1981). *Sacrifice: Its nature and functions.* University of Chicago Press.

Jacki, T., & Pamela, D. (Eds.). (2020). *Victimology: Research.* Palgrave Macmillan.

Jones, R. A. (1981). Robertson Smith Durkheim, and sacrifice: An historical context for the elementary forms of the religious life. *The History of the Behavioral Sciences, 17*(2), 184–205.

Joseph, J., & Jergenson, S. (Eds.) (2020). *An international perspective on contemporary developments in victimology.* Springer.

Josephson, A., Kilic, T., & Michler, J.D. (2021). Socioeconomic impacts of COVID-19 in low-income countries. *Nature Human Behaviour.* https://www.nature.com/articles/s41562-021-01096-7

Kahanec, M., & Zimmermann, K. F. (2016). *Labor migration, EU enlargement, and the great recession.* Springer.

Kahneman, D., & Tversky, A. (2000). *Choices, values, and frame.* Cambridge University Press.

Kawachi, I. et al. (Eds.) (2008). *Social capital and health.* Springer.

Keech, W. R., & Munger, M. C. (2015). The anatomy of government failure. *Public Choice, 164,* 1–42.

Ksiazek, T. G., et al. (2003). A novel coronavirus associated with severe acute respiratory syndrome. *The New England Journal of Medicine, 348*(20), 1953–1966.

Linnerooth-Bayer, J., & Amendola, A. (2000). Global change, natural disasters and loss-sharing: Issues of efficiency and equity. *The Geneva Papers on Risk and Insurance—Issues and Practice, 25,* 203–219.

Luhan, W. J., Poulsen, O., & Roos, M. W. W. (2019). Money or morality: fairness ideals in unstructured bargaining. *Social Choice and Welfare, 53,* 655–675.

Luhman, N. (2002). *Risk: A sociological theory.* Aldine Transaction.

Luhman, N. (1984). *Social systems.* Stanford University Press.

Meadows, D. H. et al. (2004). *The limits to growth.* Chelsea Green.

Merton, R. K. (1968). *Social theory and social structure* (1968 Enlarged Edition). Free Press.

Mizruchi, S. L. (1998). *The science of sacrifice: American literature and modern social theory.* Princeton University Press.

Murphy, A. H. (1969). On expected-utility measures in cost-loss ratio decision situations. *Journal of Applied Meteorology, 8*(6), 989–991.

Neff, S. C. (2012). *Hugo Grotius on the law of war and peace.* Cambridge University Press.

Niesiobędzka, M., & Kołodziej, S. (2020). The fair process effect in taxation: The roles of procedural fairness, outcome favorability and outcome fairness in the acceptance of tax authority decisions. *Current Psychology, 39*, 246–253.

NRC (Nuclear Regulatory Commission, U.S.A.). (2011). *The near-term task force review of insights from the Fukushima Dai-Ichi accident.* http://pbadupws.nrc.gov/docs/ML1118/ML111861807. pdf

OECD/NEA. (2016). *Five years after the Fukushima Daiichi accident: Nuclear safety improvements and lessons learnt.* https://www.oecdnea.org/nsd/pubs/2016/7284-five-years-fukushima.pdf

Pidgeon, N., Kasperson, R. E., & Slovic, P. (2003). *The social amplification of risk.* Cambridge University Press.

Rawls, J. (1971). *A theory of justice.* Harvard University Press.

Rawls, J. (2001). *Justice as fairness: A restatement.* Belknap Press.

Rodiguez-Pose, A., & Berlepsch, V. (2015) European migration, national origin and long-term economic development in the United States. *Economic Geography, 91*(4), 393–424.

Roeser, S, Hillerbrand, R., Sandin, P., & Peterson, M. (Eds.) (2012). *Handbook of risk theory: Epistemology, decision theory, ethics, and social implications of risk.* Springer.

Roth, A. E. (Ed.) (1988). *The shapley value: Essays in Honor of Lloyd S. Shapley.* Cambridge University Press.

Roth, A. E. (2008). Deferred acceptance algorithms: History, theory, practice, and open questions. *International Journal of Game Theory, 36*, 537–569.

Rothe, J. (2016). *Economics and computation: An introduction to algorithmic game theory, computational social choice, and fair division.* Springer.

Saaty, T. L. (2013). *Theory and applications of the analytic network process: Decision making with benefits, opportunities, costs, and risks.* RWS Publications.

Saltelli, A. et al. (2020). Five ways to ensure that models serve society: a manifesto. *Nature.* https://www.nature.com/articles/d41586-020-01812-9

Science Council of Japan (SCJ). (2011). *Report to the foreign academies from science council of Japan on the Fukushima Daiichi Nuclear Power Plant accident.* http://www.scj.go.jp/en/report/houkoku-110502-7.pdf

Scott, J., & Marshall, G. (Eds.) (2014). *A dictionary of sociology* (4th Ed.). Oxford University Press.

Sen, A. (1970). *Collective choice and social welfare.* Harvard University Press.

Shah, S. K., et al. (2020). Ethics of controlled human infection to address COVID-19. *Science, 368*(6493), 832–834. https://doi.org/10.1126/science.abc1076

Shapley, L. S., & Shubik, M. (1954). A method for evaluating the distribution of power in a committee system. *American Political Science Review, 48*, 787–792.

Shibuya, K. (2006). Actualities of social representation: simulation on diffusion processes of SARS representation. In C. van Dijkum, J. Blasius, & C. Durand (Eds.), *Recent developments and applications in social research methodology.* Barbara Budrich-verlag.

Shibuya, K. (2017b). Bridging between cyber politics and collective dynamics of social movement. In M. Khosrow-Pour (Ed.), *Encyclopedia of information science and technology* (4th Ed., pp. 3538–3548). IGI Global.

Shibuya, K. (2017a). An exploring study on networked market disruption and resilience. In *KAKENHI report (no. 26590105)* (pp. 1–200) (in Japanese).

Shibuya, K. (2018). A design of Fukushima simulation. In *The society for risk analysis, Asia conference 2018*

Shibuya, K. (2020b). *Identity health.* https://www.ncbi.nlm.nih.gov/pmc/articles/PMC7121317/. In K. Shibuya (Ed.). *Digital transformation of identity in the age of artificial intelligence.* Springer.

Shibuya, K. (2021a). Breaking fake news and verifying truth. In M. Khosrow-Pour (Eds.), *Encyclopedia of information science and technology* (5th Ed., pp.1469–1480). IGI Global.

Shibuya, K. (2021b). A risk management on demographic mobility of evacuees in disaster. In M. Khosrow-Pour (Eds.), *Encyclopedia of information science and technology* (5th,Ed., pp. 1612–1622). IGI Global.

Shibuya, K. (2013). Risk communication on genetically modified organisms. *Oukan, 7*(2), 125–128. (in Japanese).

Shibuya, K. (2020a). *Digital transformation of identity in the age of artificial intelligence.* Springer.

Sober, E. (2008). *Evidence and evolution: The logic behind the science.* Cambridge University Press.

Subbaraman, N. (2020). How to address the coronavirus's outsized toll on people of colour. *Nature.* https://doi.org/10.1038/d41586-020-01470-x

Thaler, T., Zischg, A., Keiler, M., & Fuchs, S. (2018). Allocation of risk and benefits–distributional justices in mountain hazard management. *Regional Environmental Change, 18*, 353–365.

Thorp, H. H., et al. (2020). Both/and problem in an either/or world. *Science, 368*(6492), 681. https://doi.org/10.1126/science.abc6859

UNISDR. (2018). *Disaster displacement: How to reduce risk, address impacts and strengthen resilience.* https://www.unisdr.org/files/58821_disasterdisplacement05a.pdf

United Nations. (2016). *Protecting humanity from future health crises: Report of the high level panel on the global response to health crises.* https://digitallibrary.un.org/record/822489?ln=en

USA (the White House). (2020). *United States strategic approach to the people's republic of China.* https://www.whitehouse.gov/wp-content/uploads/2020/05/U.S.-Strategic-Approach-to-The-Peoples-Republic-of-China-Report-5.20.20.pdf

Verweij, M. (2009). Moral principles for allocating scarce medical resources in an influenza pandemic. *Journal of Bioethical Inquiry, 6*, 159–169.

Walzer, W. (2006). *Just and unjust wars: A moral argument with historical illustrations* (4th Ed.). Basic Books.

Zack, N. (2009). The ethics of disaster planning: Preparation vs response. *Philosophy of Management, 8*, 55–66.

COVID-19 Crisis

A Crisis of COVID-19 and Its Sacrifices

1 A Crisis to Confront with the COVID-19 Pandemic

1.1 WHO Charter, International Health Regulations, and Pandemic Declaration

A cooperative system has been in place for a long time to deal with the risk of infectious diseases that are spreading worldwide regardless of the constraints of national borders or race. This is a typical example of the need to mobilize international cooperation and the best of science and technology to confront a disaster that has shaken the very foundations of human history.

First of all, the preamble of the WHO (World Health Organization) Constitution[1] reads as follows.

> THE STATES Parties to this Constitution declare, in conformity with the Charter of the United Nations, that the following principles are basic to the happiness, harmonious relations and security of all peoples:
>
> Health is a state of complete physical, mental and social well-being and not merely the absence of disease or infirmity.
>
> The enjoyment of the highest attainable standard of health is one of the fundamental rights of every human being without distinction of race, religion, political belief, economic or social condition.
>
> The health of all peoples is fundamental to the attainment of peace and security and is dependent upon the fullest co-operation of individuals and States.
>
> ....

Similarly, there is the International Health Regulations (IHR[2]) set by the WHO. These regulations used to cover only yellow fever, cholera, and plague, but were revised extensively in 2005 due to the inability to deal with health crises caused by

---

[1] https://www.who.int/governance/eb/who_constitution_en.pdf.

[2] https://www.who.int/health-topics/international-health-regulations.

K. Shibuya, *The Rise of Artificial Intelligence and Big Data in Pandemic Society*,
https://doi.org/10.1007/978-981-19-0950-4_3

emerging and re-emerging infectious diseases such as SARS and avian influenza, deficiencies in cooperation between WHO and other countries, and the need to strengthen countermeasures against terrorism (e.g., bioterrorism), which has finally been recognized as a real threat. A major revision was made in 2005.

This regulation requires the Director-General to certify international countermeasures against infectious diseases by examining the following points: [1] the seriousness of the situation, [2] whether the situation is unpredictable or unusual, [3] the risk of international spread, and [4] risk factors that impede travel and international trade. Under the current regulations, the WHO can make recommendations to member countries on entry and exit restrictions as appropriate, but it does not have the power to enforce them.

Based on this assessment of the risk and crisis of infectious diseases, countries and international organizations work together to deal with the situation. However, in the actual case of the COVID-19 pandemic in 2020 or later, although the Public Health Emergency of International Concern (PHEIC) was issued in January 2020, there was a noticeable attitude of concern towards China, and finally, on 11th March 2020, the COVID-19 pandemic was announced. Although a "pandemic declaration" was issued on March 11, 2020, the WHO (and the Chinese government) had been accused of being slow in their initial response.

1.2   Outbreak

Since an initial stage that discovered an epidemic symptom, in an official address, the leaders of China and WHO had to clearly announce that it was imperative to eradicate this disease. But they were reluctantly to recognize such situation as a pandemic or becoming a pandemic in global.

At 11th March 2020, the WHO lastly declared such global situation as a *pandemic*, and alerted this crisis and its terrible facts all over the world (Allam, 2020). But at that time, nobody believe that they were trustful and provide reliable information resources for ordinary citizens in global. Such consequences further called economic decay in global, and provoked an extraordinary and enforced daily living ways among the citizens. Namely, it multiplied the crises such as an epidemic risk itself as well as political, economics and socio-cultural facets.

Of course, as of 2021, final data on the negative impacts in markets and global economy caused by the COVID-19 pandemic as well as the number of the infected patients and deaths have not been determined (further, the total number of deaths reported by countries is estimated to be too low[3]). Therefore, it is not time to assess all

---

[3] The pandemic's true death toll: millions more than official counts: https://www.nature.com/articles/d41586-022-00104-8.

sacrifices. As of 2021, the COVID-19 *pandemic* is likely to become an *"endemic*[4]*"* (a recurring infection over a long period of time), and we will live in such a time (Shibuya, 2021). However, the fact that the situation has already become so much worse than initial optimism, and the future transformation of the world condition and our daily life have been adapting in the new normal standard. Therefore, multifaceted social scientific examinations on these issues needs to be proceeded in advance.

1.3 Causes of Infectious Diseases

There are several types of epidemic infections such as viruses, microbes, bacteria, and substances (e.g., prions). Especially, virus and microbes, and bacteria have own gene (RNA or DNA). Infectious diseases caused by viruses are ordinarily spread through each host (i.e., humankind or zoological species). The same is true for zoonotic diseases as well as the usual influenza and COVID-19. In the case of microorganisms and bacteria, they similarly require water, air, and others to transmit them. Here, it is important to note that the virus is only a capsule entity containing own RNA. In addition, it is also believed that the origin of the mitochondria in our living cells was a bacterium that entered from the outside and formed a symbiotic relationship with us. If the virus wipes out its all hosts, the virus itself will also become extinct. Therefore, the more rational solution is to have a symbiotic relationship, even if the virus infects more than a certain scale. That is, it is possible to reach mathematical equilibrium (i.e., symbiosis), but those complex systems (e.g., replicator dynamics) are likely to fluctuate and drift in real time (Fu et al., 2010; Gubar & Zhu, 2013). Namely, examining human genes and their relationship to pathogens may reveal some aspects of the history of human survival.

1.4 Symptom and Diagnosis of COVID-19

Principally, the COVID-19 was identified in a family of the corona viruses by genetic analysis of bio-informatics. An original patient seemed to be initially found in Wuhan city,[5] China at the end of 2019 (Zhu et al., 2020). SARS (Ksiazek et al., 2003; Shibuya, 2006), MERS (Nishiura et al., 2016) and the COVID-19 mutually resemble in each gene sequence, and its mutated derivations came from the unidentified wild animals. Our immunity system has been accustomed for typical corona virus itself, but many of the humankind have not immunized against the COVID-19 virus yet. Clinical cases of corona virus family can cause the humanity to be harmed severe respiratory

---

[4] The coronavirus is here to stay—here's what that means: https://www.nature.com/articles/d41586-021-00396-2.

[5] Dissecting the early COVID-19 cases in Wuhan: https://www.science.org/doi/10.1126/science.abm4454.

disease (Zhang et al., 2020), and probability of the death was generally estimated around more than several percent.

As of 2021, scientific evidences on the COVID-19 have been still ongoing matters (Ellinghaus et al., 2020; Guan et al., 2020). First, a theory that the biological weapons were artificially developed in the laboratory in Wuhan City was mainly argued by the USA, but there is not enough corroboration. Secondly, as there is no clear reason for the extremely low number of infections and deaths in Asian countries, it has been argued that a cross-immunity theory, which suggests that the immunity already gained when infected with the similar diseases of the COVID-19 virus in Asian countries. This is not well endorsed. Thirdly, a hypothesis was said that older people and those with pre-existing lung conditions were more likely to be severely affected, and of course, younger people were also observed to be infected. Fourth, the BCG (Bacillus Calmette-Guerin) vaccine was initially said to be effective, and the gene sequences of Asian peoples was said to be highly resistant to the COVID-19 virus, but no definite conclusion has been reached in both hypotheses. And fifth, because the COVID-19 virus is categorized as a type of RNA virus, it mutates quickly and was thought to have spread in Europe and the USA (Holshue et al., 2020), but the details of this process have not fully revealed yet.

On the other hand, cultural and institutional differences between Europe and Asia countries were also pointed out in the infection situations. But they were not sufficient evidences for explaining the disease and contagions. Diverse variables and detailed knowledge of the nature of the virus, droplet transmission and the process of infection in the body as well as medical instruments were quite essential (Prather et al., 2020).

In particular, at least, it could be deduced that the imbalance of infection rate among global citizens was attributable to certain socio-economic factors, mingled condition in the closed space, and quality of public health in each nation. There were the needs of appropriate, effective and timely actions for both economic and health policy at the national policy level (Weible et al., 2020). Better judgment by the government leader was still required. Differences in complex factors (such as latent variables that were still unidentified) including above them made a difference in the infection situation.

Then, researches have been conducted around the world to understand the mechanism of COVID-19 virus infection and the severity of the disease[6]. In addition to acute respiratory failure due to worsening pneumonia, it has been found that there are cases in which blood vessels become clogged due to the formation of blood clots (high risk of death due to high levels of a measure called a "*D-dimer*") and cases in which many organs fail due to an immune overload (i.e., "*cytokine storm*"). It has also been noted that there is a similarity with the symptoms of Kawasaki disease, but clinical evidence has not been endorsed enough yet.

For the purposes of distributing medicines and vaccines, globally drug manufacturers have been rushing to develop effective vaccines and treatments (Corey et al., 2020; Slifka & Amanna, 2019). DNA vaccines have shown promise, but is still now

---

[6] The quest to find genes that drive severe COVID: https://www.nature.com/articles/d41586-021-01827-w

in the process of research and development, and clinical trials (Yu et al., 2020). Each drug as a chemical substance can be referred to each database. A part of drug candidates for the COVID-19 virus are listed below. CAS[7] is an abbreviation, and it means that American Chemical Society serves Chemical Abstracts Service register number. And similarly, KEGG (Kyoto Encyclopedia of Genes and Genomes) is one of the most significant databases of bioinformatics such as genes and protein information.

Candidates of Antiviral Drug (for example, as of June, 2020):

- Remdesivir[8] (Goldman et al., 2020): CAS number:1809249-37-3, KEGG[9]: D11472
  [May 7, 2020: Approved in Japan. Review by Pharmaceuticals and Medical Devices Agency (PMDA[10])].
- Favipiravir (Avigan): CAS number:259793-96-9, KEGG: D09537
  (As of Feb. 3, 2021: Not approved in Japan. Clear efficacy against COVID-19 has not been confirmed).
  Vaccination (As of May 21, 2021).
  Start of vaccination (a part of vaccines).
  Pfizer (USA): mRNA vaccine (Feb. 14, 2021: Approved in Japan. Vaccination was permitted).
  AstraZeneca (UK): virus vector vaccine (May 21, 2021: Approved in Japan. Vaccination was permitted).
  Moderna (USA): mRNA vaccine (May 21, 2021: Approved in Japan. Vaccination was permitted).
  Biontech (Germany): mRNA vaccine.
  Sinopharm (China): Inactivated vaccine.
  And many others.

Vaccines from Pfizer, Biontech, and Moderna have all shown a preventive effect of around 95% in clinical trials involving tens of thousands of people. With regard to the Pfizer vaccine, the results of clinical trials in the USA and other countries where vaccinations were conducted earlier pointed out the risk of anaphylaxis occurring in about 1 in 100,000 people, but it is understood that the benefits greatly outweigh the risks. Besides, another vaccine collaborated by AstraZeneca and the University of Oxford has also been demonstrated to be equally effective (Voysey et al., 2021). Websites have also been established in many countries to explain risk communication in vaccination (e.g., Japan[11]).

DNA Vaccination (As of January 19, 2021):

AnGes (Japan): Conducting research and development for clinical trials of DNA vaccines.

---

[7] https://www.cas.org/.

[8] https://www.pmda.go.jp/PmdaSearch/iyakuDetail/GeneralList/62504A3.

[9] https://www.genome.jp/kegg/.

[10] https://www.pmda.go.jp/index.html.

[11] https://www.kantei.go.jp/jp/headline/kansensho/vaccine.html.

Inovio (USA): Conducting research and development for clinical trials of DNA vaccines.

And many others.

However, because the drug itself is indeed one of the subjects in the risk study, the tougher examination requires in the safety assessment. Even normally, the development of the drugs takes more than 10 years to explore drug candidates, try compounds, evaluate the safety of chemicals, and examine the effects of clinical trials and side effects and approval by the national authority. In this pandemic, therefore, the effects of taking existing drugs were being tested first, while the development and testing of new vaccines and therapeutics was recommended in a hurry. Fortunately, by early 2021, several vaccines have been approved and vaccinations have begun in Israel, the USA, the UK and other countries (In Japan, the vaccine manufactured by Pfizer was granted special approval on February 12, 2021, and vaccination was available after February 17, 2021).

Next, in general, how much does treatment for the COVID-19 case cost? In Japan, there is no co-payment for the COVID-19 case, because of one of the designated infectious diseases, both the test and treatment are paid by the nation.[12] If patients were to be borne by themselves, a huge amount of money is likely to be spent. It estimates approximately 20,000 yen (200 US dollars) at least, and an antigen test costs approximately 6000 yen (56 US dollars) with insurance points, but it can cost even more. If a patient is admitted to the hospital, it will have to be isolated, and in addition to the cost of medication, ICU (intensive care unit) costs for severe cases can be quite high. If each individual has own medical insurance and necessary investments in advance, it may be able to make up such expenditures.

1.5   Statistics

According to a report by Johns Hopkins University in the USA, the number of people worldwide who have been confirmed to be infected with the new coronavirus has exceeded 100 million (they estimated 100,063,707 infected people), as of Japan time on the 27th, January 2021. In addition, the death toll is 2.15 million, and the situation remains unpredictable even one year ever since the beginning of the pandemic.

As monitoring statistics of WHO[13] and some public data,[14] this section summarizes the "number of infected" and "number of deaths" and "number of infected and dead per population (per million)" of the COVID-19 as of 2020. Generally, the result of "number of infected and dead per population (per million)" clearly shows that New Zealand, Australia, Japan and much of Asian countries are lower than the USA and Brazil, which had suffered more (Table 1).

---

[12] https://www.mhlw.go.jp/content/000604470.pdf.

[13] https://covid19.who.int/

[14] https://graphics.reuters.com/CHINA-HEALTH-MAP-LJA/0100B5FZ3S1/index.html.

Table 1 A part of global statistics on COVID-19 (as of 30th, June 2020)

30th June 2020	Total population	Total confirmed infections	Total death	Territorial area (km²)	Total infection per Million	Total death per Million	Population density
USA	329,064,900	2,537,636	126,203	9,859,878	7711.7	383.5	33.4
China	1,441,860,300	85,227	4,648	9,753,599	59.1	3.2	147.8
Japan	126,860,300	18,593	972	377,975	146.6	7.7	335.6
UK	68,045,800	311,969	43,575	242,495	4584.7	640.4	280.6
France	67,978,400	156,930	29,730	551,500	2308.5	437.3	123.3
Germany	83,517,000	194,259	8,973	357,578	2326.0	107.4	233.6
Italy	60,550,100	240,436	34,744	301,336	3970.9	573.8	200.9
South Korea	51,225,300	12,800	282	100,033	249.9	5.5	512.1
Belgium	11,539,300	61,427	9,747	30,528	5323.3	844.7	378.0
Spain	46,736,800	248,970	28,346	505,992	5327.1	606.5	92.4
NZ	4,803,500	1,178	22	270,467	245.2	4.6	17.8
Australia	25,203,200	7,767	104	7,899,987	308.2	4.1	3.2
Brazil	211,049,500	1,344,143	57,622	8,515,767	6368.9	273.0	24.8
Russia	145,872,300	647,849	9,320	17,124,553	4441.2	63.9	8.5
India	1,366,417,800	566,840	16,893	3,287,263	414.8	12.4	415.7

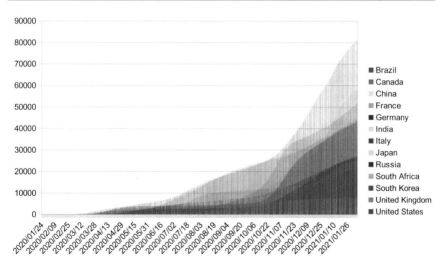

Fig. 1 Number of new cases of COVID-19 per million population in each country (as of February 3, 2021)

Table 2 Data of each epidemic disease

	Infection	Death rate
COVID-19 (data of the end of Jan., 2021)	More than 100 million people infected	More than 2,150,000 dead
SARS (2002–2003)	More than 8000 people infected	10%
MERS (2012)	More than 2494 people infected	34%
Ebola (2014–2016)	More than 28,616 people infected	40%
Spanish Flu (1918–1920)	More than 500 million people infected	17 million to 50 million
HIV	At 2016, new infection cases were approximately two million	At 2016, more than a million
Malaria	At 2015, more than 214 million cases	At 2015, more than 438,000 deaths were estimated
Tuberculosis	At 2016, more than 10.4 million cases	At 2016, more than 1.7 million deaths were estimated
Smallpox (1980: WHO certified the global eradication of the disease)	It was uncountable since the beginning of the human history	Bleeding: 100% Normal: 25–50% Mild (vaccines): less than 10%

Data from WHO

In Fig. 1, it shows the number of new infections per million population as of February 3, 2021. Of the data for major countries only, it can be seen that the USA (approximately 81,317), the UK (approximately 58,057), France (approximately 51,724), and Brazil (approximately 44,444) had the world's worst levels of infections per million people. In contrast, Japan (approximately 3195) and China (approximately 70) remained in a low state. As mentioned above, there were many theories as to the cause of such difference, and it is not still clear.

Next, regarding Table 2, human history[15] could be certainly traced as somehow endlessly survival process from lethal diseases, and it means that human history had been inscribing into the record by survivors in each time (McNeill, 1976). Each severity of the epidemic diseases should be focused. How dangerous is the COVID-19 more than other epidemics[16]? Especially, the COVID-19 appears to be characterized

---

[15] Smallpox and other viruses plagued humans much earlier than suspected. https://www.nature.com/articles/d41586-020-02083-0.

[16] Mallapaty, S (2020) How deadly is the coronavirus? Scientists are close to an answer, Nature, https://www.nature.com/articles/d41586-020-01738-2.

by its relatively high infectiousness, its susceptibility to complications, and its prominence to death in a short period of time when the disease becomes severe. The disease spread from Asia to Europe and further to the USA, because RNA virus is usually very mutable and have characteristics that make it more severe. However, there are many details of COVID-19 that has not been known yet. Therefore, the severity of this virus should have been considered higher and earlier responses in each nation and WHO.

However, the majority of infections of the COVID-19 did not result in serious injury, and a small number of cases had been observed to worsen and result in death. For example, Lavezzo et al. (2020) had examined statistics such as medical capacities and infections from contagion to suppression process in the Italian rural city. Furthermore, in global, the actual number of deaths in large cities and rural areas of developing countries was not fully known.[17]

1.6 Number of Hospitals, Beds, Doctors, and the Status as a System

According to OECD data,[18] furthermore, an index of *hospital beds* (Total per 1000 inhabitants, 2018 or latest available) indicates that a top of the nations is Japan (13.1, at 2017) and second place is South Korea (12.3, at 2017). And successively, Germany (8.0), France (6.0), Italy (3.2), the USA (2.8) and the UK (2.5). In such point, these statistic data appear ordinary beds as total capacity, but it clearly shows the strength of each nation for medical and public health before the pandemic.

Next, and index of *medical doctors*[19] per 1000 inhabitants indicates that top of the nations is Austria (5.18), Germany (4.25), Italy (3.99), Spain (3.8),France (3.37), Belgium (3.08), the UK (2.85), the USA (2.61), Japan (2.43),South Korea (2.34) and others. It seems ironic that countries with more deaths have more medical doctors than countries with less deaths. It's not just the number of doctors, it should be laid weight on the overall medical efficiency and support system of the whole country, and the public health seems to be too important in ordinary. Of course, the COVID-19 required more specific medical devices and instruments (e.g., ventilator), but it seemed that such differences have apparently strong evidences to explain total amount of infection and death among the nations.

Similarly, according to a comprehensive report that conducted by the SDG index (2020), the most effective nations that were prevented from the pandemic were respectively enumerated South Korea, Latvia, Australia, Lithuania, Estonia, Japan, Slovenia, Slovak Republic, New Zealand, Norway and the others. As of 2020, this report said that the top developed countries in the OECD such as the USA and Western Europe had suffered significant economic damage as well as mortality rates, while

---

[17] https://www.nature.com/articles/d41586-020-02497-w.

[18] https://data.oecd.org/healtheqt/, https://data.oecd.org/healtheqt/hospital-beds.htm.

[19] https://data.oecd.org/healthres/doctors.htm.

smaller countries in Asia and Europe including Japan and South Korea were able to contain the damage to a lesser extent. Those nations almost relatively succeeded their policies for prevention control, public health and economic regulations.

However, there was something to keep in mind. In the case of Japan, national and public hospitals account for about 20% of the total number of beds in Japan, and the remaining 80% is the figure for small hospitals run by medical doctors. Because of the large number of them, the number of doctors who can provide full-time care for the COVID-19 patients became insufficient, and during the third peak in late 2020 and early 2021, it became overwhelmed, resulting in a large number of people receiving home treatment (more than 30,000 people nationwide). It was reported that some of them died without receiving adequate medical care. Therefore, it is important to consider not only the numerical status of the medical system, but also whether it has been designed to function as a system. The pandemic revealed the degree of effectiveness of this function. The inadequacy of the system certainly imposed heavy sacrifices on the medical personnel who responded to the corona, not only for more than two years, but also for the many infected, seriously ill, and dead citizens. This also resulted in a stagnation of economic activities, a decline in GDP, and the bankruptcy of many small and medium-sized enterprises.

1.7 What Are Definitive Factors of the COVID-19?

What are the major causes of the spread of COVID-19 virus? As of 2021, the causes of COVID-19 and the detailed process of its spread are still under verification. First, it should be verified whether the cause of the worsening of the pandemic is anthropogenic or not. Secondly, which did factors interact with each other? Candidate variables are, at least, viral mutant, human errors, each government leader's failures, globalism, international conflicts before the pandemic (e.g., the USA and China), a degree of economic inequality among the citizens, population density, urbanization, the national border landlocked or not, the number of immigrants, public health policy and its quality, and accessibility for health insurance (Bempong et al., 2019). And consequentially, principal variables of the cause that invoked this pandemic should be more crucially investigated by scientific manners (Haug et al., 2020; Hsiang et al., 2020). Further, it should clarify how complicated dynamics multiplied by such risk factors finally induced sacrifices as a result.

Firstly, since the population heterogeneity increases the variability in the process of acquiring immunity, by mathematical models (Britton et al., 2020), the homogeneous acquisition of immunity with a target of 60% or more noted to be difficultly. Of course, statistical assumptions (e.g., BRR (Basic Reproduction Rate), contact probability) will lead to different results for the estimates, so that this viewpoint should be regarded as a single indicator. Additionally, the actual cases based on key variables such as population distribution, contact patterns, symptom severity, quality of care and capacity of health care institutions have been already examining data on infection and deaths by age and generation (Walker et al., 2020). These findings are only valid in countries where the infection has not reached an advanced stage yet, and thus immediate action is still required.

Further, even after the pandemic in early 2020, COVID-19 had not easily ended, and many people around the world had been repeatedly re-infected. One of the reasons is that the COVID-19 is highly infectious, this virus mutates frequently (with a short period of our effective immunity) and is not highly lethal, so then, it will be taking longer to terminate. On the other hand, in the case of infections that are as same as COVID-19 but as more highly lethal, the disease will gradually decrease, because people who are infected are all dying. This is because the infected person dies and there are no more people left to infect. Therefore, the COVID-19 pandemic does not easily end because it is less lethal (Hence, it tends to be called an *"endemic"* condition).

Some big-data and useful contents online on the statistics of the COVID-19 have provided (please also see "further references" of the end of this book). But each country has not opened own raw data to do statistically multivariate analysis (perhaps, the reason is not to open data of infected persons' privacy such race, gender, and other personal profiles). Alternatively, the author hypothetically laid weight on below factors.

- *Population Density in specific area* (e.g., city, inner-building, etc.): especially, urban area and limited spaces have potentials to infect more seriously around the mingled citizens.
- *Critical Physical Contacts with Infected Persons*: when someone has critically each history to contact with infected persons, its infected possibility will raise.
- *Viral mutability*: an EU-type and other variant types (as a mutated type of a RNA virus) that could lead to more severe damages (variant of concern[20]: e.g., alpha (the UK), beta (South Africa), gamma (Brazil), delta (India), lambda (Peru), omicron (South Africa), etc.).
- *Ease of crossing the national borders*: it is likely that the more landlocked a country is, the more likely it is to spread. Island territorial are less likely to spread (e.g., Japan, Taiwan, New Zealand, Australia). The UK is connected to the EU by a tunnel in the Strait of Dover, and then it's a nearly landlocked country. If such nation perfectly abandons both aircraft and ship navigation, the island nation is equal to an entire nation on lockdown. Since mobility restrictions of the citizens encourage the spread of infection between multiple cities in the country as well, using ICT and GIS data to monitor the behavior of citizens, each nation had proceeded its policies.
- *Initial measures of each government*: whether effective and scientific based measures have been taken promptly and effectively by the ruling government leader or not.
- *A Degree of Economic Inequality*: the view is to consider an economic status among the citizens, and higher degree of economic based stratification (or social

---

[20] https://www.who.int/en/activities/tracking-SARS-CoV-2-variants/.

class) ought to be unjustified in term of equability for achievement of each individual in democratic society.

- *Number and density of immigrants and low-wage workers*: Many of the nations are reluctant to serve them well public health. Larger numbers of immigrants, refugees, and undocumented foreign workers have been staying in poor health condition. And there are the cases that some of the racial prejudices also has been constraining its preventing such supports. Was it able to provide both necessary treatments of public health and daily living status for them comparing with equal level of the ordinary citizens?

- *Public health and welfare policies*: Is the public system providing equal access and treatment of health care for everyone (including immigrants, low-wage workers, and homeless peoples)? For example, according to Fair and Johnson (2021), during the COVID-19 pandemic, it has been found to be unfair to certain racial groups worldwide. In particular, a study of more than 17 million adult patients in the United Kingdom concluded that the risk of death from COVID-19 among blacks and Asians was about twice that of whites, likely due to racial discrimination and inequity in access to various health care resources and opportunities.

- *Public health awareness and behavior*: Before the pandemic, whether ordinary citizens were wearing masks or not, whether they were eagerly to be clean or not, and sustaining the quality of their living environment (e.g., air, water, house, city) (Chernozhukov et al., 2021).

A part of those statistics is accumulated by OECD,[21] for example. Especially, labor migration (ADB Institute, 2021), the high number of collective immigrants, illegal undocumented workers, low-wage foreign workers and homeless, the high population density in urban areas, the constant poverty problem, and the lack of public health and medical services for these people can lead to epidemic clusters in urban areas, which can lead to further contagion in the megalopolis. If proper treatments are not accessible for them, some of them might easily become super-spreaders in the urban area.

Namely, dependent variable of *world infection rank* was obviously classified by the parameters such as ease of crossing national borders, number of immigrants and low-wage foreign workers, economic inequality in each nation, and public health for the people. And then, the results seem to be likely to interpret that lower data of Japan' total death can be mainly depended on the national border (the isolated islands) and higher quality of public health policy, whereas higher data of the USA's total death may be relied on larger population of immigrants, lower-wage workers and undocumented workers (it also suggests mobility frequency of the peoples crossing the borders), economic inequality, and poor policy of public health.

---

[21] https://www.oecd.org/els/mig/keystat.htm, https://www.oecd.org/tokyo/statistics/.

Actually, at 5th July, 2020, New York Times[22] published the evidence in the USA, which endorsed above results by the author. They reported that black and Hispanic people have more triple vulnerability against infecting the COVID-19 than white citizens. And it was observed that elderly people have also sensibility of infection.

1.8   What Corona Brought?

1.8.1   General

This emerging corona virus was a threat to the survival of the humanity, an irrevocable loss about global economics, and a catalyst for various social changes and international conflicts. Around the world, the COVID-19 pandemic has overhauled traditional production processes, economic transactions, administrative decision-making, health care, and education styles. This infection has also damaged the traditional economic system that depends on physical contact and population density that constitutes human society, and has transformed human behavior to the behavioral principles firmly based on the digital transformed one. It can be said that it proceeds the clock of the human history in the direction of expanding the benefits of the AI driven services and big data through digital transformation.

It should also reconsider the world structure since the post-Cold War that had been built around the framework of international cooperation which leading by the United Nations. However, due to the COVID-19 pandemic, it has allowed the rise of China while making cooperation between Western countries led by the USA vulnerable. The COVID-19 pandemic, from which originated in China, was suggested to be a sign of the coming of the age for China hegemony. The vigilance against the world structure that divides the world by the two great powers between the USA and China, namely, Westness and Eastness, will bring about a major turning point in the human history.

Here, since the outbreak of the COVID-19, what the pandemic brought can be summarized as follows.

[Nation]

- A discussion on the superiority of democratic political system or despotism
- China Hate
- Strengthening state control in emergency situations
- Management and monitoring of medical care, health, and citizen behavior using AI and big data
- Many megalopolises fortified by locking down to prevent further epidemic contagion.

[International Relations]

---

[22] https://www.nytimes.com/interactive/2020/07/05/us/coronavirus-latinos-african-americans-cdc-data.html.

- China's hegemonic age?
- The end of American-centrism?
- Rising discussions on differences between Westness and Eastness
- Vigilance of decoupling the states and alliances
- Shift to nationalism and centrism of the own nation
- Emerging severe risk of international tension between China and the USA.

[Prejudice and Discrimination]

- The manifestation of prejudice and discrimination against medical professionals in relation to the COVID-19
- The manifestation of prejudice and discrimination against those infected with the COVID-19
- The manifestation of prejudice and discrimination against Asians
- The race riots in reaction to the life unrest caused by the pandemic.

[Social and economic aspects]

- Postponement of the 2020 Olympics at Tokyo
- Many of the academic and industrial conferences were suspended, postponed, or alternatively held online
- Stagnating the global economy, and disruptions in trade and supply chains (Verschuur et al., 2021)
- Rapid increase in the number of unemployed people
- Tightness in welfare and medical facilities for the elder people
- Focusing on the trust of government leaders (Shibuya, 2017, 2020a; Yang & Tsai, 2020)
- The recessions of globalism and its economic idealism
- A move to strengthen the country's economy
- Reconstructing remote-workplaces and distributed working styles
- Income support for citizens and companies
- Redesigning supply chain management.

[Citizen]

- Forcing to keep social distancing (physical distancing) among them
- Halting acceptance and expulsion of immigrants and refugees
- Alienationism and nationalistic movements
- The reduction of the working environment and the rise of remote work
- People hoarded foods and daily necessaries in anticipation of an impending crisis.
- Heightening social anxiety and mental stress against the COVID-19 and other factors
- Increasing domestic violence
- Bringing learning opportunities online and difficulties in accessing low-income populations.

[ICT aspects]

- Providing medical care and health monitoring

- Analyzing big-data such as online activities and search queries
- Analyzing collective dynamics of citizens behavior using smart-phone and GIS (Boulos & Geraghty, 2020)
- AI driven monitoring tools and smart-phone applications
- Demands on epidemic and socioeconomic models for simulation.

1.8.2   GDP

Globally, it checks initial economic impacts caused by the pandemic. Table 3 comparatively shows GDP of each nation (as of May 2020). It clearly displays declining economic growth and many economists predict pessimistic pathway to be recovered from the pandemic. In early 2021, the United Nations also published an outlook on the impact of the COVID-19 pandemic on the socioeconomic aspects of the world and economic recovery, based on statistical data (United Nations, 2021). It was assumed that the economy will recover once the corona scourge was contained by the end of 2021, but this was contingent on global progress in vaccination.

In the case of Japan immediately after the outbreak of the pandemic (at the end of May 2020[23,24]), the number of peoples absent from work in April was 5.97 million, an increase of 4.2 million from the same month 2019. Immediately after the Lehman shock, the number of peoples who took a leave of absence was around 1 million as a high growth rate. Here, an absent worker refers to a person who was not unemployed, but was out of work. The seasonally adjusted full unemployment rate in April was estimated as 2.6%, and 0.1 percentage point worse than the previous month. The number of fully unemployed people was similarly estimated as 1.78 million, and it means an increase of 60,000. The number of workers (mainly casual and temporal workers) fell by 1.07 million from the previous month, it denotes a first decline since January 1963 in Japan. The longer the economic downturn lasts, the harder it will be for companies to keep hiring such workers. Namely, these economic indexes warn about that the unemployment rate, which has remained at 2.6%, could jump in the future. Similarly, a number of workers in the other developed countries has been suffering resemble conditions or worse than Japan cases (Kimura et al., 2021). However, except for some industries (e.g., aviation, tourism, restaurants, etc.), the Japanese domestic economy itself turned out to be steady with certain demands (e.g., medical products, foods, transports, ICT-related).

1.8.3   Markets

During the pandemic (from 2020 and early 2021), some industries, especially restaurants and small and medium-sized enterprises (SMEs) were affected by the pandemic,

---

[23] https://www.jil.go.jp/kokunai/statistics/covid-19/c03.html.

[24] https://vdata.nikkei.com/newsgraphics/coronavirus-economy/.

Table 3 GDP of each nation at 2019 and 2020 (at the present of May, 2020) (Data from IMF[25])

GDP	China (%)	Japan (%)	Brazil (%)	Russia (%)	USA (%)	Canada (%)	UK (%)	Germany (%)	France (%)	Italy (%)
2019	6.1	0.7	1.1	1.3	2.3	1.6	1.4	0.6	1.3	0.3
2020	1.2	–5.2	–5.3	–5.5	–5.9	–6.2	–6.5	–7.0	–7.2	–9.1

meanwhile the ICT industry was generally doing well worldwide due to the growing demand for remote work, online gaming, and online education.

Economic trends were also evident in the global marketplace. Soon after the pandemic, the USA chart[26] of the Dow Jones the Industrial Average (DJIA) stock market index had shown that the price was down to approximately $21,763 at March 2020, but the price soon recovered to about $25,605 and has remained steady at June 2020. Similarly, Japan's Nikkei index fell below 20,000 yen only in March 2020, but since April, this index has recovered to over 20,000 yen. The comparison of GDP suggests that many workers in other developed countries were also suffering from the same situation as in the Japanese case.

Additionally, the IMF (international monetary fund) had issued their economic outlook[27] in April 2020, and it offers a pessimistic outlook in global. They said *"Growth is expected to be above trend in 2021, with a partial recovery, but the level of GDP. It will remain below the "Before Corona"'s trend. Moreover, great uncertainty exists about the strength of the recovery. Growth rates could be even lower, and that is even more likely. Prolonged global epidemic and spread prevention measures, with emerging market and developing economies hit even more severely"*. And further *"The current crisis requires two-phase responses. It means a phase of containment and stabilization of the infectious disease, and a subsequent phase of recovery. In both phases, both public health and economic policies will play a crucial role. Quarantine, urban containment, and social distancing can delay not only the spread of infection, but give researchers efficient time to address the surge in demand and to develop treatments and vaccines. These measures will prevent more prolonged decline in economic activity and help the economy for recovery"*.

However, in February 2021, the Nikkei average of the Japanese Tokyo stock market reached the 30,000 yen level for the first time in 30 years, pointing to signs of a bubble. Since it was a global trend, the divergence from the real economy and trends in interest rates and other factors during the course of the pandemic were being closely watched.

1.8.4 Public Supports

Actually, according to a report by Teikoku Databank[28] (the largest credit research company in Japan), the number of "corona-related bankruptcies" in Japan was approximately 985 as of February 3, 2021, from February 2020, after the outbreak of the pandemic. In particular, it is known that restaurants and small and medium-sized businesses were forced to go bankrupt, suspend operations, or go into legal liquidation due to lack of cash flow or difficulties in continuing their business.

Furthermore, more than 20,000 people committed suicide in Japan in 2020 due to the hardships of life, with about 13,000 men and about 7000 women, with the rate of

---

[26] https://www.macrotrends.net/1319/dow-jones-100-year-historical-chart.

[27] https://www.imf.org/en/Topics/imf-and-covid19.

[28] https://www.tdb.co.jp/tosan/covid19/pdf/tosan.pdf.

increase for women being higher than the previous year. Most of these are conceived to be due to reasons such as layoffs and difficulties in continuing business due to the corona pandemic, which has led to the first increase since the Lehman shock in 2009 (Tanaka & Okamoto, 2021).

Of course, public supports (e.g., METI in Japan)[29] were enhanced for business continuity, employment stability, and livelihood security, but financial support for students, part-time employees, and fixed-term employees was not sufficient.

On the other hands, in the case of the USA, on April 6, the FRB (The Federal Reserve Board) announced that, as part of the pandemic policy, the USA federal government had begun on April 3 to implement the preferential loans for small business, and recovery rebate and other programs for citizens in the USA. And then, the federal government and state governments were necessary to bear the financial burden. Furthermore, the results of positivist economic analysis revealed significantly that income disparities among residents in each infected area of the USA show a clear difference in the impact of the COVID-19 pandemic and subsequent recovery. Using available data, they found the mechanisms through which the COVID-19 affected the economy by analyzing heterogeneity in its impacts (Chetty et al., 2020).

In the EU, there are also already some estimation of economic burden in each country (Salje et al., 2020), and further studies will reveal more critical evidences against domestic and international impacts on economics (Bartik et al., 2020). At 2020, the EU has a plan to allocate a total of 750 billion euros in new funds to help rebuild economy after the pandemic. It finally reached consensus among the member countries, even though there is no obligation to repay. But there were concerns that such a frame of support itself could lead to other increased burdens and the question was who will bear such a burden.

2   Discussion Based on the Distribution of Sacrifice

2.1   Quid Pro Quo

"*Quid pro quo*". The world always requires sacrifices to achieve the goals. As repeating once again, the history of the humanity is not a litany of visible "glories", rather a more realistic history built on numerous negative consequences, namely it can concise them as the "*sacrifices*" corresponding with the crisis. *Fortuna*, the goddess of fate, is blind, as a proverb says. The fate of sacrificed person is determined accidentally and blindly, however, the many cases are actually intended by the human, especially by political leaders.

In ordinary, the human history only told time-periodical causal process and its factual results (McNeill, 1976, 1999), or an epic story on autobiography of each renown personage. However, the author focuses on rather sacrifices in each crisis

---

[29] https://www.meti.go.jp/covid-19/pdf/pamphlet.pdf.

and emergency situations during those historical sequential process. For example, uncountable events such as warfare, pandemics, environmental disasters, international conflicts, political revolutions, and so forth. Namely, such sacrifices almost implied the total cost to achieve and solve something historical events. In other words, the author frequently wonders how to equalize the total value of sacrifices and final consequences as achievements. Perhaps, both of them could not be evaluated as equal values, because those are not always insured an equivalent exchange between total sacrifices and final results. If so, who and what such *difference* (absolute value: results minus sacrifices) would be undertaken? The author considers that such inevitable difference can be also called as real sacrifices.

For example, how many peoples did participate as well as die during developing a pyramid? Total amount of them suggested a part of total sacrifices for pursing the goal, at least. Such reality of the historical event should be shed light on an unaware fact. As compensation by the loser countries (i.e., Germany (Habermas, 1987)) to the winner countries would become larger after the Second World War, their' sacrifices were probably overburdened than they initially expected (Inoki, 2014), not equivalent one. French revolution and the "Arab Spring" could not only obtain the benefits (Shibuya, 2020a), but someone forced to be undertaken severe sacrifices. Namely, the most important was who and what sacrifices did undertake it.

As denoted at earlier chapter, the author named such perspective as the "*distribution of sacrifice*". It is often ignored whether decision making is properly done or not, the "*distribution of sacrifice*" becomes particularly salient when "*both individuals and the society face a crisis of the existence if it is not properly and promptly executed*". Rather it is naturally present in everyday life.

2.2 Political Attitudes

In times of emergency, the significance of the decision based on "*distribution of sacrifice*" can be inevitably recognized. However, it means neither any criticism nor any irony. It confronts directly with the actual problem. Kennedy clearly said "*There are risks and costs to a program of action, but they are far less than the long-range risks and costs of comfortable inaction*". And he encouraged the citizens "*When written in Chinese, the word 'crisis' is composed of two characters. One represents danger and the other represents opportunity*". At another scene, he said "*Efforts and courage are not enough without purpose and direction*".

It is important to note that there is usually no correct answer to any of the alternatives in decision-making in each crisis. In other words, sacrifices must be required in any alternatives. Political leaders often face a "*trolley problem*" that chooses who will die and how much. If they have a policy that emphasizes medical care, more people may commit suicide due to economic deprivation. If they enforce a policy that focuses on the economy, more people will die without access to medical care. Of course, if the political policy stays ambiguous, even more people may die without both access to medical care and economic support.

It will not be enough to merely adopt an alternative that meets a slogan *"greatest happiness of the greatest number"* by *Utilitarianism.* No one knows the "correct answer". Every alternative has the sacrifice in politics.

In other words, from the "distribution of sacrifice" perspective, rationality and empathy are quite necessary among the stakeholders, but these factors never guarantee the elimination of sacrifice itself. The most prioritized disposition of political leaders is whether they are devote supporting for all of the people (for not specific person).

2.3   Policy Priority in the COVID-19 Pandemic

2.3.1   Choosing Sacrifice

Let me take some examples in terms of sacrifices for further discussions.

(1)   If you are a movie fan, please imagine a movie (*"Deux jours, une nuit"*) at 2014. *A story of a woman who will be certainly fired by her boss in order to serve employee bonus except for her, and then struggles to take back her own dismissal. Just as Sandra is about to return to work, she is told on Friday that she will be fired, because it serves a bonus for other employees. However, her colleagues have arranged for the staff to vote at the beginning of the week, and if more than half of them accept saving for Sandra, she can avoid being fired. Next weekend, Sandra scrambles to convince her colleagues to go to the polls on Monday...* If you are an employed worker in that company, should you cooperate her or company-side? Or, if you are appointed by company as an alternative candidate to be fired, can you undertake its sacrifices?

(2)   WHO has announced their concerns about development and its R&D patents and intellectual properties for vaccines and medicines, and sufficient policy of medicine allocation for everyone in global. But major pharmacy companies in developed countries almost occupied those patents and well founded to produce vaccines and medical tools, and then there are the problems how to prioritize and allocate among the global citizens and who will undertake the enormous financial burden. WHO[30] called it *"Making the response to COVID-19 a public common good"*.

And actually, an international collaboration called the COVAX Facility[31] has been established and has begun hands-on support activities. According to their action plan, *"The 75 countries, which would finance the vaccines from their own public finance budgets, partner with up to 90 lower-income countries that could be supported*

---

[30] https://www.who.int/emergencies/diseases/novel-coronavirus-2019/global-research-on-novel-coronavirus-2019-ncov/covid-19-technology-access-pool/solidarity-call-to-action/.

[31] https://www.who.int/news/item/15-07-2020-more-than-150-countries-engaged-in-covid-19-vaccine-global-access-facility.

through voluntary donations to Gavi's COVAX Advance Market Commitment (AMC).
Together, this group of up to 165 countries represents more than 60% of the world's
population. Among the group are representatives from every continent and more than
half of the world's G20 economies". In essence, it is a framework for cooperation for
the common good of humanity in global, with donations and good will for the optimal
distribution of vaccines from wealthy developed countries to developing countries.

At the same time, however, there were also issues of what is known as "vaccine
nationalism", in which people were fighting for their own vaccines. In the EU, for
example, some countries imposed restrictions on the export of vaccines from their
own countries, which became a worldwide problem.

(3) Taking an example from a series of psychological experiments
     (Kahneman&Tversky, 2000), let me consider a balance between an insurance
     merit and its cost. *"We are interested in your reaction to a new kind of*
     insurance. Imagine that you are offered a new kind of health insurance that
     supplements your major medical insurance. This insurance policy your if
     you are hospitalized for any disease (followed by any accident)....(hereafter
     omitted)". Namely it is a kind of the problematic situation, which estimates
     future risk and intends to optimize own proper costs. Such frame of psycho-
     logical experiments has repeatedly indicated our bounded rationality for risk
     estimation between profits and costs, and revealed our mental vulnerability of
     decision making and dispositions for risk aversion.

(4) Disclaimer issues should be focused in terms of sacrifices. It often prioritizes
     the larger corporations and organizations rather than the individual by public
     policy. The tort liability related to the COVID-19 has been worried about
     legal matters with complaints from the customers and employees. Actually,
     in some developed countries (e.g., the USA), according to Hunton Andrews
     Kurth LLP,[32] complaints related to the COVID-19 has been increasing in the
     USA, and the national government prepares to exempt or reduce the risk of
     litigation for companies if an employee is infected with corona.

Above all, the point is that whichever the political leaders choose, or none at all,
they will eventually make a sacrifice. Decision making in politics in emergencies is
about who and at what sacrifices. Above dilemma stories respectively exemplified
and depicted a part of the *"distribution of sacrifice"*, which introduced by the author
at earlier chapter. The first case means when one alternative will be selected, another
alternative must be aborted. It is a zero-sum game situation. Both alternatives cannot
be simultaneously coexisted in same context. Otherwise, second case indicates a
problem on allocation with fair costs, and it also means a disparity between rich and
poor countries. Third case is to exhibit our mental accounting and vulnerability based
on bounded rationality, and consequentially our decision making will not always
guarantee risk evaluation with higher accurateness and precision, and it strongly
imply that we cannot often handle with sacrifices at the crisis. And fourth is to pay
attention to legal risks at the crisis. In the end, the policy focuses on the global aspects

---

[32] https://www.huntonak.com/en/covid-19-tracker.html.

and not the individual circumstances. In other words, the relative sacrifice is almost imposed to each individual, unless each individual brings a lawsuit to justice with sacrificing much cost and efforts.

2.3.2   AHP Modeling

The author exemplifies a result of AHP assuming for the COVID-19 initial policy. As mentioned earlier chapter, AHP is a mathematical method based on pairwise comparisons and quantifies multiple alternatives in a preference order (Saaty, 2013). Detail calculation process refers to Saaty's book.

Now, it considers which of the three alternatives of initial responses should be the most important measure to implement. The first step is to consider which of the three policy candidates for initial response should be implemented. In Fig. 2, this case (it is a hypothetical model for discussion) firstly configures weights of *medical care, economic, closing border, public health*, and *global cooperation* in the second layer. Based on the weighting settings for the elements of *medical oriented, public health,* and *global cooperation*, the third tier is used to select the highest policy among the three candidates. In each case, the elements and its each weight differ and need to be considered. For example, in the following situation, it can use AHP modeling to determine own policy "medical oriented" (0.451) > "global cooperation oriented" (0.348) > "economic oriented" (0.198) in descending order. Thus, the policy of "medical oriented" is used to counter the pandemic. It can reasonably decide what to prioritize first. In this case, specific considerations will include the number of medical doctors and hospitals, the amount of drugs, and the preparation of a system for conducting PCR tests and other necessary preparations (Badano, 2018). On the other hand, for the time being, it means sacrificing elements pertaining to domestic economy, and then immediate supports to businesses and individuals will be considered. Although there will be a need to collect and analyze more information from WHO and other international organizations, these will not take precedence over domestic health policy.

2nd Layer	1	2	3	4	5	λ	5.355
	Medical Care	Economic	Closing Border	Public Health	Global Coop.	CI	0.089
	0.343	0.116	0.309	0.143	0.090	Effective	Under 0.1

Factors	3					
3rd Layer		1	2	3		
		Medical Oriented	Economic Oriented	Global Oriented	λ	C.I.
Medical Care		0.611	0.113	0.276	3.094	0.047
Economic		0.120	0.453	0.426	2.123	−0.438
Closing Border		0.424	0.137	0.439	3.001	0.001
Public Health		0.448	0.271	0.281	6.209	1.605
Global Coop.		0.374	0.307	0.318	4.925	0.963
TOTAL		0.452	0.200	0.349		

Fig. 2 An example of AHP to compare with three alternatives for initial policy of COVID-19

In actual, the pandemic has shown a fact that each nation should switch affordable policies and regulate the citizens are very important for prevention of the pandemic. The frequent asked question was at when the pandemic would cease and which timing it should be going to switch from medical to economic policy or not. Few countries can bring their economies to a halt until full convergence. There are many countries (e.g., Brazil) that have been keeping economic policy and have suffered many casualties. It may be true that the switch from medical to economic has been relatively successful in Japan, Taiwan, South Korea and China, for example. It seemed likely that the necessary requirements for switching policy were in advance. Global cooperation policies were also important, and all citizens were obliged and required to comply thoroughly with those policies.

2.3.3 Concurrent Processes, Not Single Process

In actual cases, it will not be a single entity that responds to these crises, rather there will be more multiple entities. This does not mean that only an option can be chosen and several alternatives cannot proceed simultaneously. In practice, it may be possible to execute multiple scenarios in parallel based on the common goal.

For example, there is not just decisions made by the national executive such as prime minister and the president. The Red Cross and other NGOs will also voluntarily participate in these crises and operate under their own policies. In addition, scholars, research institutes, universities, and other organizations will be also working from their own perspectives and with their own authority to contribute. Such activities may lead to minimizing the sacrifices and costs to the society and the global.

However, it cannot forget that there is the issue of their legal responsibility. If there is any trouble or defect due to the presence or absence of constitutional or legal authority, some of the cases will claim for compensation and legal disputes. Administrators and others would be protected in many countries if they were conducted in their official capacity, but NGOs and other stakeholders are likely to face considerable litigation risks. There will be also difficulties in continuing to operate under the constraints of lack of personnel, equipment and budget.

Furthermore, where is the equilibrium point between economically-oriented policies and healthcare-oriented policies? The question is whether or not a compromise can be reached where economic interests and medical and public health are compatible. Concurrent processes with multiple entities make it difficult to find an equilibrium point, and there may not even be a guarantee that there will be one in the first place.

And the importance of the global citizens' ability to cooperate with each other taking such circumstances into account is a key factor in the COVID-19 pandemic. There were many business corporations confronted with a dilemma whether it should be maximizing their economic interests or the humanity as a whole. All of the citizens now could be aware of the structure of distribution of sacrifice.

2.3.4   Comparing Policies and Scenarios in Actual Cases

Next, we turn to actual policies of the countries in the COVID-19 pandemic (Yano et al., 2022). It has forced us to similarly prioritize one of the policy alternatives (Bach et al., 2020). As Table 4 shows, all countries can be broadly divided into either *"health care oriented strategies"* or *"economic oriented strategies"*.

The former was to enforce epidemiological measures to prevent the coronavirus and limit the spread of infection. This had led to a spate of layoffs and bankruptcies at the great expense of economic growth and GDP. The medical organizations were also forced to make life choices in the face of significant shortages of both medical and necessary supplies, even though they were proactive in providing medical care.

Table 4   Initial differences between medical care oriented strategy and economic oriented strategy

	Medical care oriented	Economics oriented
Basic principle	Medical Care Intensives	Economics Intensives
countries	China, Japan, India, the USA, the EU, the UK, South Korea, Taiwan, Cuba, etc.	Brazil, Sweden, Poorer countries in Africa, etc.
Basic ways for medical cares and public health	Lock-down, Containment, Intensive PCR checks, Expanding ICU, beds and hospitals, etc.	Promotion of natural immunity acquisition among the citizens. Or abandonment of aggressive medical cares
Economic status	Stagnation, Shortage of Cash Flow, Exhaustion	Gradual Rescission due to Increasing Infectious Peoples
Civilian status	Behavior Changes for the New Normal Daily Living, Social Distancing, Loss of Income, Dismissal, Bankruptcy, Suspension of Trade and Travel, Social Unrest, Frustration, Prejudice, etc.	Threaten Mentality against Infection and Death Suspension of Trade and Travel, etc
Sacrifices of the people	Medical doctors and health care workers died on duty, elder citizens, citizens with physical vulnerabilities, citizens in lower economic status, immigrants, illegal undocumented workers, homeless, etc.	Especially, poorer citizens, citizens in lower economic status, etc.
Sacrifices of resources	Additional National Budget for Medical care, economic supports for the needy citizens and companies, and R&D for medicines (e.g., Vaccine) GDP Human Resources Common Wealth, etc.	Human resources Many of the countries are too poor to invest more strategical policies for medical and economic developments, etc.

The latter, on the other hand, took steps as if to dare to abandon health care and social welfare policies in order to maintain economic activity. Some of developing countries gave up promotions of medical cares for their citizens, and rather they continued liberalized economic trades and activities [e.g., Brazil (De Souza et al., 2020)]. In many cases, even effective policy for preventing a seasonal flu is practically neglected, and they do not proceed special medical measures. Some countries were even willing to promote aggressive economic policies while encouraging the acquisition of *herd immunity*[33] among the citizens (e.g., Sweden). These were relatively poor countries (López-Roldán & Fachelli, 2020), and the national governance system might be disrupted when the poor people would recognize it even more difficult to live their daily lives due to the suspension of economic activity.

Indeed, in the USA, there were differences in public health measures depending on the political party that supported the state governor. It was often said that states with strong conservative parties had a clear emphasis on economic policy. Mixed state-by-state policies made uniform measures across the USA difficult.

Additionally, epidemic regulation usually requires strict mobility control of the citizens. NY state has launched own contact tracing program, and their efforts are quite necessary for prevention. But national economy is hardly unstoppable (Bonaccorsi et al., 2020). So then, to what extent is the sacrifice acceptable? Do citizens prioritize health care, the economy or both simultaneously? How can the government steer the opinions of stakeholders? This pandemic obliterated larger extent of predominantly indigenous daily living styles. In terms of medical cares, such policy has been acceptable, but the benefits of a policy to its beneficiaries may outweigh its costs as sacrifices in the economic oriented side. In risk management studies, there are ideas such as *risk acceptance* and *risk financing* in such context, especially, financial insurances for any risks and its consequences have possibility to be fulfilled shortage of the budgets. Well organized sets of the policies have to be prepared before the crisis.

Certainly, the common goal among all nations was to overcome the COVID-19 pandemic. But because the underlying strategic goals were different, and the people and resources that ultimately were sacrificed were quite different. A complexity of various factors including the content of their policies and whether they responded quickly, the extent of their public health and welfare policies, the level of acceptance of immigrants and other groups, and their living conditions determined the final outcome for each country. In the former, Taiwan, South Korea, Japan, and Germany were the only countries that accomplished this efficiently and effectively, while in the latter, due to the emphasis on economic policy, many of the deaths occurred in succession, mainly among the poor. It also placed the USA in the former position, but despite its high medical technology and ample budget, the USA had one of the world's worst situations due to the president's emphasis on economic policy in the early stages of the infection. It could be said that there was a confusion between economic

---

[33] The false promise of herd immunity for COVID-19: https://www.nature.com/articles/d41586-020-02948-4.

orientation on the president side and medical orientation on the administration and medical staffs.

It ventures to say, the tragedy of the USA was caused by rulers and economists alike, who pursued only short-term interests. Even if they obtained economic benefits, was it equivalent to the lives of those sacrificed? On the other hand, the Chinese government, which permitted the cause of the COVID-19 pandemic was able to finally minimize the many casualties in the country by focusing on thorough repression and medical welfare, as they asserted.

2.3.5   Inequality During COVID-19

The victims who caused by the COVID-19 were directly independent on a market efficiency or randomness, but many cases were inevitably depended on their attributes and external factors of the victims. It especially suggested an inequality of socioeconomic reasons (McNutt, 2020). At 2020, UNDP (United Nations Development Programme) has warned about problems on the inequality among the peoples in developing countries confronting with the pandemic (UNDP, 2020). There could be an argument for fairness principles (Rawls, 1971, 2001), how much should be done to them. Even before the pandemic, developed countries have been spending much investment on developing countries to offset the disparity in economic benefits (Kaltenborn et al., 2020). However, the situation in developing and developed countries has been staying to be too divergent, even if each person's conditions (e.g., living efforts, labor time, calories) equalize to the same, both are completely qualitative differences between the peoples living in developed or other countries. It is too difficult to compare with the peoples of both countries, where conditions are too different. Certainly, the alternatives such as a subjective index of wealth, which is not GDP, and the coexistence of a variety of values still have significant meaning (Josephson et al., 2021).

Similarly, at the bottom of all democratic nations, there will be always a working class with low wages. In modern societies, it has been supported by legally vulnerable workers such as immigrants, refugees, or even illegal workers. For example, in the USA, EU countries, and the UK, there is a strict "social class" (or social stratification), though it is rarely denounced overtly. What is problematic is, for example, the *immigration crisis* in the EU in 2015, which resulted in a massive influx of immigrants beyond the scope of state control (Bauboeck, 2019; Kahanec & Zimmermann, 2016; Rodiguez-Pose & Berlepsch, 2015; United Nations, 2015, 2016; UNHCR, 2016), resulting in the rise of nationalism and populism, or the influx of terrorists causing security problems (Shibuya, 2017, 2020a).

This fact seems to be true during the COVID-19 pandemic as well. As one of the hypotheses for such conditions, first, immigrants and undocumented low-wage workers[34] are less likely to receive benefits related to health and welfare because they are less likely to be well paid, and more likely to engage in simple job. In addition to a lack of housing, many immigrants are often overcrowded in their homes, so their chances of getting infected are likely to be higher than in the rest of the population. Some areas such as New York State have more homeless people than any other state in the USA with more than 70,000 unidentified homeless people in the state (They don't have enough shelters, and then they're overcrowded and easily infected). In addition to being vulnerable to survival in the first place, they tend to be placed outside the confines of health care policy (Azose et al., 2016; Brader et al., 2008; Buzby, 2016). Vulnerable people who are put outside the covering of health care and insurance are sacrificed every year.

In the past, Kennedy warned about *"If a free society cannot help the many who are poor, it cannot save the few who are rich"*. Whoever recall this phrase in the USA? Certainly, this pandemic had revealed the nature of national governance that has been deceiving some of them as a slogan of "democracy and freedom" (Pande, 2020). It eagerly argues for individual right of the freedom, but it is clear that it is a discourse for "ordinary citizens" (who could be said to have a certain amount of annual income and prestige) and not including "people living at the bottom of the society". The pandemic exhibited the working class that is forced to do essential labor (Lohmeyer & Taylor, 2020). Some people may say they are offering opportunities or the American Dream, but the probability of success is low to begin with. Even if refugees fleeing from other countries actively counter that they are also being received from a humanitarian standpoint, it will only offer jobs for the *lower* positions of their own. In some cases, they hire foreigners with Ph.D. degree and other high-educated skills (Weinar & Koppenfels, 2020), but even then, they put restrictions on their work visas and other requirements (Ganning & McCall, 2012; Powell, 2015). In short, those who live at the bottom are victims of a capitalist economy, and is massively squandered and discarded.

Essentially, even as far as globalism is concerned, it is nothing more than an economic network to enrich some of the richest people in the world (Schumpeter, 2008). The contradictions in the countries upholding democracy (e.g., the freedom of mobility between nations, working, private rights) have all had negative consequences in these aspects. Perceptions of the deprivation of domestic job opportunities led to the election of the USA president and to the synchronized movements based on extreme xenophobia and nationalism in the EU. The issue of identity is also, in essence, caused by their deception of purported democracy.

On the other hand, the lower number of deaths from the COVID-19 pandemic in Asian countries and Japan might have relatively advantages of sustaining their cultural heritages. There are still relatively lower number of immigrants, refugees, undocumented workers, and homeless people than other developed countries.

---

[34] Facing COVID-19 as an Undocumented Essential Worker:https://www.sapiens.org/culture/covid-undocumented-essential-workers/?utm_source=SAPIENS.org+Subscribers&utm_campaign=4a00cb95ea-EMAIL_CAMPAIGN_2020_4_12&utm_medium=email&utm_term=0_18b7e41cd8-4a00cb95ea-227310007&ct=t().

Nuclear families are becoming more common, and the number of people living together is not larger than other developing countries. The relatively stable form of the statehood by ethnic majority makes it less likely to be disrupted by uncertainty and social unrest (Johri et al., 2012).

2.3.6 Deeper into the Problem

All politics whether it is democracy or not, in the end, is about enforcing the citizens to bind with the "*distribution of sacrifice*". In order to achieve the goal, there will be the elements that must be "sacrificed" on many occasions. And such sacrifices are either unrewarding or inadequate. It is up to the political decision-making process to decide who will bear such sacrifices and to whom they will be imposed. What often tends to happen in democracies is that "*everyone equally undertakes own sacrifices*". This is particularly true in Japan, where the discourse of everyone putting up with each other and supporting each other speaks for itself. Political decision is willing to sacrifice someone in particular, but therefore always makes late and poor decisions.

Even when they don't seem to sacrifice anyone, for example., they do sacrifice people outside the country (in short, part-time workers who come to study in Japan,[35] and foreigners aren't considered as domestic people, and they don't realize that such people in foreign countries work subcontractually for low wages) and future generations (and it sacrifices the biosphere of the earth and other species considerably). And the lower birthrate and higher aging population are depended on women's unwillingness of childcare and take on childbirth, and then there are politicians who envision the development of the nation at the sacrifices of their labor force. Even in this pandemic, the sacrifice of many medical personnel (died on duty) has saved the national crisis. These issues are either unrewarding or inadequate. Lastly, no one may want to carry it, but not securing the sacrifices won't solve a lot of the problems. Hence, democracy is often slow to build consensus and judgment, whereas tyranny is quick to judge, but sacrifice is not even an issue.

During the COVID-19 pandemic, it was often noted that there were differences in the number of people infected and the number of deaths due to differences in national governance systems. Significant differences by the races such as blacks, Hispanics, and those at the bottom of the economic pyramid statistically observed exposing a much higher risk of infection and death (Price-Haywood et al., 2020; Subbaraman, 2020). And, at 28th, May 2020, UNICEF[36] warned about living status on needy children in global, and they predicted that the number of those who suffer severe poverty caused by this pandemic will boost to more than 800 million. Indeed, as the author mentioned above, the negative aspects of democracy and the capitalist economy have been highlighted because they are able to prosper at the *sacrifice* of immigrants and refugees, wasting those human resources and continuing to accept the shortfall every new year. If medical care is to be provided to immigrants and

---

[35] https://www.nikkei.com/article/DGXMZO56835690W0A310C2000000/.

[36] https://www.unicef.org/press-releases/covid-19-number-children-living-household-poverty-soar-86-million-end-year.

refugees, there is also the question of who will *undertake* the high cost of medical insurance premiums. Donations alone have their limits, unless there is something extraordinary, the activities of the United Nations and other projects will be derailed (then, it won't be resolved for a long time). Even states with strict state control (e.g., China) have achieved high growth at the expense of their citizens and rural peasants. The economic situation of the middle class in China is also gradually becoming more and more contradictory (Zhou & Hu, 2021).

The sacrifices can be assumed to be inevitable no matter what decision is made. The *"Arrow's Impossibility Theorem"* which brought the Nobel Prize in Economics means that there is a mutual similarity between the ultimate form of democracy and a kind of dictatorship (Arrow, 1963; Brandt et al., 2016; Sen, 1970). As a result, the COVID-19 pandemic response demonstrated that these differences will ultimately have the same effect on the operation of state governance. To be sure, democracies emphasize the aggregation of opinions through a deliberative consensus-building process in which independent individuals participate, but this is a process in which the independent preferences of individuals are coordinated by the society as a whole. This is analogous to a dictator making decisions based on his values and preferences from the beginning, since the result is that the preferences of society as a whole are largely set in one. Therefore, although some say that the open participation for the citizens, the right for expressions of their opinions and civic engagement in politics are the basis of liberal democracy. However, as much as it is the essence of state governance to set a common direction for the whole society, unless the social preference is consistent with the preference of individuals, its citizens and their values are sacrificed. It could be also said that the maximization of individual freedom only leads to individual irresponsibility. Any action that deviates significantly from the policies and preferences (i.e., common values) determined in the society, even if it is not illegal, should be said to be irresponsible rather than freedom.

3 Social Theory on Crisis

3.1 How Differences Are Between Westness and Eastness?

At 2020, Munich Security Conference (MSC) had been held in Munich, Germany. Those who participated in this conference discussed a theme on *Westnessless*. This concept clearly was very impressive for us. This implied a symmetrical concept, and it articulates the conflicts between "Westness" and "Eastness" societies (Root, 2020). It reflects not only these cultural differences, but also their attitudes about how to deal with the rise of China in particular. In fact, the world history had given an important meaning for China's civilization and its cultural influence (McNeill, 1999), and moreover, Morgenthau (1978) had warned about increasing China's presence in global at the cold war after the Second World War.

National governance and cultural differences were also noted in the pandemic. Until recently, the definitions to discern between *Westness* and *Eastness* are merely conceptual categories based primarily on the differences between the constituent cultures and the leading ethnicity (Bell, 2000). The former is based on social institutions and cultures that the western European countries had primarily cultivated, while the latter is mainly based on the institutions and cultures that the Asian peoples had histrionically inherited. Past studies in cultural psychology, archaeology, and anthropology had shown also a similar framework (Triandis, 1995).

However, likewise Hellenism, the differences between Westness and Eastness will mutually provoke not only crises and conflicts among those heterogeneous members, but innovative products interweaving with each of them. For example, Hofstede (1980) reported an empirical fact that culture can be systematized by four categorical axes from survey results. Namely, these axes were *power, uncertainty avoidance, individualism-collectivism,* and *masculinity-femininity.* In particular, uncertainty avoidance composed with minimization of unstructured situations, adaptations, and innovation in groups. Diversity of cultural backgrounds and multipurposes in each company, organization, and nations have been vigorously facilitated (Aoki, 2010), he noted one of the organization principles in differences such as "*rule-based*" or "*self-governing*". Of course, the former may be frequently identified as Eastern culture, whereas the latter is characterized as Western culture. But those adoptions in business partnerships ought to be collaborative and inclusive memberships unless there are any rational reasons.

Here, as below category shown, the author hypothesizes the major differences between Westness and Eastness. However, due to globalization, these clear differences cannot discern clearly which element is Westness or Eastness. Especially, Japan has been committing in both two sources between them, and namely political and socioeconomic foundations are depended on capitalism of Westness since the Era of the Meiji, while their cultural history has inherited from wider Asian area.

As basically, Japan's ruling regime had been continually leading by the emperor through the Japanese history, but in some part periods of the entire history of Japanese, there were several times of ruling government that conducted by a clan of the Samurai General, who was assigned a hereditary status by an emperor. Historically, Japanese regime had been related to religious thought by Buddhism and Shinto, and their background of the regime could be founded to save own nation by religious relief among the citizens in the context of Asian cultural heritages. For example, Japan's "*imperial era name*" (or *Japanese Era*) means not only the historical time as a period but somewhat hope for the nation and citizens. Current imperial era name "*Reiwa*" is the 248th (it had changed from "*Heisei*" to "*Reiwa*" from 1st, May, 2019). Historically, changing era name had definitive relation with proliferation of the plague, and the most common reason for changing was a religious intention and a revival from the "cataclysm" or "disease" dozens of times.

Certainly, Japan has also a tradition of honoring such noble family lineage, but after the Second World War, the aristocratic system was abolished by the new constitution based on western democracy, and this disparity disappeared on the surface. However, before the Second World War, in the Meiji Era of Japan, at 1875, when

Table 5 Differences between Westness and Eastness

	Eastness (Mainly Cultural Bases)	Westness (Mainly Cultural Bases)
Basic principle	Collectivism	Individualism
Countries	China, Japan (Cultural, genetic features), India, Asian Countries, etc.	The USA, EU nations, the UK, etc.
Religions	Buddhism, Shinto, Islam, etc.	Christianity, etc.
Economics	Communism, Capitalism	Capitalism
Political Regime	Communism, Democracy	Democracy
Languages	Chinese, Japanese, etc.	English, French, German, etc.
Genetic Races	Mongoloid, etc.	Caucasoid, etc.
Permanent Five Members of The United Nations	China (and Russia?)	The USA, the UK, France
Bloc Economics, or Trades	ASEAN, TPP, OBOR (One Belt, One Road Initiative), AIIB, etc.	The EU, TPP, ADB (Asia Development Bank), etc.
Total population (2019)	Total 2,935,138,400: China (1,441,860,300), Japan (126,860,300), India (1,366,417,800)	Total 632,234,800: the USA (329,064,900), the UK (68,045,800),France (67,978,400), Germany (83,517,000), Italy (60,550,100), Spain (46,736,800), Belgium(11,539,300)
GDP (2018: Million US$)	Total 22,779,061: China (13,368,073), Japan (4,971,767), India (2,718,732), South Korea (1,720,489)	Total 33,643,964: the USA (20,580,250), Germany (3,951,340), France (2,780,152), the UK(2,828,833), Italy (2,075,856), Spain (1,427,533)

Fukuzawa published his book "*An Outline of a Theory of Civilization*", he theorized Japan's own national governance of the legitimation as an independent and westernized nation. And his main intention of this book was to theorize Japan's civilization about Asian cultural aspects that inherited from China and own historical backgrounds and Japan's modernization imported from the latest essences about scientific technology and knowledge in the Western countries (Table 5).

3.2 The Sacrifices Conjoined With Habermas's Social Theory on Crises

As Habermas (1976) noted, *crises* in each aspect (political, economic, and cultural) in the social system are different, and there are the necessary measures and responses for each aspect. Based on above discussions, it can be seen that the direction of the responses to various crises can be characterized in each occasion. Tables 6, 7 and 8 show ordinary condition of each nation, and these patterns of the social systems suggest either each advantage or vulnerability for each crisis and risk.

Next is the author's own perspective on the "*distribution of sacrifice*" with Habermas' social system. The purpose is to make it clear to examine whether such risks can be minimized by scrutinizing the crises and sacrifices in each political, economic, and socio-cultural category. Many of the factors are intricately intertwined and interacted with each other as a system. In other words, it should manage an entire process such as politics, economics, and social culture as a kind of system dynamics. Based on these validations, it will be possible to build necessary system models. It is essential in the political process that the problems are not dealt with as it occurs, but that the risks and crises that may occur before it occurs are detailed, and that policies are prepared to avoid or minimize those sacrifices. This is the essence of the political process, whether democratic or tyrannical which is commonly required.

China's system of governance is often said that it is counterposed to liberal democracy (Bell, 2016), but is it really so bad? Is it a political system hostile to liberal democracy? For the individual, the system of governance should be more preferable to the system of governance that enjoys the higher benefits. So then, which is larger

Table 6 Social systems in China

China	Westness	Eastness
Politics	× (Slight against democratic election)	○ (Ruling Governance conducted by Communism)
Economics	○ (Capitalism, but Restricted by the Central Governance)	○ (Marxism, Communism)
Culture	○ (Restricted by the Central Governance)	○ (Traditional Eastern Culture)

Table 7 Social systems in the USA

USA	Westness	Eastness
Politics	○ (A President, and the States)	×
Economics	○ (Capitalism)	×
Culture	○ (Traditional Western Culture)	○ (But, melting-pot as heterogeneous culture)

Table 8 Social systems in Japan

Japan	Westness	Eastness
Politics	○ (Constitutional democracy)	○ (Historical Emperor System as a Symbol)
Economics	○ (Capitalism)	×
Culture	○ (Imported Western Culture)	○ (Traditional Eastern and Own Culture)

the outcome of freedom and private rights (e.g., the democracy of Western societies) or the outcome of enjoying coordinated freedoms in society as a whole (e.g., a system of governance like the Asian cultures and China)? Traditionally, many of the citizens undoubtedly supported the former. In this manner, maximizing individual freedom through democracy has led to the formation of ubiquitous online networks, but in the end, rather than maximizing individual freedom, it has led to the formation of opinion clusters, new divides among the citizens, and the spread of inconvenience lacking harmony and responsibility throughout society. On the other hand, now, digital transformation using AI and big data technologies has the potential to lead us globally towards the latter. A system that coordinates mutual interests across the network, with inducements achieved through AI and big data technologies, could ultimately be more beneficial to both individuals and society as a whole. When we take into account not only the immediate benefits, but also the unseen benefits to be gained by ensuring safety, environmental considerations (e.g., SDGs), and social institutions, the maximization of individual liberty would rather have only maximized social unrest and made coexistence with the others more difficult. Moreover, which is the system of governance that people really find desirable in the pandemic crisis?

On the other hands, an organization often faces a matter of power and responsibility. Particularly in times of crisis, when it is not clearly determined in advance how that authority will be exercised and whether or not sacrifices will be allocated, the crisis is often even worse. As sociologist Bell (2016) already pointed out, China relies on an inherent "*meritocracy*", and its electoral and power structures appear unique to Western societies. "*political leaders in meritocracies such as China are meant to exercise political judgment in a wide range of domains. They hold the ultimate power in the political community (including control over the instruments of violence), like elected leaders in democracies. And there is no clear institutional distinction between civil servants and political leaders in a political meritocracy. In short, meritocratically selected public servants in democratic countries are not meant to be political, whereas meritocratically selected public servants in political meritocracies are meant to exercise political power*".

Actually, the crises triggered by the case of the COVID-19 pandemic were, of course, different in the policies called for in response to each risk. In addition, each country had different facets between Westness and Eastness, and then further the insights were also different. For example, in China, the central government forcefully suppressed the pandemic through a collective leadership system, whereas in the USA, because their respect for individual human rights and autonomy is a national policy and it is based on a multicultural society with many heterogeneous backgrounds, those factors had caused the world's worst infection as a result. Japan, on the other hand, had survived the terrible risk of infection nevertheless their government was not effective, but because it is a territorial area surrounded by the sea, their national safety was still defended.

As categorized below, principal sacrifices and the crises in each element can be summarized as follows.

Politics:

Because of failures caused by preexisting international tensions between the USA and China, along with WHO, international cooperation against preventing pandemic could not be coordinated among the nations. The existing global conflict structure was also seen to have influenced the outcome of the COVID-19 pandemic. Complicated factors were multiplied to interlink with various preexisting and ongoing problems, and further it would not reach appropriate solutions for the pandemic.

Economics:

Globalism and its economic networks and migration routes have bridged countries as a route of infection. Due to economic interconnectability and physical portability beyond the borders between China and other nations, when in particular some nations may play either a hub position for further contagion, an economic damage could seem to be became exponentially.

Culture:

Because of failures that conducting improperly domestic policies on public health and medical cares, even many of the developed countries suffered collapse situations at the multiple layers in medical and public foundations. These consequences in each nation could be crucially dominated by whether public and medical services could be well prepared rationally or not (Tables 9 and 10).

Table 9 Crises on each aspect of the social system at the time of the COVID-19 Pandemic

	China	Japan	USA
Politics	A Decaying Risk against Central Governance System Minimizing the impact of the loss of the prides of the central government Avoiding pressure on China in diplomatic matters Avoiding the collapse of the medical policies Refinement of public health policy	Confusions in Ruling Party The negative effects of domestic governance due to considerations for both the USA and China The collapse of the medical policies Refinement of public health policy	A negative effect against President Election Campaign Intensifying Accusing Attitudes against China Government Avoiding the collapse of the medical policies Refinement of public health policy
Economics	Industrial and Trade Halt Downturn of GDP Reorganizing the budget, spending the corporate support budget	Industrial and Trade Halt Downturn of GDP Reorganizing the budget, spending the corporate support budget	Industrial and Trade Halt Downturn of GDP Reorganizing the budget, spending the corporate support budget
Culture	Strict and Severe Regulation and Imperatives for The New Normal	Regulation and Imperatives for The New Normal	Regulation and Imperatives for The New Normal

Table 10 Sacrifices on each aspect of the social system at the time of the COVID-19 Pandemic

	China	Japan	USA
Politics	Permanent decline in China's international prestige Personnel Purge on Wuhan City's lockdown Containment of criticism against domestic and international dissent Containment of facts that cannot be scientifically verified	Citizens' resentment of the government, declining approval ratings, and declining popularity with the prime minister Skepticism about Japan's response to domestic and international issues	Citizens' resentment of the government, declining approval ratings, and declining popularity with the president Abandonment of the strategy of dominance in presidential elections and refinement of electoral measures International criticism and loss of prestige to American politics that has caused the world's worst damage
Economics	Economic downturn and huge budget for recovery	Declared a state of emergency, but many economic blows due to the delay in timing A number of companies have gone out of business due to delay of economic support The daily lives of citizens are exhausted	The demolition of Republican-promoted economic policy The economic blow to the poor and the devastation of civic life Justification for criticism of immigration and refugee policies that brought about large-scale infections
Culture	Quality of life of citizens Decaying the economic situation of the poor citizens Rising international skepticism of Chinese culture	More people are not benefiting from the shift to an online culture Lack of support for students to attend school has left many students in distress Extensive damage caused by the cancellation of various scheduled events	Increased racial conflicts Increased discrimination, prejudice and crime against Asian immigrants The social divide between citizens who can live online and those who can't The "social distance" between the rich and the poor

In the event of a pandemic, medical, epidemiological, and public health experts shared the goal to protect the health of the total population. It's about saving the humanity not individuals. In other words, it is the utilitarian pursuit of "*the greatest happiness of the greatest majority*" that should be the top priority, not confusing meaning and purpose with saving each individual. Thus, they handle detail variables to explain, predict, and control variables such as the number of infected, dead, and recovering people, and constitute a plan that serves the goal by controlling those operable variables.

At the same time, in the case of the COVID-19 pandemic, the severe concerns about the negative impact on the economy and industry were raised. To be sure, there is no opposition to restricting their economic activities for infectious disease control and preventive measures, but there was a backlash in all countries against the perceived length of time that was too long or more restrictive than necessary.

At the brink of such emergency situation, the quality of politicians' decision-making is called into question. Do we prioritize the public health oriented policies, do we emphasize the economic oriented policies, or rather try to have it both ways if possible? It is not easy to find a general solution because the actual situation in each country and region is different. However, from the point of the former view of "the whole of the humanity", regardless of differences in political systems, regimes or ideologies, it is clear that only the former should be pursued. The survival of "humanity as a species" through the implementation of medical policy is the key to the survival of civilized society and the survival of the economic and industrial structure. Immediate sacrifices and financial losses are irrelevant to the achievement of this "goal". This is because taking sufficient proactive and prompt preventive measures as soon as possible will minimize civilian casualties and loss of national wealth.

In terms of the conflicts between Westness and Eastness, even though there is a common understanding of how to achieve the "goal", there are clear differences in the "ways" to achieve the "goal". According to system theory, it meant bottle-necks or unavoidable constraints for responses of the COVID-19 policy in each nation. Namely, the USA has the fundamental cause (i.e., severe constraints) on the president's personality and its direction (meanwhile CDC imperatives always matched with scientific manners), Japan government frequently delays their responses but many of the citizens obey the rules in scientific manners, and China performed the policy conducted by the central government's directions. In short, the political leaders themselves had exposed whether or not they were the bottleneck in the system for prevention, and their quality determined total quality and performances of the national system. Moreover, Western society values individual human rights, it is difficult to take forceful and tyrannical policies even in emergency situations. On the other hand, tyrannic countries like China, which are traditionally collectivist and tend to prioritize the interests of the whole over the interests of the individual, tend to adopt "survival as a whole" strategies and tactics in these emergency situations. From that aspect alone, at glance, Eastness' despotism is superior.

Indeed, Rawls' (1971, 2001) fairness principle itself could not save any life in the pandemic. It was more practical difference to tackle the actuality rather than the country's welfare oriented policies or liberalism. Infected people and deaths were concentrated among the economically vulnerable and elderly, especially in overpopulated urban areas. It seemed that Sweden's herd immunity strategy had completely failed. Even though they had followed a welfare-oriented policy in the Nordic countries, the damage was enormous, especially for the elderly. Germany and France also had good social welfare policies, but the economically vulnerable and the elderly peoples were sacrificed. Even in these democratic and liberal-minded industrialized countries, the effects of the pandemic they had not been able to escape.

Further, as one of the other necessary viewpoints, rethinking the usefulness of common wealth and social common capital is quite necessary for sustaining our lives with equality and accessibility of public health (Uzawa, 2005). In addition to the enhancement of public medical institutions and welfare, and the conservation of the sound environment are still more important as quality of life (QOL). By intersecting them, it can ensure that the priority of individual liberty is superseded by lasting values. It is to focus on not only utility of each individual but values for the society as well as for the all generations of humanity.

4   Conclusion: What We Need to Learn from Misrule

What the author intended to consider was how above theoretical differences could explain actual differences which were observed in the pandemic. Some countries have already begun to scrutinize the effectiveness of a COVID-19 pandemic policies.

According to Asia Pacific Initiative Japan,[37] they had published their comprehensive report interviewed details from politicians, academic experts, epidemic specialists and public officers in charge. This report said the "Japanese model", which has been described as successful worldwide, was simply a fake image. The Japanese government's approach was defined as "*an approach aimed at both deterring the spread of infection and limiting economic damage through a combination of behavioral change measures, mainly the tracking of individual cases through cluster measures and requests for voluntary restraint and absence from work without penalties, without legally enforceable behavioral restrictions*". This report revealed the serial responses to the Diamond Princess' arrival in port, difficulties in improving the country's quarantine and PCR testing systems, poor decision-making and insufficient policy implementation.

On the other hand, in the USA as well, despite having their CDC and medical technology, which are considered to be among the best in the world, there have been reports that the government has not been able to respond effectively and enforce its policies, and that there have been systemic problems (Piller, 2020).

In contrast, the Chinese government's intensive and effective implementation of city blockades and quarantines was clearly more excellent domestic responses than other countries.

Therefore, it can be seen that the implementation of effective decision-making and thorough public health policies, not to mention the governance system of each country itself, had an impact on the situation of infection and the subsequent stability of each socio-economic system. If government leaders wait until after a risk or crisis occurs to think about how to respond, it is too late. There is always the possibility that what could happen could be worse than expected, and that governments will have to deal with the possibility that they did not anticipate. Certainly, in a crisis, a totalitarian system of the governance would be best if the axis of evaluation is to minimize sacrifice. Of course, it is a prerequisite that the supreme authority is more intelligent and judgmental, and able to properly exercise initiative and leadership

---

[37] https://apinitiative.org/en/project/covid19/

(Choi & Mai-Dalton, 1998). And while certain cultural and ethnic differences will need to be taken into account somewhat, it will also be very important to see if the entire population could be unitedly attuned to the goal in line with scientific manners.

Finally, terminating COVID-19 pandemic will be impossible through nationalism and can only be achieved through international collaboration. It will involve not only weighing lives and economics, but also weighing both the interests of one's own country and the interests of the world as a whole. However, during the pandemic, international tensions between the USA and China had risen (Thorp et al., 2020). How to examine the numerous secondary future events caused by the casualties of this pandemic will be the focus in the further steps.

References

ADB Institute. (2021). *Labor migration in Asia: Impacts of the COVID-19 crisis and the post-pandemic future*. https://www.adb.org/sites/default/files/publication/690751/adbi-book-labor-migration-asia-impacts-covid-19-crisis-post-pandemic-future.pdf#page=69

Allam, Z. (2020). *Surveying the covid-19 pandemic and its implications urban health*. Elsevier.

Aoki, M. (2010). *Corporations in evolving diversity: Cognition, governance, and institutions*. Oxford University Press.

Arrow, K. J. (1963). *Social choice and individual values*. Yale University Press.

Azose, J. J., et al. (2016). Probabilistic population projections with migration uncertainty. *PNAS, 113*(23), 6460–6465.

Bach, K., et al. (2020). Ten considerations for effectively managing the COVID-19 transition. *Nature Human Behaviour, 4*, 677–687.

Badano, G. (2018). If you're a Rawlsian, how come you're so close to utilitarianism and intuitionism? *A Critique of Daniels's Accountability for Reasonableness, Health Care Analysis, 26*, 1–16.

Bartik, A. W., et al. (2020). The impact of COVID-19 on small business outcomes and expectations. *PNAS*. https://doi.org/10.1073/pnas.2006991117

Bauboeck, R. (2019). *Debating European citizenship*. Springer.

Bell,D.A (2000) *East Meets West: Human Rights and Democracy in East Asia*, Princeton University Press

Bell, D. A. (2016). *The China model: Political meritocracy and the limits of democracy*. Princeton University Press.

Bempong, N.-E. et al. (2019). Critical reflections, challenges and solutions for migrant and refugee health: 2nd M8 Alliance Expert Meeting. *Public Health Reviews, 40*(Article number: 3). https://doi.org/10.1186/s40985-019-0113-3

Bonaccorsi, G., et al. (2020). Economic and social consequences of human mobility restrictions under COVID-19. *PNAS, 117*(27), 15530–15535. https://doi.org/10.1073/pnas.2007658117

Boulos, M. N. K., & Geraghty, E. M. (2020). Geographical tracking and mapping of coronavirus disease COVID-19/severe acute respiratory syndrome coronavirus 2 (SARS-CoV-2) epidemic and associated events around the world: how 21st century GIS technologies are supporting the global fight against outbreaks and epidemics. *International Journal of Health Geographics,19*(Article number 8). https://doi.org/10.1186/s12942-020-00202-8

Brader, T., Valentino, N. A., & Suhay, E. (2008). What triggers public opposition to immigration? Anxiety, group cues, and immigration threat. *American Journal of Political Science, 52*(4), 959–978

Brandt, F., et al. (2016). *Handbook of computational social choice*. Cambridge University Press.

Britton, T., Ball, F., & Trapman, P. (2020). A mathematical model reveals the influence of population heterogeneity on herd immunity to SARS-CoV-2. *Science*. https://doi.org/10.1126/science.abc6810

Buzby, A. (2016). Locking the borders: Exclusion in the theory and practice of immigration in America. *International Migration Review, 52*(1), 273–298. https://doi.org/10.1111/imre.12291

Chernozhukov, V., Kasahara, H., & Schrimpf, P. (2021). Causal impact of masks, policies, behavior on early covid-19 pandemic in the U.S.. *Journal of Econometrics, 220*(1), 23–62.

Chetty, R. et al. (2020). *The economic impacts of COVID-19: Evidence from a new public database built from private sector data*. https://opportunityinsights.org/wp-content/uploads/2020/05/tracker_paper.pdf

Choi, Y., & Mai-Dalton, R. R. (1998). On the leadership function of self-sacrifice. *The Leadership Quarterly, 9*(4), 475–501.

Corey, L., et al. (2020). A strategic approach to COVID-19 vaccine R&D. *Science, 368*(6494), 948–950.

De Souza, W. M., et al. (2020). Epidemiological and clinical characteristics of the COVID-19 epidemic in Brazil. *Nature Human Behaviour*. https://doi.org/10.1038/s41562-020-0928-4

Ellinghaus, D., et al. (2020). Genomewide association study of severe covid-19 with respiratory failure. *The New England Journal of Medicine*. https://doi.org/10.1056/NEJMoa2020283

Fair, M. A., & Johnson, S. B. (2021). Addressing racial inequities in medicine. *Science, 372*(6540), 348–349. https://doi.org/10.1126/science.abf6738

Fu, F., et al. (2010). Imitation dynamics of vaccination behaviour on social networks. *Proceedings of the Royal Society B*. https://doi.org/10.1098/rspb.2010.1107

Ganning, J. P., & McCall, B. D. (2012). The spatial heterogeneity and geographic extent of population deconcentration: Measurement and policy implications. In L. J. Kulcsar, & K. J. Curtis (Eds.), *International handbook of rural demography*. Springer.

Goldman, J. D., et al. (2020). Remdesivir for 5 or 10 days in patients with severe covid-19. *The New England Journal of Medicine*. https://doi.org/10.1056/NEJMoa2015301

Guan, W.-J., et al. (2020). Clinical characteristics of coronavirus disease 2019 in China. *The New England Journal of Medicine*. https://doi.org/10.1056/NEJMoa2002032

Gubar, E., & Zhu, Q. (2013). Optimal control of influenza epidemic model with virus mutations. In *European Control Conference (ECC)*, Zurich, pp. 3125–3130. https://doi.org/10.23919/ECC.2013.6669732

Habermas, J. (1976). *Legitimation crisis*. Heinemann

Habermas, J. (1987). *Eine Art Schadensabwicklung*. Nachdruck.

Haug, N. et al.(2020) Ranking the effectiveness of worldwide COVID-19 government interventions. *Nature Human Behaviour, 4*(12), 1303–1312. https://doi.org/10.1038/s41562-020-01009-0

Hofstede, G. (1980) *Culture's consequences: International differences in work-related values*. SAGE.

Holshue, M. L., et al. (2020). First case of 2019 novel coronavirus in the United States. *The New England Journal of Medicine*. https://doi.org/10.1056/NEJMoa2001191

Hsiang, S., et al. (2020). The effect of large-scale anti-contagion policies on the COVID-19 pandemic. *Nature, 584*, 262–267. https://doi.org/10.1038/s41586-020-2404-8

Inoki, T. (2014). *A history of economics*. Chuokhoron-shinsha (in Japanese).

Johri, M., Chung, R., Dawson, A., & Schrecker, T. (2012). Global health and national borders: the ethics of foreign aid in a time of financial crisis. *Globalization and Health, 8*(Article number: 19). https://doi.org/10.1186/1744-8603-8-19

Josephson, A., Kilic, T., & Michler, J. D. (2021). Socioeconomic impacts of COVID-19 in low-income countries. *Nature Human Behaviour*. https://www.nature.com/articles/s41562-021-01096-7

Kahanec, M., & Zimmermann, K. F. (2016). *Labor migration, EU enlargement, and the great recession*. Springer.

Kahneman, D., & Tversky, A. (2000). *Choices, values, and frame*. Cambridge University Press.

Kaltenborn, M., Krajewski, M., & Kuhn, H. (2020). *Sustainable development goals and human rights.* Springer.

Kimura, F., et al. (2021). Pandemic (COVID-19) policy, regional cooperation and the emerging global production network. *Asian Economic Journal, 34*(1), 3–27.

Ksiazek, T. G., et al. (2003). A novel coronavirus associated with severe acute respiratory syndrome. *The New England Journal of Medicine, 348*(20), 1953–1966.

Lohmeyer, B. A., & Taylor, N. (2020). War, heroes and sacrifice: Masking neoliberal violence during the COVID-19 pandemic. *Critical Sociology, 47*(4–5), 625–639.

López-Roldán, P., & Fachelli, S. (2021). *Towards a comparative analysis of social inequalities between Europe and Latin America.* Springer.

Lavezzo, E., et al. (2020). *Suppression of a SARS-CoV-2 outbreak in the Italian municipality of Vo'*, Nature, https://doi.org/10.1038/s41586-020-2488-1

McNeill, W. H. (1976). *Plagues and peoples.* Anchor Press.

McNeill, W. H. (1999). *The world history.* Oxford University Press.

McNutt, M. (2020). Science and equality of opportunity. *PNAS, 117*(28), 16090–16091. https://doi.org/10.1073/pnas.2011794117

Morgenthau, H. J. (1978). *Politics among nations: The struggle for power and peace.* Knopf.

Nishiura, H., et al. (2016). Identifying determinants of heterogeneous transmission dynamics of the Middle East respiratory syndrome (MERS) outbreak in the Republic of Korea, 2015: A retrospective epidemiological analysis. *British Medical Journal Open, 6*, e009936. https://doi.org/10.1136/bmjopen-2015-009936

Pande, R. (2020). Can democracy work for the poor? *Science, 369*(6508), 1188–1192.

Piller, C. (2020). The inside story of how Trump's COVID-19 coordinator undermined the world's top health agency. *Science.* https://doi.org/10.1126/science.abf2632

Powell, B. (2015). *The economics of immigration.* Oxford University.

Prather, K.A., Wang, C.C., & Schooley, R.T. (2020). *Reducing transmission of SARS-CoV-2*, Science, Vol. *368*, Issue. 6498, 1422–1424, https://doi.org/10.1126/science.abc6197

Price-Haywood, E. G., et al. (2020). Hospitalization and mortality among black patients and white patients with Covid-19. *The New England Journal of Medicine.* https://doi.org/10.1056/NEJMsa2011686

Rawls, J. (1971). *A theory of justice.* Harvard University Press.

Rawls, J. (2001). *Justice as fairness a restatement.* Harvard University Press.

Rodiguez-Pose, A., & Berlepsch, V. (2015). European migration, national origin and long-term economic development in the United States. *Economic Geography, 91*(4), 393–424.

Root, H. L. (2020). *Network origins of the global economy. East Vs. West in a complex systems perspective.* Cambridge University Press.

Saaty, T. L. (2013). *Theory and applications of the analytic network process: Decision making with benefits, opportunities, costs, and risks.* RWS Publications.

Salje, H., et al. (2020). Estimating the burden of SARS-CoV-2 in France. *Science.* https://doi.org/10.1126/science.abc3517

Schumpeter, J. A. (2008). *Capitalism, socialism, and democracy* (3rd Ed.). Harper Perennial Modern Classics.

SDG Index. (2020). *Sustainable development report 2020.* Cambridge University Press. https://s3.amazonaws.com/sustainabledevelopment.report/2020/2020_sustainable_development_report.pdf

Sen, A. (1970). *Collective choice and social welfare.* Harvard University Press.

Shibuya, K. (2006). Actualities of social representation: Simulation on diffusion processes of SARS representation. In C. van Dijkum, J. Blasius, & C. Durand (Eds.), *Recent developments and applications in social research methodology.* Barbara Budrich-verlag.

Shibuya, K. (2017). Bridging between cyber politics and collective dynamics of social movement. In M. Khosrow-Pour (Ed.), *Encyclopedia of information science and technology* (4th Ed., pp. 3538–3548). IGI Global.

Shibuya, K. (2021). A spatial model on COVID-19 pandemic. In *The 44th Southeast Asia Seminar, The Covid-19 Pandemic in Japanese and Southeast Asian Perspective: Histories, States, Markets, Societies*. Kyoto University.

Shibuya, K. (2020a). *Digital transformation of identity in the age of artificial intelligence*. Springer.

Slifka, M. K., & Amanna, I. J. (2019). Role of multivalency and antigenic threshold in generating protective antibody responses. *Frontiers in Immunology*. https://doi.org/10.3389/fimmu.2019.00956

Subbaraman, N. (2020). How to address the coronavirus's outsized toll on people of colour. *Nature*. https://doi.org/10.1038/d41586-020-01470-x

Tanaka, T., & Okamoto, S. (2021). Increase in suicide following an initial decline during the COVID-19 pandemic in Japan. *Nature Human Behaviour*. https://doi.org/10.1038/s41562-020-01042-z

Thorp, H. H., et al. (2020). Both/and problem in an either/or world. *Science, 368*(6492), 681. https://doi.org/10.1126/science.abc6859

Triandis, H. S. (1995). *Individualism and collectivism*. Westview Press.

UNDP. (2020). *COVID-19 AND HUMAN DEVELOPMENT: Assessing the crisis, envisioning the recovery*. http://hdr.undp.org/sites/default/files/covid-19_and_human_development_0.pdf

UNHCR. (2016). *Syria Regional Refugee Response*. http://data.unhcr.org/syrianrefugees/regional.php

United Nations. (2015). *Global trends forced displacement in 2015*. http://www.unhcr.org/576408cd7

United Nations. (2016). *Addressing large movements of refugees and migrants: Draft for adoption by The President of the General Assembly*. United Nations. http://www.un.org/pga/70/wp-content/uploads/sites/10/2015/08/HLM-on-addressing-large-movements-of-refugees-andmigrants-Draft-Declaration-5-August-2016.pdf

United Nations. (2021). *World economic and situation prospects*. https://www.un.org/development/desa/dpad/wp-content/uploads/sites/45/WESP2021_FullReport.pdf

Uzawa, H. (2005). *Economic analysis of social common capital*. Cambridge University Press.

Verschuur, J., Koks, E. E., & Hall, J. W. (2021). Observed impacts of the COVID-19 pandemic on global trade. *Nature Human Behavior*. https://doi.org/10.1038/s41562-021-01060-5

Voysey, M., et al. (2021). Safety and efficacy of the ChAdOx1 nCoV-19 vaccine (AZD1222) against SARS-CoV-2: An interim analysis of four randomised controlled trials in Brazil South Africa, and the UK. *The Lancet, 397*(10269), 99–111.

Walker, P. G. T., et al. (2020). The impact of COVID-19 and strategies for mitigation and suppression in low- and middle-income countries. *Science*. https://doi.org/10.1126/science.abc0035

Weible, C. M., et al. (2020). COVID-19 and the policy sciences: Initial reactions and perspectives. *Policy Sciences, 53*, 225–241.

Weinar, A., & Koppenfels, A. K. (2020). *Highly-skilled migration: Between settlement and mobility*. Springer.

Yang, W.-Y., & Tsai, C.-H. (2020). Democratic values, collective security, and privacy: Taiwan people's response to COVID-19. *Asian Journal for Public Opinion Research, 8*(3), 222–245.

Yano, M. et al. (Eds.) (2022). *Socio-life science and the COVID-19 outbreak public health and public policy*. Springer.

Yu, J., et al. (2020). DNA vaccine protection against SARS-CoV-2 in rhesus macaques. *Science*. https://doi.org/10.1126/science.abc6284

Zhang, X., et al. (2020). Viral and host factors related to the clinical outcome of COVID-19. *Nature*. https://doi.org/10.1038/s41586-020-2355-0

Zhou, S., & Hu, A. (2021). *China: Surpassing the "Middle Income Trap"*. Springer.

Zhu, N. et al. (2020). A novel coronavirus from patients with pneumonia in China. *The New England Journal of Medicine*. https://doi.org/10.1056/NEJMoa2001017

Formalizing Models on COVID-19 Pandemic

1 Introduction

Exponentially rising concerns over the intolerable risk have provoked much severe interest in how prevention policies might mitigate the pressures in growth of the COVID-19 within and across the cities. In many countries, preventing policies of the epidemic diseases focuses on regulating the daily mobility of the citizens. These policies are adopting in an immediate response to the predicted growth in the COVID-19.

In these concerns, at March 2020, one of the most prestigious scientific journals *Nature*[1] has posed some significant questions for the global pandemic caused by COVID-19. "*Has the coronavirus that causes COVID-19 been spreading undetected in some populations?*", "*Can scientists estimate the size of an outbreak without widespread testing?*", and "*With so many cases undetected, how can the WHO make claims about how many countries have sustained transmission?*". These questions explicitly reflect our holding questions.

In these days, both data on human mobility and its geospatial information in the urban areas can also be utilized (Shibuya, 2004a). With such big data, real-time prediction and analysis is possible. However, there is no definitive model including spatial information. In this pandemic, what was often used was big data (e.g., mobility flow data of the peoples) which provided by mobile phone operators. But an increase in the number of people does not mean an immediate increase in concentrated contacts, and there is no guarantee that the assumption of an increase in infection will hold. Japan government authority defined "contact frequency" as an indicator of contact reduction. This is the total number of concentrated human-to-human contacts in the urban areas, and is estimated as "population multiplied by the number of contacts per each individual". Namely, such model intends to define a better model in which the number of contacts simply does not depend on population density.

---

[1] https://www.nature.com/articles/d41586-020-00760-8.

Regarding these points, this chapter intends to contribute to clarifying some parts of overriding issues. It can paraphrase such situation as identifying mingled condition in geometric closed space, dynamic estimations of infectious growth, transmission control conducted by inhibiting mobility, and a proposal to handle clustering patterns. For such purposes, at first, the author gets in touch mathematical principles measuring necessary real-time data in practice. And of course, the author examines own models and simulations on the bases of such contexts.

2   Public Health by Digitized Solution

Evasive policy of the government against such pandemic must be advised by scientific manners, and it should conduct to plan necessary stringent policies led by their responsibility. Each president and government leader in each nation made an emergency declaration on corona virus. Ideally speaking, it should conduct effective prevention policy before the crisis, and it ought to detect such emerging sign in advance. In public health studies, at least, there are needs to prioritize *identifying the disease itself, checking health conditions against suspicious case, surveying actual cases in regional area, containment of infectious damage, purifying infectious area, isolating patients, providing nursing services, pursuing the infectious routes* and *verifying both explicit and implicit risks of the disease.*

Actually, there were many criticisms against policy makers and governmental leaders. For example, the UK government laid weight on *social distancing strategy*[2] for infectious control, but some of the UK scientists on the contrary worried about its effects for delaying and prevention against the COVID-19 progress. Avoiding physical contact among all citizens is usually very crucial policy for prevention through epidemic process, they faithfully weighted on their concerns about small and closed kinship such as home and pubs rather than larger public space in the city. Otherwise, the EU nations almost agreed with each other to temporarily suspend and prohibit free mobility of the citizens and visitors across the EU borders and migrating within the EU nations. Further, in some of the EU nations (e.g., France, Italy), their governments conducted strict imperatives for the prevention of pandemic progress in their homeland. Particularly, their set of the policies was to keep a social distance among each individual (Fig. 1), avoid to participate massive meeting in public space, restrict going outside, and recommended working in each home using online tools.

In this point, National Security Commission on Artificial Intelligence (NSCAI[3]) in the USA recommended a set of the COVID-19 responses using the AI and its technologies. Certainly, estimation on larger population and tracking its dynamics in public space might be hardly handled by only our sensitive recognition. There are now affordable solutions (Munzert et al., 2021).

---

[2] https://www.gov.uk/government/publications/coronavirus-action-plan.

[3] https://www.nscai.gov/home.

Fig. 1 A public rule of social distancing strategy in Italy (it means "*It is recommended to keep the interpersonal distance of at least 1 m.*" in English)

For example, below services for smart-phone application has been launched to measure two meters centered around each individual using AR (augmented reality).

- Google Sodar: https://sodar.withgoogle.com/
- Keep Distance Ruler:https://keepdistanceruler.prty.jp/.

In other cases where smartphones were used, many countries took steps to acquire and use necessary data such as GIS data and trajectory data, while taking privacy into consideration. Each government[4] recommended the use of applications that capture personal health information using only a unique ID to identify each user.

Indeed, all of the global citizens live in digital transformed age. In practical solutions, the sensing system driven by the AI can automatically monitor continual patterns, measure their mingled conditions, and alert to the authority if the criteria exceed or expect such patterns in advance (e.g., Johns Hopkins University: Boulos & Geraghty, 2020). Nowadays, such smarter services enable to enhance identifying, positioning, measuring and mining real-timely dynamic distribution patterns of objects and humans in urban space (Jiang & Yao, 2010; Pan et al., 2013; Rossi et al., 2015; Qi et al., 2019; Shibuya, 2020a, 2020b; Dai et al., 2020).

Namely, computational and mathematical supports can be applied in epidemic and public health policy (Squazzoni et al., 2020; Aleta et al., 2020). As a first example, using super-computers,[5] it is possible to solve more strict models. And taking with smart-phone, geospatial informatics applied to measure distribution of people and its statistical meanings in arbitrary space (Haining, 1990; Shibuya, 2004a, 2004b). Otherwise, autonomous face recognition system driven by the AI can be already working around the clock, and these mechanisms can be also able to analyze health condition based on each facial information. Any symptoms often represent subtle

---

[4] Cocoa (Japan): https://www.mhlw.go.jp/stf/seisakunitsuite/bunya/cocoa_00138.html.

[5] https://www.riken.jp/medialibrary/riken/pr/news/2020/20200424_1/t_20200424_1_p.pdf, https://www.r-ccs.riken.jp/en/topics/fugaku-coronavirus.html, https://www.youtube.com/watch?v=JhWSKwG9pdw.

difference visually, and then AI based medical services will detect latent sign by deep-learning mechanism. Further, it accumulates location-based data on citizens' mobility, and trajectory (Gonzalez et al., 2008; Renso et al., 2013) as well as clustering patterns of citizens.

3   Modeling for the Solution

To quantify human population and its dynamics in a particular space or city, it is not enough to simply acquire and analyze geo-location data. Above all, such population density needs to be defined and modeled by more measurable ways (Britton et al., 2020). Since population density merely indicates the ratio of individual people in the area of a particular space, it is not possible to divide density from social distance and is therefore inadequate as a pandemic countermeasure.

1.   Single space: inner city and closed space (e.g., cruise ship, music club, concert hall)
2.   Mobility dynamics across multiple spaces (e.g., commuters and visitors).

To be sure, big-data and social media resources have much capability to provide necessary statistics online. For example, Google reports[6] exhibited their own data, statistics and trajectories in each nation, and Tableau COVID-19 project[7] has been launched in line with their voluntary objectives. Otherwise, in some countries, the government and some internet ventures have eagerly offered such detail information for citizens using smart services and ubiquitous solutions, and in Japan, the national institute provided a part of actual data.[8]

3.1   Estimation of Mingled Condition: Solving the Geometric Covering Problem

3.1.1   Formalizing Mingled Condition and Keeping Distance

According to past studies in social physics (Pentland, 2014), he also introduced a fact that HIV infection clusters would be correlational and interdependent with higher population density in the mega-city.

However, such pseudo-correlation could imply just a general possibility of infection within the city. Although population density in each city can be prepared for the consideration (Table 1), it is clear that these variables cannot explain and predict epidemic consequences on infections and deaths caused by the COVID-19 in each

---

[6] https://www.google.com/covid19/mobility/, https://arxiv.org/pdf/2004.04145.pdf.

[7] https://www.tableau.com/covid-19-coronavirus-data-resources.

[8] https://v-resas.go.jp/.

Table 1 Actual data of cities (at the end of March, 2020)

Cities	m^2 (T)	Population (N)	Population density (N/T)	Infected
Wuhan	8,494,000	11,080,000	1.30	75,815
NY	783,800	8,623,000	11.00	44,635
Paris	105,400	2,148,000	20.38	5793
London	1,572,000	8,900,000	5.66	3400
Milano	181,800	1,352,000	7.44	27,206
Tokyo (23 wards)	619,000	9,210,000	14.80	299

Both values 'm^2' and 'population' in each city were cited from actual statistics

city. As a megalopolis always permits to build higher buildings, for example, if it estimates strict solutions, real estate data on total floor spaces should be substituted to explain positive correlation between infection and population density in each city.

Rather, the COVID-19 cases seem likely to be more important to be considered both closed distance among citizens and mingled degree in enclosed space than total city area itself. Certainly, further contagion is deeply depended on total amounts of the other peoples, but it must put more weight on physical distance with each other in the space. Thus, as it simulates later, the author contemplated that the higher extent of closed contacts with the infectious patients (*super spreaders*) and clusters consisted of infectious human relationships could accelerate secondary and more contagion among sound citizens in their proximity.

Here, the most important is to primarily measure and detect both latent and actual risk of the *mingled* condition, which causes to provoke infectious diseases in the public or closed spaces. In brief, Japan government recommends below three factors for preventing the COVID-19. It is called *social distancing strategy* for preventing and delaying further epidemic growth in the city.

1. *Mingled Risk*: at least, keeping from 1 to 2 m between the people in public and private space.
2. *Closed-Physical Risk*: prohibiting physical contacts frequently and inter-communication directly in whether it is larger public or not (Prather et al., 2020). Not clustering each other.
3. *Airborne Risk*: venting room air and opening the windows to breath fresh air (Doremalen et al., 2020). Taking care of own hand-wash, gargling, and healthy conditions.

At July 2020, Japan government[9] and WHO similarly redefined *3C* for preventing more infection as follows.

1. Crowded places
2. Close-contact settings
3. Confined and enclosed spaces

---

[9] https://www.mhlw.go.jp/content/10900000/000615287.pdf.

However, how to define a degree of the crowding condition for preventing emerging diseases? How to discern between the *clusters* consisted of infectious people and sound people? And how can it predict an arbitrary scenario of epidemic progresses based on somewhat sensing data real-timely? Namely, the necessary parameters are measuring physical distance among people (mingled factor), estimating total infectious population in the spatial area (infectious cluster factor), and simulate some models based on the scientific scenarios.

Regarding this matter, there are two important facts related to mathematics and geometry. These are a covering problem in two dimensions and percolation in geometric space. Especially, as the former focus on the physical constraints of our daily life space, the author touches a solution to apply these facts for own modeling.

3.1.2 Tóth's Theorem

First, the author picks up the Fejes Tóth's theorem in geometry (Tóth et al., 1940, 1964). And it is a theorem which typically solves a pattern of the covering problems (Eq. 1, 2, and Fig. 2). The covering problem is a problem in geometry in which a two-dimensional space is filled with a specific figure without any vacancy and gap. Above theorem assumes the total number of N points (for example, people and objects), assuming a hypothetical square D^2 which is one side D surrounds each individual (N), and each N is not overlapped by each other. Where, D is the most important variable which manages the shortest distance among all N points in the specific space T. The area T covered with D^2 that multiplied by N (Overlap of each square D^2 can be permitted), which is not larger than T plus 15.4%. Therefore, at least two points are required. On this theorem, where T is the area of the specific space, and the area multiplied by 1.154 to T is the theoretical maximum area based on Tóth's theorem. If it calculates directly from geological measured information such as google map, simply substitutes such data for variable T. When it estimates the lower limit of N and the upper limit (plus 15.4%), the actual N can be inferred between the ranges.

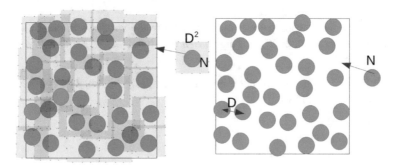

Fig. 2 Visualizing the Theorem (right pattern depicts distributions of N in space T, D is found by the shortest distance among all pair of N, and left pattern exhibits covering T with ND^2)

$$\lim_{n\to\infty} nd_n^2 \leq \frac{2\sqrt{3}}{3}T \tag{1}$$

[Eq. 1: Original Formula (Tóth's theorem, 1940)].

$$N \leq \frac{1.154T}{D^2} \tag{2}$$

(Eq. 2: Estimating the total population).

The most important of this theorem underlies on a precision for estimating both total population N and the shortest distance D in the space with the upper limit. Mathematically, there is more decisive meaning which will not exceed over 1.154 T. And it should also discern the difference between implication of this theorem and population density (= number of people/area). This theorem can handle with a mingled degree at the time$_t$ in dynamic continual time process measured by D, in contrast, population density cannot be altered its static meaning if total population and spatial area sizes are regularly stable. In actual, when each point denotes a human, because the shoulder width of Japanese adult (male) is 40 cm on average, excluding the extreme congestion situation like a crowded train, a square of 40 cm can be theoretically defined as a minimum. It would be reasonable to hypothesize such model in actual space.

For example, if three persons (a, b, and c) simultaneously coexist in the area T and their mutual distance among them (a–b, a–c, and b–c) are 1.2, 1.5, and 1.1 m respectively, variable D is determined as the shortest distance 1.1 m among these data. In such case, because T must be satisfactorily constrained by necessary N keeping the shortest distance D and its square (D^2), without any vacancy space to cover T, and then N in T will be increased or decreased by fitting those parameters. In another case, given that area T (= 10 m^2) and people (N = 4) are known, variable D must be calculated around 2.0 m between each of the citizens for satisfying perfectly covering constraints of this theorem within T. On the contrary, given that area T (= 10 m^2) and people (N = 6) are observed, variable D becomes around 1.3 m (to adjust each person's position and mutual distance without any gaps). Namely, the latter case is more mingled condition than the former, and then variable D can clearly implicate such crowding. In this point, population density lose subtle information of such differences and it cannot manage physical distance among the citizens in those meanings.

Let me firstly exemplify a simple thought experiment. Please imagine an actual case (e.g., EU cities). Government leaders of the EU nations and the UK led strictly conducting *Social Distancing Strategy* against severe proliferation of the COVID-19 m virus in the public space. They often requested that each citizen should keep from 1.0 to 2.0 from each other in the cities and it should not progressively go outside. In such situation, Tóth's theorem and geospatial informatics can answer below consequences, for example. Table 2 indicates an example calculation based on above model. If covering constraints are satisfied by the shortest physical distance (D) among the citizens in public space (T), it can also determine the upper limit of

Table 2 An example which calculated by configurations based on the model

D	D^2	T (m^2)	Theoretical maximum T (m^2)	Estimated N	Population density
0.5	0.25	100	115	400	4.00
1	1	100	115	100	1.00
1.2	1.44	100	115	69	0.69
1.5	2.25	100	115	44	0.44
1.8	3.24	100	115	31	0.31
2	4	100	115	25	0.25

estimating number of people (N) in T. As each citizen can be allowed distributively staying in T by keeping D with covering constraints, it possibly deals with spatial dynamics of the humans in the geometrical field.

Thus, the advantage of this theorem is that both total number of N and the shortest distance D can be estimated in arbitrary space by assuring a square (D^2) centered around each person with the geometrical constraint of covering area T.

4   Simulations

4.1   Single Space: City and Closed Space

A model on the single space by the author represents two impressive cases of the COVID-19 such as single city[10] (e.g., Wuhan City, EU cities) and inner cruise ship as an enclosed space. The former cases were to regulate infection progress in urban area (Hao et al., 2020), but in contrast, the latter cases were unexpectedly faced the difficulty of infectious control in the enclosed space. Such cruise ships (e.g., Diamond Princess)[11] called further international disputes for the prevention of emerging diseases.

In many cases, as it seems probably that the COVID-19 has capable of infecting to the next noninfectious hosts as naive citizens at the stage of clinical latency, efficient policy of each nation could not be activated. In the earlier stage of infection, social distancing strategy for delaying the growth rate of new infections would be almost more effective than the later stage of total infectious process.

Basic configuration of this model exhibits below parameters for both constants and variables. The author configures that it can omit somewhat clinical latency in this simulation process. Renown SIR model and its other derived models (e.g., SIER model (Chiu et al., 2020) in mathematical studies presumes a clinical latency of the disease, namely these patterns are *Susceptible, Infected* and *Recovered*. Otherwise,

---

[10] Cities—try to predict superspreading hotspots for COVID-19: https://www.nature.com/articles/d41586-020-02072-3.

[11] https://en.wikipedia.org/wiki/COVID-19_pandemic_on_Diamond_Princess.

COVID-19 virus has a risk that it can infect uninfected citizens before the onset of the disease (pre-symptomatic) or even in an asymptomatic state (asymptomatic). And the COVID-19 has higher rate to get worse than other diseases (it seems that there are several genetic difference of virus), the author did not put clinical latency (e.g., Susceptible) and immunized factor (e.g., Recovered) into necessary parameters below model.

Parameters

1. D: the shortest distance by dynamically measuring or estimating among all pairs of the citizens at the time$_t$. Social distance can be assumed.
2. D^2: A hypothetical square surrounding each person in space T_i.
3. T_i: Regional area. For example, it is a public space in the city, a room in the building, or the whole city.
4. N_i: Estimated total number of people in T_i.
5. BRR[12] (Basic Reproduction Rate): A constant to calculate a new reproductive rate per an infectious patient. WHO officially assumed that BRR of the COVID-19 could be ranged around 1.5–2.0, but BRR was more severe around 5.0 within Wuhan City and some countries (Xu et al., 2020). Otherwise in the Japan cases, many of the patients could be regulated around 1.0. Here, actually, BRR represents the number of people in the vicinity of the infected person at the time of infection. Because ordinary finite space assumes the Moore neighborhood, at each time, each person is maximally surrounding by other eight persons in physical environment. Then, a default constant of BRR ranges from 1.0 to 2.0, and this maximum score is theoretically up to 8.0 in this model. Furthermore, in Japan, national government policy had been requesting every citizens to reduce own interpersonal contact by "*80 percents*". In this case, it is quite necessary to maintain a BRR from 2.0 ($2/8 = 0.25$, it means $1 - 0.25 = 0.75$) to 1.0 ($1/8 = 0.125$, it means $1 - 0.125 = 0.875$).
6. n: Before starting simulation, initial patient(s) can be configured around 1–n.
7. E: Environmental Risk Constant. It totally exemplifies polluted air condition, effectiveness of air conditioners, transmission risk on infection without venting facility, air deterioration caused by congested proximity, risk on closed contacts with the others in the limited space, and cleanness of the environment (Doremalen et al., 2020). If E is 0, such case can be assumed as a perfect clean room in the lab experiment.

Furthermore, there are also measurable indexes in the simulation process.

1. Occupying rate of infected patients in space (P_r): A_t/N_i
2. Percolation threshold in square lattice (Malarz & Galam, 2005; O'Sullivan & Perry, 2013) (P_c): $((A_t D^2)/1.154)/T_i$. Percolation phenomena can be understood as growing larger cluster occupied by infectious people in space. When this criterion will be exceeded, *phase transition* occurs. In this case, medical cares

---

[12] A guide to R -the pandemic's misunderstood metric: What the reproduction number can and can't tell us about managing COVID-19, https://www.nature.com/articles/d41586-020-02009-w.

and prevention for further infection will be unable to lastly regulate epidemic situation, and such efforts itself becomes meaningless.

3. Entropy (H_i): It was defined in Shannon's Information Theory. H_i can be determined by this formula: $- p \log p - (1 - p) \log (1 - p)$, and here, variable p is inputted above P_c. In this condition, maximum of H_i means the most watchful conditions whether it is infectious risky or not in space ($P_c = 0.5$), whereas minimum of H_i can be decisively discerned as infectious risky state or sound state in the space ($P_c = 1.0$ or 0.0).
4. Probability for each person who will be directly infected from the closed infectious patients and neighborhoods (R_d): P_r E (BRR/8.0)
5. New infectious numbers (I_t): $R_d S_t$
6. Accumulated infectious numbers (A_t): $A_{t-1} + I_t$
7. Sound citizens (S_t): $N_i - A_t$.

Now, it exemplifies a simulated result in continual progress (Fig. 3 and 4: please see an appendix of this chapter). The author inputs parameters into variables $D^2 = 1.5$ m^2 and $T = 500,000$ m^2. And initial infectious patient $= 1$ (i.e., initially, n = 1, A_0 = 1), BRR = 2, and E = 3. Estimated population can seamlessly cover all over the geometrical area. In actual case of Wuhan city, China, this area is approximately 8494 km^2 and gross of total population was approximately 11,080,000. Further, it presumes that each citizen does not actively move around space by the regulation, and then D, D^2 and total number of the N do not vary (in other words, this model treats only the infected state, not the patient's death).

At Fig. 4, in this model, percolation threshold (P_c) and its process decisively denotes the evidence on the growing larger infectious cluster in the finite and enclosed space. And then it gradually rises population of total infections in the city (Liu et al., 2020; Lavezzo et al., 2020; Stechlinski, 2017). As such threshold of site percolation in the two-dimensional finite space (i.e., square lattice) has already well known

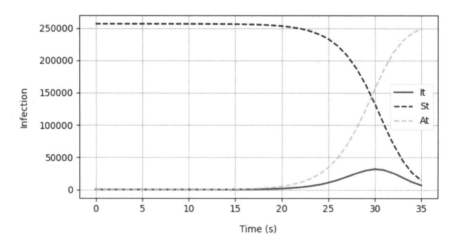

Fig. 3 Simulated results in the public space (Y axis denotes population of each variable)

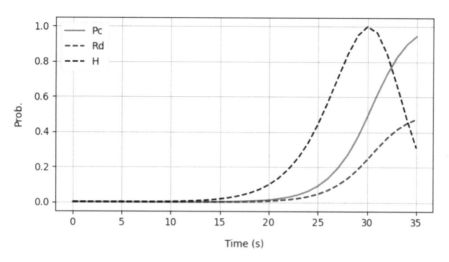

Fig. 4 Simulated results in the public space (Y axis denotes rate of each variable)

(Malarz & Galam, 2005; O'Sullivan & Perry, 2013), it is definitively 0.59. Here, if it exceeds such threshold at once, it is called *phase transition* in percolation model and it is also described *overshoot* in epidemic studies (Ksiazek et al., 2003; Nishiura et al., 2016). Such condition will reach the worst case, as it suggests that nobody would be able to no longer manage the catastrophe by conducting the infection control and prevention. Moreover, entropy H_i can sensitively monitor such watchful trend as its growing risk since the earlier stage of infection process. Before exceeding these criteria, public health officers in charge and government leaders must tackle all set of the problems for the prevention and controls. Therefore, at the brink of catastrophic ending, losing effective ways to conquer pandemic will be faster tragedy than we initially expect.

Thus, contact between citizens can easily generate further infected clusters.[13] The same is true for COVID-19. Infected clusters are ordinarily spread in crowded places, often by the presence of super spreaders. This means that intensive patient care while isolating healthy, uninfected citizens from infected clusters is a more effective solution (TTI policy in epidemics: Test-Trace-Isolate).

As illustrated from another perspective in Fig. 5, it can be mathematically modeled as a percolation model in geometrical space, and it is an example of epidemic simulation on sphere (left sphere depicted an initial stage of the outbreak, whereas right proceeded its process). In this case, the infecting process with parameters such as initial clients = 3 and BRR = 2 was shown. This model can say that if not stopped in the early stages, the number of infected people will exponentially increase, finally leading to the collapse of the medical field and the economy.

Additionally, tracing infectious path and detecting a cluster among the human-relationship is very regular way to examine suspicious clinical case by public health

---

[13] The Science of Super spreading: https://vis.sciencemag.org/covid-clusters/.

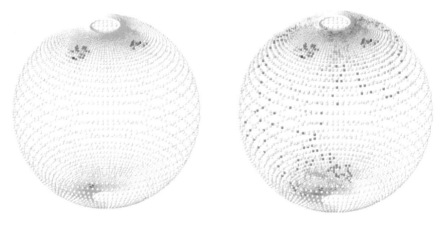

Fig. 5 As a percolation model, an example of epidemic simulation on three dimension sphere (left sphere depicted an initial stage of the outbreak, whereas right sphere proceeded its process)

investigations. Exactly, regarding this necessity, network science had been delving into the frontier of disease diffusion models. Watts (1999) had already clarified small world effect on epidemic contagion. And until recently, this topic attracted various researchers, and then there seems to be many computational and mathematical studies. For example, Newman (2002) also clarified physical and mathematical modeling on spreading disease in social networks. Similarly, Moore et al. (2000) investigated a process of spreading and transmission of disease in small world network, and Kuperman and Abramson (2001) further verified an epidemiological model using small world paradigm. As results, their models are common fundamentals which are mathematical feature of small world and simple rules such as "susceptible", "infectious" and "removed" in structured populations.

Next, the results of the above simulations (Figs. 3 and 4) are well with the actual data on the gradual increase in the spread of infection in Japan shown in Fig. 6. Here, it is important to note that the essence of the social phenomenon of the spread of infection is the cumulative effect. Even if the number of initial infections is small and the number of new infections increases only slightly each day, the cumulative total of these infections will eventually exceed the number of people who can receive medical care. This will lead to a situation where the infected cannot receive adequate medical care (so-called medical collapse). In other words, the probability of not being able to receive adequate medical care increases in the case of late infection than the benefit that a patient can receive in the case of early infection, and the patient suffers a disadvantage. The same risk of infection will have a different risk of disparity in access to health care. Triage is the doctor's decision to prioritize patients, but it can also occur as to whether or not a patient can receive adequate medical care in practice, depending on the timing of infection. In fact, as mention later, there were many such cases in Japan when the infection spread in early 2021.

In the first place, it seems that many Japanese characteristics and daily customs have advantages of keeping cleanness around surroundings for their health at the

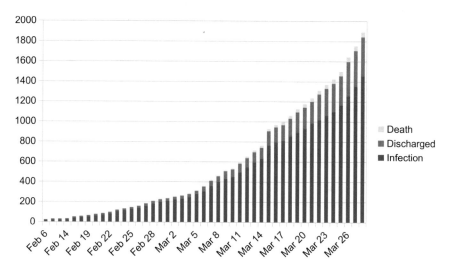

Fig. 6 Actual data of infection, discharged, and death in Japan from Feb 6th, 2020 to Mar 28th, 2020 (only in Japan homeland)[14]

time of this pandemic. Their customs are frequently and spontaneously to wear on facial masks, wash their hands, and keep clean in public space. National medical care services led by a set of public policies could also primarily take care of their health. Regardless governing leaders did not effectively control this disease in the homeland, as WHO's data and guidelines[15] also indicated, growth rate of the infectious progress and death rate in Japan homeland could be still comparatively regulated by much effort committed by each citizen than other countries.

4.2 Mobility Dynamics Across Multiple Spaces

4.2.1 Mobility Flow

Mobility analysis on collective dynamics of human behavior already became one of the hot topics in computational social science (Renso et al., 2013). This case of the COVID-19 impressively indicates the difficulty of regulating mobility of the citizens. In particular, it was rechargeable to restrict movement based on appropriate analysis based on anonymous location and behavioral history information of commuters and visitors who traveled between multiple cities (e.g., crossing between the USA cities) (Venkatramanan et al., 2021).

---

[14] https://www.mhlw.go.jp/english/.

[15] https://experience.arcgis.com/experience/685d0ace521648f8a5beeeee1b9125cd,   https://www.who.int/emergencies/diseases/novel-coronavirus-2019/technical-guidance.

First, let me review previous studies on mobility dynamics. Mathematically, mobility flow based on the mutual distance between two sites have a good reason to be selected for recent studies on human behavior. For example, there is a *potential model* on moving tendencies can be formalized as following equations (AIST, 2013). It is often applied when mobility flow of numerous people seems to be related to the nature of space and physical distance between two sites (S_i and S_j).

$$P_{ij} = \frac{V_{ij}}{\sum_{j-1}^{m} V_{ij}} \tag{3}$$

$$V_{ij} = \frac{E_j}{D_{ij}^a} \tag{4}$$

where,

S_i, S_j	Site$_i$ and Site$_j$
P_{ij}	Probability of moving dynamics of people from S_i to S_j
m	Total numbers adjoining S_i
V_{ij}	Potentials on moving tendency of people from S_i to S_j
E_j	Size of population in S_j
D_{ij}	Geospatial distance (e.g., cartesian distance)
a	Constant (it is usually inputted two)

Among above potential model, variable E_j can be actually regarded as the scalability of facility and capacity. And the distance D_{ij} will not be treated as inversely proportional and rather it will be better to be defined as square (a = 2). In such way, it can be refined as moving patterns between two geospatial sites, and nearer distance can be supposed to be an influential factor for explaining and predicting social phenoma on collective dynamics of behavior. As another model, the *gravity model* is based on a similar idea (Anderson, 2011).

On the other hands, some measuring indexes on the population statistics can also help to understand the dynamic mobility flows. For example, preference index (PI) can be formalized as below (Shryock et al., 1973).

$$PI = \frac{M_{ij}}{\left(\frac{P_i}{P_t} \cdot \frac{P_j}{P_t} \right) \cdot \sum M_{ij}} \tag{5}$$

where,

M_{ij}	mobility from area$_i$ to area$_j$
P_i	total population in area$_i$
P_j	total population in area$_j$
P_t	total population in whole country
$\sum M_{ij}$	gross migrants within whole country

4.2.2 Simulating Dynamics Between Multiple Spaces with Mobility Flow

Here, the author commits next simulation. And an expanded model from previous single space similarly exhibits mobility dynamics across more than two cities, and it prepared below additional parameters with aforementioned parameters of single space.

1. Total Cities (X_n): It configures multiple spaces. Default is two.
2. Migrating Citizens (M_n): Many of them can be regarded as commuters and visitors in ordinary cases. Every time, this model can permit some parts of them to move from X_i to X_j and vice versa. All commuters and visitors must go back to the own living city at the next $time_{t+1}$, and they repeat commuting by the end of simulation. In this simulation, it configures to restrict 10% among citizens in each city. And then, total population itself in each city varies in each $time_t$. It nearly denotes a difference between daytime and night population.

And extra index can be formalized as below.

1. PI_n: Preference Index (PI) in each city, as noted above. In this simulation, each city aims to controlling mobility flows every $time_t$. Regulation of mobility flows between cities for the prevention in epidemic process asymptotically approaches and converges to zero in terms of this index.

In an example of this simulation, city area X_1 was configured $D = 1.5$, 100,000 m^2, $BRR = 2$ and starting infectious patient $= 1$ (i.e., $n = 1$, $A_0 = 1$), $BRR = 2$ and $E = 5$. Otherwise X_2 was $D = 1.2$, 250,000 m^2, $BRR = 2$, initial infectious patient $= 0$ (i.e., $n = 0$, $A_0 = 0$) and $E = 2$. Additionally, initial estimated populations of each city (X_1 and X_2) were $N_{1} = 44,444$ and $N_2 = 173,611$ (e.g., Eq. 2), but as mentioned above, those parameters such as populations, D and D^2 were dynamically altered its value because of permitting the mobility such as commuting and visiting between X_1 and X_2 in each $time_t$.

As a result, scaling difference between two cities did not normally engender different consequences except for each convergence speed reaching the catastrophe. Even if initial infectious patient was perfectly absent in city X_2, when it freely permitted mobility flows between the cities at each $time_t$, PI of each city (X_1 and X_2) were around 1.26 and 4.91 respectively. In many actual cases (e.g., China (Liu et al., 2020), EU cities), mobility flows including infectious patients and clinical latency across multiple cities would stir and spread more virus than this configuration. Lockdown tactics against the infectious city and mobility regulation across the cities could be approved in this context.

5   For Balancing Health Care and Economics

5.1   *Spatial Zoning by Mathematical Methods*

The COVID-19 pandemic raised the issue of how to achieve a balance between economy and health care. The author further proposes an idea, which is mathematically related to urban planning with the *"Maximum facility covering problem"*, *"Location Problem"* (Farahani & Hamacher, 2002) and the *"Partitioning algorithm"* (Andoni et al., 2018). Principally, it is based on the *optimal allocation of facilities* of multiple cities and compartmentalization through *zoning*. And it must be relying on the containment procedures of the epidemics (Lemon et al., 2007; WHO, 2018).

In the first place, politicians must anticipate crisis management for the civic life of the city in advance. And even if there are a few areas that stop functioning, the city need to have a resilient urban design that allows it to maintain a minimum amount of functionality and recover from the crisis.

The author exemplifies a basic procedure[16] in line with epidemics. For example, suppose there are N cities in a country, and infection is confirmed in one of them. First, this infected city must be locked-down from the outside, and all citizens who live in this city are not allowed to travel to and from the outside city and all routes also shuts down. These infected cities will focus on treating the infected patients within the city. The remaining cities, on the other hand, will either provide medical support to the infected cities, or focus on economic actions and mass production. Here, the problems are how to allocate necessary facilities for the remaining cities, and how to coordinate with those functions (medical supports or economic relation), and how to conduct zoning in across the cities such as safety zones (uninfected city areas) or danger zones (suspicious infection and infected areas).

In this way, Batty (1976) practically introduced algorithms and mathematical formulas in his urban modeling. Urban models for the epidemic prevention would be applied in various purposes. Quan et al. (2012) has similarly introduced their models based on graph theory and network structure for determining the location and management of affiliated hotels and stores in a particular region. The paper is also analyzed by algorithms by machine learning (semi-supervised learning).

In such algorithms, those models can design optimal location, zoning of hospitals and production sites. For other resources, it can also be applied to the problems how to install surveillance cameras, IoT and sensor networks to detect people's medical conditions and collective behavior in the city. The question of where and how much to install is optimal can be solved in this way.

To summarize a proposal of the author as following.

(1)   Infected city and its area must be locked-down, and it must separate necessary facilities and functions of remaining other cities from infected city and its area.

---

[16] What the data say about border closures and COVID spread, https://www.nature.com/articles/d41586-020-03605-6.

- It is more preferable that all city is well designed and constructed in such a concept that urban functions can be compartmentalized in anticipation of the pandemics and natural disasters in advance.
- Zoning *Green Zone* (uninfected safety area) and *Red Zone* (suspicious and infected area). And it also separates all routes green or red. Red zones and its adjoined routes must be blocked.
- Complete separation of the economic production area from the medical intensive area.
- In the area of economic production, it will continue to product mass produce foods, and other necessary goods. The same applies to water, gas and electric energy.
- In the medical intensive functioning area, the focus will be on medical care for infected patients. No one but patients and medical personnel will be allowed to enter. And as necessary, beds, tools and hospitals must be immediately expanded and produced.
- Uninfected areas can support for infected areas to transport necessary goods with strict health regulations.

(2)   Dynamic monitoring for citizens' health and tracking of citizens' behavior

- To minimize and optimize medical staff bottlenecks by strictly zoning, urban blockades and mobility regulation.[17]
- The ordinary citizens will be allowed to stay only in the green zone.
- Promoting PCR tests and other necessary medical surveillance for all citizens.
- Dynamic monitoring and tracking of the citizens using ICT, GIS and ubiquitous technology (Shibuya, 2004a, 2004b; Cybenko & Huntsman, 2019).
- To constantly manage the health of all citizens, and to build distributed database with the help of blockchain technology.
- Infected patients are immediately transferred to the medical intensive area, quarantined and isolated. In case of suspected infection, such person must be checked immediately.

Figure 7 denotes to do zoning green zones (uninfected areas and routes) or red zones (suspicious and infected areas and routes) in the city, and reallocate necessary facilities for the needs such as hospitals, industrial equipment, business and residences. And mobility restriction and monitoring health of the citizens as well as locking-down must be required without saying. A right pattern implies to be both dysfunction of economic and medical functions in the city as whole, meanwhile left pattern minimizes to both economic loss and medical burdens if the city as whole.

However, there are no examples of urban design that takes into account preventing mechanisms for the natural disasters and infectious diseases (e.g., COVID-19,

---

[17] Coronavirus: share lessons on lifting lockdowns: https://www.nature.com/articles/d41586-020-01311-x.

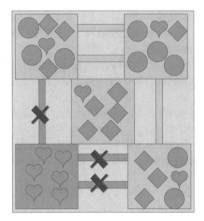 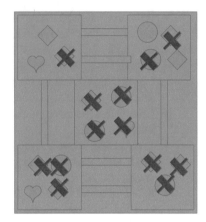

Fig. 7 Conceptualization on partitioning and zoning areas (green or red zone) in the city (left: a success case) or not (right: a failure case)

Ebora). Even in Japan, where natural disasters are common, there are no such examples.

In the actual case of the USA, there are 50 states, it seemed that they could not totally organize and conduct how to balance health care and the economy in the country. There seemed to be no design and leadership. A suggestion of the author would be to have certain states specialize in economic production, while isolating the badly infected states from the outside world. Infected states had to concentrate on health care. These measures would prevent infection while maintaining economic and productive activity throughout the country. Not to mention enforcing the necessary restrictions on mobility and zoning.

On the other hands, China's response was appropriate in this regard (Chen & Yuan, 2020). After completely locking down Wuhan city, China central government reallocated necessary production to other cities. All transportation to and from Wuhan city was blocked with the exception of medical staffs and supplies. And large new hospitals and beds were built by the direction from the central authority. For example, many cases of containment in China have been reviewed and reported (Maier & Brockmann, 2020; Zhang et al., 2020; Salzberger et al., 2020; Gatto et al., 2020). Questions have been raised as to whether all the correct data was being released, but at least within China, as of the end of June 2020, large-scale infections are not believed to have occurred outside of the Wuhan City area.

5.2   M/M/S Model on Public Health Issues in System Science

Furthermore, from a systems science point of view, the bottleneck in all socio-economic activities is the maintenance of a socio-economic system that must depend

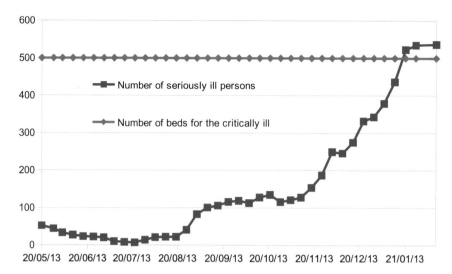

Fig. 8 Number of critical care and critical care beds in Tokyo (as of May 2020–February 3, 2021)

on the number of medical professionals (i.e., doctors, nurses, various therapists, pharmacists). This means that the daily life of the citizens and the socio-economic activities are highly dependent on the number of health care workers and their working time. For example, in Japan, after the second declaration of the state of emergency (after 7th, Jan., 2021), six indicators such as "hospital bed utilization rate", "positive rate", "number of newly infected patients" and others have been presented, but the most important indicator was missing. It is the "number of patients per medical doctor against COVID-19" and the "number of dedicated beds for COVID-19 patients". When these variables will exceed its criteria, no matter how many beds there are, there will be a medical collapse. In other words, all citizens must conduct their daily lives and economic activities in accordance with the "amount of medical services that medical doctors can provide per person" and not exceed it. This does not mean that health care workers should conform to daily life and economic activities.

In fact, working medical doctors in Japan are extremely busy. It has long been pointed out that their actual working hours and overtime hours are the worst level in the world. Since 37.8% of working physicians in hospitals (e.g., university hospitals, public hospitals) work more than 960 h per year, there have been proposals to not only shorten working hours but also to improve their work-life balance by FY2024 (the Ministry of Health, Labor and Welfare[18] in Japan). Faced with the COVID-19 pandemic, many medical doctors continued to treat patients for more than a year in the medical field under the added burden of medical work and the high risk of infection and death. However, when that limit eventually arrives, it will become impossible to support the medical field, and the country will be turned into the hell.

---

[18] https://www.mhlw.go.jp/content/10800000/000516867.pdf.

Figure 8 shows the number of critically ill people and the number of beds for the critically ill in Tokyo from May 2020 to February 3, 2021 (based on data from the Ministry of Health, Labor and Welfare, Japan). The number of beds for the critically ill in Tokyo was only 500, and the number of critically ill patients was within that range, but the beginning of 2021 was a period when the number of critically ill patients was the highest in Japan, and the number of deaths was so high that it exceeded that range. As shown in Figs. 9 and 10, there were a series of deaths of severely ill people, mainly the elderly, in early 2021, it was called the third wave (as of February 3, 2021, the cumulative number of deaths in Japan was about 6400). Similarly, those indicators in Tokyo have also recorded their worst (as of February 3, 2021, the cumulative number of deaths in Tokyo was about 1000). It is also clear that these figures had weighed heavily on the health care workers, placing an invisible burden on them.

In order to resolve such a situation, mathematical planning and proactive measures are essential. Here, various problems related to public health and infection prevention, such as the number of hospitals that can handle corona, the number of beds dedicated to corona patients, the shortage of medical personnel, and the "vaccination problem" can be considered mathematically as the same M/M/S model, or "*queueing problem*". These are studies that have been devised in operations research, management engineering, and computational social science (Shibuya, 2020a, 2020b). The queueing problem in infectious disease control and public health can be envisioned in the following situations. This is because it formulates a mathematical model of a situation in which patients and citizens continuously visit and stay (wait for services) in order to receive services such as medical care.

(1)   **[Vaccination issue]** How can we efficiently vaccinate all those who wish to be vaccinated while minimizing the waiting time for vaccination?

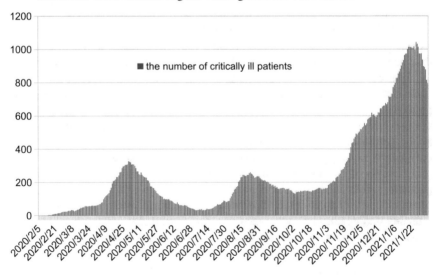

Fig. 9 The number of critically ill patients in Japan (as of May 2020–February 3, 2021)

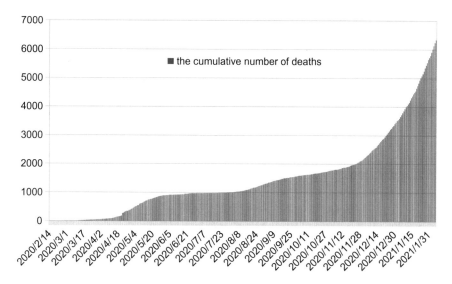

Fig. 10   Cumulative number of deaths in Japan (as of May 2020–February 3, 2021)

(2)   **[Medical service provision issue]** How much medical service can be provided to the patients who rush to the hospital every day in response to the availability of dedicated beds?

(3)   **[Resource distribution problem]** How can we effectively distribute various resources (e.g., masks, cash) in line with the social distance policy (i.e., limited arrival interval)?

Here, the event of a patient or citizen's visit is a contingent event that follows a Poisson distribution (discrete probability distribution), while the event of service provision time, such as medical care, is considered to be a system defined by statistical fluctuations that follow an exponential distribution (continuous probability distribution). The constraint of "serviceable medical care" acts on the entire system as the bottleneck mentioned above, and since the quality and quantity of the bottleneck determine the performance of the entire system, it is necessary to grasp these numbers dynamically as much as possible. In the above example, the "number of available contact points" is likely to be the first bottleneck. For example, in (1), the number of doctors to be vaccinated, in (2), the set of the number of available beds and the number of dedicated medical personnel (it is not possible to have an excess of either), and in (3), the variable representing the number of reception desks to distribute necessary resources for the citizens. If these variables are not sufficient to meet the demand (patients, citizens, etc.), then (1) "major delays in vaccination" ($\rightarrow$ vaccine spoilage, piles of vaccines in stock, etc.), (2) "collapse of medical care" and "increase in the number of waiting people receiving treatment at home" ($\rightarrow$ increase in the number of deaths due to lack of access to medical care), and (3) "claims of not being able to receive necessary support when needed" ($\rightarrow$ financial support may not be provided in time for needy households) will occur.

In vaccination case, suppose that citizens come to "one doctor" at a vaccination site at a rate of one every five minutes (Poisson distribution), and that the average time required to vaccinate each citizen is three minutes (exponential distribution).

(1)   The probability that one doctor is vaccinating a citizen, ρ (= utilization)

Define the average arrival rate λ (person/minute) and service rate μ (person/minute).

$$\lambda = 1/5, \mu = 1/3$$

Therefore, the utilization rate ρ is as follows

$$\rho = \lambda/\mu = 0.60$$

(2)   Average waiting time

According to the following equation, input into parameters such as ρ, and μ.

$$W = \frac{1}{\mu} \frac{\rho}{1-\rho} \tag{6}$$

$W = 4.5$, and the average waiting time is 4.5 min.

In fact, the response to the COVID-19 pandemic was a test of the accuracy of predictions and planning based on such mathematical models in each country. For example, in Japan, from the beginning to the middle of 2021, (1) the "vaccination problem" became an actual issue of concern. In particular, securing medical doctors as well as vaccine delivery plans became a serious matter. In other words, the bottleneck was the number of doctors in this case, and it would determine the number of people who could be vaccinated per day, which in turn would determine not only the progress of vaccination in the country as a whole, but also whether or not the plan to prevent the spread of infection would succeed. Furthermore, since the beginning of 2021 corresponded to the worst period of infection spread in Japan, the number of beds available and the number of full-time medical staff in (2) was so tight, and the number of medical doctors who could help with vaccination was limited. As a result, there was an extremely high risk of a vicious cycle in which infection would spread further, serious infectious cases would occur one after another, beds would not be available, and medical doctors would not be able to cooperate in vaccination.

5.3   The Importance of Modeling for Public Health and Infection Prevention Planning

Thus, it is necessary to take all possible measures to achieve the goal of ensuring public health, preventing infections and preventing the spread of infections, and it is worthwhile to do so even if it means making reasonable sacrifices to citizens and businesses, as long as the benefits of achieving the goal are relatively greater.

However, the reality was that in many countries around the world, the neglect of these bottlenecks had led to health care providers overstepping their bounds, the reckless timing of economic stimulus measures without regard for these bounds, and the lack of scenario-based forecasting and preparation for what to do in the face of the worst-case scenario had led to confusion in policy and medical practice, resulting in a collapse of the balance between profit and loss for the country as a whole.

When the global COVID-19 pandemic escalated in early 2021, it seemed reasonable at first glance to increase the number of full-time medical doctors and build more medical facilities and beds, the idea being to allocate surplus personnel not working on COVID-19 to its response. However, this is unacceptable from a systems science perspective. The total number of medical personnel is a valuable human resource in each country, and it is difficult to train them easily. Allocating a large number of personnel for a specific disease would sacrifice medical care for other diseases. Furthermore, it will be difficult to dispatch medical personnel in the event of the unexpected natural disasters (in fact, in February 2021, a major earthquake occurred in Japan during the pandemic). In the end, it can understand that the number of medical personnel is the bottleneck of such the crises, and further they need to proceed with vaccination at the same time. In other words, in order to save one person in the immediate, other people who can be saved is thrown away. Rather, from a systems science point of view, the existence of redundancy is essential to cope with such unpredictable statistical fluctuations and randomness. In order to cope with such crises, it should not be fixated only on the simple events that it can see. We need to look at the system as a whole, the statistical variability, the dependent events, and the complexity of these factors. The only way to achieve this is to maintain the socio-economy in line with the number of healthcare workers, and to reduce the burden on healthcare workers by strictly enforcing the necessary legal systems such as restrictions on citizens' behavior and lockdowns in cities.

During the pandemic, some prefectures in Japan have made a sharp distinction between hospitals for the critically ill and those for the mildly ill, and have taken the idea of transferring critically ill patients to hospitals for the mildly ill once they have recovered, leaving beds for the critically ill vacant to receive the next critically ill patients. However, this is also impossible without a well-developed mathematical plan and information management on overall availability and logistics. In another study, the author introduced a simulation to match refugees with acceptable cities using the Deferred Acceptance Algorithm (DAA: Roth, 2008) and genetic algorithm optimization (Shibuya, 2020a). It can also be applied to the problem of matching hospitals.

In 2021, the Japanese government finally began to detail policy scenarios according to a systematic mathematical model and to verify their effectiveness before they are implemented. If it can secure medical personnel for vaccination and provide citizens who wish to be vaccinated, while reducing the burden on the medical field, then it will be able to find solutions to explore scenarios for economic recovery. In actual, at the National Diet in session on February 9, 2021, an estimate was presented that if vaccination of the elderly was completed by the end of June 2021, the number of deaths could be reduced by more than 6,000 and economic losses could be curbed. If vaccination proceeds at a pace of 4 million people per week starting from the first week of April 2021, the cumulative number of deaths by the end of 2021 could be reduced by more than 6,000 compared to the pace of 1.6 million people per week in the USA and other countries where vaccination was more advanced, and GDP in Japan was expected to improve by about 0.08%, as it depended on the labor productivity of workers. However, in another scenario, if the vaccination rate was 1 million per week, the number of deaths would increase by more than 10,000.

In the age of AI, thus, the issue of "distribution of sacrifice" can be significantly improved in policy decision making, if it takes into account appropriate policy planning and analysis of the prevention of disease rather than the dichotomy of medical care versus economics.

6 Discussion

Generally, as everyone already experienced in global, the COVID-19 pandemic suggested the accessible level to public medical services, effectiveness of the governmental imperatives for prevention control, quality of public health cares, and citizens' consciousness for their daily health in each country. In this point, the USA and the EU states were comparatively the worst cases than other developed countries and regions (e.g., Taiwan, Japan). Then, many of the governments and WHO could not be effectively, the global world was rife with anxiety about the infection and its spreading risks.

Putting it simply to consider, the last resort against COVID-19 was to delay its contagion by manually inhibiting mobility of the citizens. Nevertheless, in the early 2020s, when not all citizens could be vaccinated, natural immunization policies were not enough. In many countries, the spread of the disease has spiraled out of control, leading to increased deaths and medical collapse in the field. Many governments failed to have a concrete plan to curb the rapid spread of the disease from its early stages. Thus, since the outbreak of COVID-19, citizens and governments lost a lot of lives, time, and national wealth, apart from valuable lessons learned.

In all actual cases, the policy of inhibiting against mobility across multiple cities would be effectively if such imperatives were committed since the earlier stage of infectious process. For such purposes, as mentioned before, it should be conducted firstly by not only medical checks but somewhat AI based technologies in the digitized age. Furthermore, aforementioned modeling and forecasting on epidemic risk will also contribute to evaluating emergent possibilities on financial systemic risk and demands on the insurance against the global pandemic (Bedson et al., 2021).

Exactly, as simulation results of this chapter denoted, imperative policies in each country simultaneously must conduct quarantining suspicious cases and achieving isolation of detected patients as necessary and fast as possible. Vulnerability based on the level of medical and public health care is not only the degree of provision of an effective medical system against a pandemic, but also causes socio-economic systemic risk worldwide. That is a reason why it is necessary to institutionalize high quality normal medical care and public health policies in each country on a regular basis.

Therefore, monitoring and collaboration based on data science (Allam, 2020) and AI technology should be organized internationally to detect signs of a pandemic before it happens. The WHO has certainly been coordinating with other countries and medical institutions, but it is difficult to take countermeasures simply by restricting the movement of citizens as before. Therefore, an automatic detection and infection prevention system using smart sensing technology is quite needed.

As mentioned before, our digitized society has achieved smart-city and smart-intelligent services such as wearable devices, location-based services and ubiquitous sensor network (Shibuya, 2004a, 2020a; Clifton, 2016; Geng, 2017; Marin et al., 2016; Pentland, 2014). For example, each citizen can wear on one of the arbitrary digital tools which equipped with AI driven smart-sensors, and those computational mechanisms can periodically monitor user's health data (e.g., Heart rate, body temperature, breathing state, kinetic motion, walking, daily energy consumption of calorie) and transmit those big-data to the cloud data storage. In addition, each user's communication log and its contents may be reflected a personalized medical care for their latent symptom (Ginsberg, 2009; Bakker et al., 2016; Broniatowski et al., 2013). If there is close contact in a public space, the AI system can quickly warn citizens (e.g., Pan-European Privacy-Preserving Proximity Tracing[19]). Certainly, consideration must be given to the respective privacy and prejudice against infected people (Mello & Wang, 2020).

In actual cases of China, pandemic of the COVID-19 facilitated their computational health cares and advanced services using big-data in their homeland. Digital transformation based on the AI driven health services has a strong potential to enhance promoting our quality of life and detecting anomaly condition such as suspicious case of the epidemics in advance. At the earlier stage, such smart-space driven by the AI will regulate and control infectious growth for prevention (Zhu et al., 2019).

Similarly, in the UK, the National Health Service (NHS[20]) reported the effectiveness of the COVID-19 contact verification application (using an API co-developed by Google and Apple). This application was launched in September 2020 in the UK and estimates that over 21 million people have downloaded it and 1.7 million people have been notified to voluntarily quarantine themselves, preventing about 600,000 infections by the end of December, 2020 (Wymant et al., 2021).

In Israel, on the contrary, the Health Maintenance Organization (HMO)[21] has been centrally managing personal information including health and medical data of citizens for some time. In the case of the COVID-19 pandemic, the prime minister himself negotiated directly to Pfizer to set up a vaccination system in the country by

---

[19] https://www.pepp-pt.org/.

[20] https://www.nhs.uk/.

[21] https://govextra.gov.il/ministry-of-health/corona/corona-virus-en/.

the end of 2020, and consequently it had been realizing an early vaccination policy for citizens.

But in Japan, the Ministry of Health, Labor and Welfare (MHLW) and other government agencies had been seriously lagging behind in digitization, which led to a number of problems. With regard to vaccination, developing medical information database[22] and the high level of checks on safety evaluation and risk assessment were sufficient, as Japan has experienced drug-related injuries in the past. However, the delay in the digitalization also caused serious matters on both the vaccine supply network and the ongoing management of the vaccination system. In this way, it can be said that the thoroughness with which the government has not been promoting information management and state governance through the promotion of digitalization has been exposed in this respect as well.

Appendix

As described in this chapter, one of the examples written in Python can be simulated as follows.

```
#----------------------------
# COVID-19 Toth Model of Percolation Phenomenon (ver.2021)
# by Kazuhiko Shibuya, Ph.D.
# Python ver.3.9
#----------------------------
import math
import numpy as np
from matplotlib import pyplot as pyt
from scipy.stats import entropy

T = 100 * 100 # 100 m * 100
D = 0.5 # 50 cm
D2 = Math.Pow (D, 2)
N = (1.154 * T)/(D2)

C = 30 #Max Loop

BRR = 2 # 8 = Max of Moore neighborhood
n = 1 # initial infectious client(s)
E = 2 # environmental risk constant

At = 0
It = 0
Rd = 0
St = N
```

---

[22] https://www.pmda.go.jp/safety/surveillance-analysis/0018.html.

```
Pr = 0
Pc = 0
H = 0
Rd = 0

a = np.zeros (C + 1)
b = np.zeros (C + 1)
c = np.zeros (C + 1)
x = np.zeros (C + 1)
y = np.zeros (C + 1)
z = np.zeros (C + 1)
t = np.zeros (C + 1)
#---------------------------
# 【Simulation】
#---------------------------
At = n
lp = 0
while (lp ≤ C)

    try:

        St = N – At # Healthy citizens
        Pr = At/N # Occupancy of the infected person in the space
        Pc = ((At * D2) / 1.154) / T #Percolation threshold
        p = Pc
        H = entropy([p,1.0 - p], base = 2) #entropy
        #Probability of people being directly infected by infected people and their
        neighbors
        Rd = Pr * E * (BRR/8)
        It = Rd * St #Number of newly infected people
        At = At + It #Cumulative number of infected people

        a [lp] = Pc
        b [lp] = Rd
        c [lp] = H
        x [lp] = It
        y [lp] = St
        z [lp] = At
        t [lp] = lp
        lp = lp + 1

    except Exception as e:

        print (e)

fig = pyt.figure()

ax1 = fig.add_subplot(2, 2, 1)
```

```
ax1.set_title('COVID-19 Model')
ax1.set_xlabel('Time (s)')
ax1.set_ylabel('Infection')
ax1.grid(True)
ax1.plot(t, x, color = "r",label = "It", linestyle = "solid")
ax1.plot(t, y, color = "b",label = "St", linestyle = "dashed")
ax1.plot(t, z, color = "y",label = "At", linestyle = "dashed")
ax1.legend()

ax2 = fig.add_subplot(2, 2, 2)
ax2.set_xlabel('Time (s)')
ax2.set_ylabel('Prob.')
ax2.grid(True)
ax2.plot(t, a, color = "g",label = "Pc", linestyle = "solid")
ax2.plot(t, b, color = "m",label = "Rd", linestyle = "dashed")
ax2.plot(t, c, color = "k",label = "H", linestyle = "dashed")
ax2.legend()

ax3 = fig.add_subplot(2, 2, 3)
ax3.set_xlabel('Time (s)')
ax3.set_ylabel('Infection')
ax3.grid(True)
ax3.bar(t, y, color = "b",label = "St", linestyle = "dashed")
ax3.bar(t, z, color = "y",label = "At", linestyle = "dashed")
ax3.bar(t, x, color = "r",label = "It", linestyle = "solid")
ax3.legend()

ax4 = fig.add_subplot(2, 2, 4)
ax4.set_xlabel('Time (s)')
ax4.set_ylabel('Prob.')
ax4.grid(True)
ax4.bar(t, a, color = "g",label = "Pc", linestyle = "solid")
ax4.bar(t, b, color = "m",label = "Rd", linestyle = "dashed")
ax4.bar(t, c, color = "k",label = "H", linestyle = "dashed")
ax4.legend()

pyt.show()
```

References

AIST (Ed.) (2013). *Handbook of measuring human behavior*. Asakura Press. (in Japanese).
Allam, Z. (2020). *Surveying the Covid-19 pandemic and its implications urban health*. Elsevier.

Aleta, A., et al. (2020). Modelling the impact of testing, contact tracing and household quarantine on second waves of COVID-19. *Nature Human Behaviour*. https://doi.org/10.1038/s41562-020-0931-9

Anderson, J. E. (2011). The gravity model. *Annual Review of Economics, 3*(1), 133–160.

Andoni, A., et al. (2018). Data-dependent hashing via nonlinear spectral gaps. In *Proceedings of 50th annual ACM SIGACT symposium on the theory of computing (STOC'18)* (pp. 787–800).

Bakker, D., Kazantzis, N., Rickard, D., & Rickard, N. (2016). Mental health smartphone apps: Review and evidence-based recommendations for future developments. *JMIR Mental Health, 3*(1), e7.https://doi.org/10.2196/mental.4984

Batty, M. (1976). *Urban modeling: Algorithms*. Predictions, Cambridge University Press.

Bedson, J., et al. (2021). A review and agenda for integrated disease models including social and behavioural factors. *Nature Human Behaviour, 5*, 834–846.

Britton, T., Ball, F., & Trapman, P. (2020). A mathematical model reveals the influence of population heterogeneity on herd immunity to SARS-CoV-2. *Science*. https://doi.org/10.1126/science.abc6810

Broniatowski, D. A., Paul, M. J., & Dredze, M. (2013). National and local influenza surveillance through twitter: An analysis of the 2012–2013 influenza epidemic. *PLoS One, 8*(12), e83672.https://doi.org/10.1371/journal.pone.0083672

Boulos, M. N. K., & Geraghty, E. M. (2020). Geographical tracking and mapping of coronavirus disease COVID-19/severe acute respiratory syndrome coronavirus 2 (SARS-CoV-2) epidemic and associated events around the world: How 21st century GIS technologies are supporting the global fight against outbreaks and epidemics. *International Journal of Health Geographics, 19*, 8. https://doi.org/10.1186/s12942-020-00202-8

Chen, L., & Yuan, X. (2020). China's ongoing battle against the coronavirus: Why did the lockdown strategy work well? *Socio-Ecological Practice Research, 2*, 175–180.

Chiu, W. A., Fischer, R., & Ndeffo-Mbah, M. L. (2020). State-level needs for social distancing and contact tracing to contain COVID-19 in the United States. *Nature Human Behaviour, 4*, 1080–1090.

Clifton, D. A. (Ed.) (2016) *Machine learning for healthcare technologies*. Oxford University Press.

Cobey, S., et al. (2020). Modeling infectious disease dynamics. *Science, 368*(6492), 713–714. https://doi.org/10.1126/science.abb5659

Cybenko, G., & Huntsman, S. (2019). Analytics for directed contact networks. *Applied Network Science, 4*(106). https://doi.org/10.1007/s41109-019-0209-1

Dai, S., et al. (2020). Temporal social network reconstruction using wireless proximity sensors: model selection and consequences. *EPJ Data Science, 9*(19). https://doi.org/10.1140/epjds/s13688-020-00237-8

Doremalen, D. H., et al. (2020). Aerosol and surface stability of SARS-CoV-2 as compared with SARS-CoV-1. *The New England of Journal of Medicine*. https://doi.org/10.1056/NEJMc2004973

Farahani, R. Z., & Hamacher, M. (Eds.) (2002). *Facility location: Application and theory*. Springer.

Gatto, M., et al. (2020). Spread and dynamics of the COVID-19 epidemic in Italy: Effects of emergency containment measures. *PNAS, 117*(19), 10484–10491.

Geng, H. (Ed.) (2017). *Internet of things and data analytics handbook*. Wiley.

Ginsberg, J., et al. (2009). Detecting influenza epidemics using search engine query data. *Nature, 457*, 1012–1014.

Girvan, M., & Newman, M. E. J. (2002). Community structure in social and biological networks. *Proceedings of National Academy Science, 99*(12), 7821–7826.

Gonzalez, M. C., Hidalgo, C. A., & Barabasi, A. (2008). Understanding individual human mobility patterns. *Nature, 453*, 779–782.

Haining, R. (1990). *Spatial data analysis in the social and environmental sciences*. Cambridge University Press.

Hao, X., et al. (2020). Reconstruction of the full transmission dynamics of COVID-19 in Wuhan. *Nature*. https://doi.org/10.1038/s41586-020-2554-8

Jiang, B., & Yao, X. (2010). *Geospatial analysis and modelling of urban structure and dynamics.* Springer.

Jin, et al. (2020). A rapid advice guideline for the diagnosis and treatment of 2019 novel coronavirus (2019-nCoV) infected pneumonia (standard version). *Military Medical Research, 7*(4). https://doi.org/10.1186/s40779-020-0233-6

Ksiazek, T. G., et al. (2003). A novel coronavirus associated with severe acute respiratory syndrome. *The New England Journal of Medicine, 348*(20), 1953–1966.

Kuperman, M., & Abramson, G. (2001). Small world effect in an epidemiological model. *Physical Review Letters, 86*(13), 2909–2912.

Lavezzo, E., et al. (2020). Suppression of a SARS-CoV-2 outbreak in the Italian municipality of Vo. *Nature.* https://doi.org/10.1038/s41586-020-2488-1

Lemon, S. M., et al. (2007). *Ethical and legal considerations in mitigating pandemic disease: Workshop summary.* The National Academic Press. https://www.ncbi.nlm.nih.gov/books/NBK54163/

Liu, X., & Stechlinski, P. (Ed.) (2017). *Infectious disease modeling.* Springer.

Liu, Z. et al. (2020). Modeling the trend of coronavirus disease 2019 and restoration of operational capability of metropolitan medical service in China: A machine learning and mathematical model-based analysis. *Global Health Research and Policy, 5,* 20.

Lopez-Garcia, M., et al. (2019). A multicompartment SIS stochastic model with zonal ventilation for the spread of nosocomial infections: Detection outbreak management, and infection control. *Risk Analysis, 39*(8), 1825–1842.

Maier, B. F., & Brockmann, D. (2020). Effective containment explains subexponential growth in recent confirmed COVID-19 cases in China. *Science, 368*(6492), 742–746. https://doi.org/10.1126/science.abb4557

Malarz, K., & Galam, M. (2005). Square-lattice site percolation at increasing ranges of neighbor-bonds. *Physical Review E, 71,* 016125.

Mallapaty, S. (2020). Why does the coronavirus spread so easily between people? *Nature.* https://doi.org/10.1038/d41586-020-00660-x

Marin, H., et al. (Eds.) (2016). *Global health informatics (1st Ed.): How information technology can change our lives in a globalized world.* Oxford University Press.

Mello, M. M., & Wang, C. J. (2020). Ethics and governance for digital disease surveillance. *Science, 368*(6494), 951–954.

Moore, C., & Newman, M. E. J. (2000). Epidemics and percolation in small-world networks. http://www.santafe.edu/sfi/publications/Working-Papers/00-01-002.pdf

Munzert, S., Selb, P., Gohdes, A., et al. (2021). Tracking and promoting the usage of a COVID-19 contact tracing app. *Nature Human Behaviour.* https://doi.org/10.1038/s41562-020-01044-x

Newman, M. E. J. (2002). The spread of epidemic disease on networks. http://www.santafe.edu/sfi/publications/Working-Papers/02-04-020.pdf

Nishiura, H., et al. (2016). (2016) Identifying determinants of heterogeneous transmission dynamics of the middle east respiratory syndrome (MERS) outbreak in the Republic of Korea, 2015: A retrospective epidemiological analysis. *British Medical Journal Open, 6,* e009936. https://doi.org/10.1136/bmjopen-2015-009936

O'Sullivan, D., & Perry, G. L. W. (2013). *Spatial simulation: Exploring pattern and process.* Wiley.

Pan, et al. (2013). Urban characteristics attributable to density-driven tie formation. *Nature Communications, 4.*https://doi.org/10.1038/ncomms2961

Pentland, A. (2014). *Social physics.* Penguin Press.

Prather, K. A., Wang, C. C., & Schooley, R. T. (2020). Reducing transmission of SARS-CoV-2. *Science, 368*(6498), 1422–1424. https://doi.org/10.1126/science.abc6197

Qi, M., Wang, Z., He, Z., & Shao, Z. (2019). User Identification across asynchronous mobility trajectories. *Sensors, 19*(9), 2102. https://doi.org/10.3390/s19092102

Quan, X., et al. (2012). Rink graph analysis for business site selection. *IEEE Computer, 45*(3), 64–69.

Renso, C., et al. (2013). *Mobility data.* Cambridge University Press.

Rivers, C., et al. (2019). Using "outbreak science" to strengthen the use of models during epidemics. *Nature Communications, 10*(3102). https://doi.org/10.1038/s41467-019-11067-2

Rossi, L., et al. (2015). Spatio-temporal techniques for user identification by means of GPS mobility data. *EPJ Data Science, 4*(11). https://doi.org/10.1140/epjds/s13688-015-0049-x

Roth, A. E. (2008). Deferred acceptance algorithms: History, theory, practice, and open questions. *International Journal of Game Theory, 36*, 537–569.

Salzberger, B., Glück, T., & Ehrenstein, B. (2020). Successful containment of COVID-19: The WHO-Report on the COVID-19 outbreak in China. *Infection, 48*, 151–153.

Shibuya, K. (2004). A framework of multi-agent based modeling, simulation and computational assistance in an ubiquitous environment. *SIMULATION: Transactions of the Society for Modeling and Simulation International, 80*(7), 367–380.

Shibuya, K. (2004). Perspectives on social psychological research using agent based systems. *Studies in Simulation & Gaming, 14*(1), 11–18. (in Japanese).

Shibuya, K. (2006). Actualities of social representation: Simulation on diffusion processes of SARS representation. In C. van Dijkum, J. Blasius & C. Durand (Eds.), *Recent developments and applications in social research methodology*. Barbara Budrich-Verlag.

Shibuya, K. (2020). *Digital transformation of identity in the age of artificial intelligence.* Springer.

Shibuya, K. (2020b). *Identity health.* https://www.ncbi.nlm.nih.gov/pmc/articles/PMC7121317/; *Digital transformation of identity in the age of artificial intelligence.* Springer.

Shibuya, K. (2021). A spatial model on COVID-19 pandemic. In *The 44th Southeast Asia seminar, the Covid-19 pandemic in Japanese and Southeast Asian perspective: Histories, states, markets, societies.* Kyoto University.

Shryock, H. S., & Siegel, J. S. (1973). *The methods and materials of demography* (Vol. 2, U.S. Bureau of Census).

Squazzoni et al. (2020). Computational models that matter during a global pandemic outbreak: A call to action. *Journal of Artificial Societies and Social Simulation, 23*(2), 10. http://jasss.soc.surrey.ac.uk/23/2/10.html

Starnini, M., et al. (2013). Modeling human dynamics of face-to-face interaction networks. *Physical Review Letters.* https://doi.org/10.1103/PhysRevLett.110.168701

Tóth, L. F. (1940). Über einen geometrischen Satz. *Mathematische Zeitschrift, 46*, 83–85. (in Germany).

Tóth, L. F., Sneddon, I. N., Ulam, S., & Stark, M. (1964). *Regular figures: International series of monographs on pure and applied mathematics.* Pergamon.

Venkatramanan, S., et al. (2021). Forecasting influenza activity using machine-learned mobility map. *Nature Communications, 12*, 726. https://doi.org/10.1038/s41467-021-21018-5

Watts, D. J. (1999). *Small worlds.* Princeton University Press.

WHO (2018). *Managing epidemics.* https://www.who.int/emergencies/diseases/managing-epidemics-interactive.pdf

WHO (2020). Report of the WHO-China joint mission on coronavirus disease 2019 (COVID-19).

Wymant, C., et al. (2021). The Epidemilogical impact of the NHS COVID-19 App. https://github.com/BDI-pathogens/covid-19_instant_tracing/blob/master/Epidemiological_Impact_of_the_NHS_COVID_19_App_Public_Release_V1.pdf

Xu, C., et al. (2020). Estimation of reproduction numbers of COVID-19 in typical countries and epidemic trends under different prevention and control scenarios. *Frontiers of Medicine.* https://doi.org/10.1007/s11684-020-0787-4

Zhang, S., et al. (2020). COVID-19 containment: China provides important lessons for global response. *Frontiers of Medicine, 14*, 215–219.

Zhu, T. T., Pimentel, M. A. F., Clifford, G. D., & Clifton, D. A. (2019). Unsupervised Bayesian inference to fuse biosignal sensory estimates for personalizing care. *IEEE Journal of Biomedical and Health Informatics, 23*(1), 47–58.

AI and Our Society

AI Driven Scoring System and "Reward or Punish" Based on a Theory of Han Fei

1 Human Resources for Economic Development

The COVID-19 pandemic has posed the challenge of how to sustain economic growth, human and material capital, and democracy. This is why the differences between Western society and China's political and economic system had attracted so much attention (Bell, 2000, 2016).

According to Inoki (2014), one view is *the institutional view*: "a government with limited power such as by principally establishing democracy and guaranteeing property rights should be firstly established, and then investment in human and material capital should be promoted, which will lead to economic growth". In contrast, the second hypothesis is *the development view*: "First, leaders who start investing in human and material capital and emphasize the market will allow private ownership in their policies, and as the level of education increases and wealth accumulates, democracy and other political systems will be improved".

The former may be closer to Western society, meanwhile the latter is to the policies of the modern Chinese government. In addition, the two hypotheses have opposite causal relationships. Inoki posed further questions: Which of these two opposing hypotheses is more useful for viewing historical developments or contemporary international political developments? He also said that these hypotheses have not been tested in a satisfactory manner and no conclusions have been reached yet. But at present, human resources (i.e., the intellectual and moral quality of the citizens) seems to be the most important factor for national growth and democratization. Differences in political institutions are only a secondary effect on the performance of the economy, with the enhancement of human and material capital being the most important factor.

In order to pursue this issue, education and human resource development, including the cultivation of social morality, are quite essential for further discussions. For example, in Western society, Bentham (1789), a pioneer of utilitarianism, wrote "An Introduction to the Principles of Morals and Legislation". The preface is a masterpiece. "*Nature has placed mankind under the governance of two sovereign*

© The Author(s), under exclusive license to Springer Nature Singapore Pte Ltd. 2022
K. Shibuya, *The Rise of Artificial Intelligence and Big Data in Pandemic Society*,
https://doi.org/10.1007/978-981-19-0950-4_5

matters, pain and pleasure. it is for them alone to point out what we ought to do,
as well as to determine what we shall do. On the other hand the standard of right
and wrong, on the other chain of causes and effects, are sastend to their throne".
Since merit is, after all, the principle that guides people toward profit and encourages
them to avoid damage, it is too natural that the standards of good and evil values
will follow, and the chain of cause and effect related to them will also need a system
based on it. This also shows that it is possible to combine morality and the legal
system to educate ordinary people.

Similarly, this perspective has existed for some time in Asia, particularly in China.
Then, the author pays attention to the *"reward or punishment"* based on a theory of
Han Fei in the ancient China.

2   A Theory of Han Fei and AI Scoring Services

2.1   *Han Fei*

Han Fei, who was an ancient theorist in political governance (280BC–233BC), was
invited to see an employer who was a first emperor of the Qín dynasty[1] of the ancient
China. And he proposed his written book for this emperor which compiled his idea to
govern the nation and retainers. A set of elaborated ideas which inscribed his theoretical backbone drastically characterized the nature of the human as maliciousness and
untrustworthy. If a ruler will stand on the top of the hierarchy for governing nations,
he must handle own power to keep his dynasty and control against the retainers and
ordinary citizens.

Han Fei further told one of his secrets for the effective ruling governance (e.g.,
despotism) in his book, for example, it was a rule of *reward or punish*. It concretely
means *"rewarding for the goodness of retainers or punishment against the badness
of them"*. This rule was conducted by only an emperor, and he can exclusively decide
and legitimate what better or worse it is. Namely, only an emperor decides what is
valuable in his preferential order, and then he evaluates something and someone by
this value standard. He must always exhibit an own source of the highest power for
all retainers and citizens, and hence he can let strictly them to be controlled by the
directions what an emperor desire. Of course, for such purposes, anyone seeks to
pursue merits and avoid punishment given by an emperor. Moreover, he also crucially
said that even if clever scholars cannot be induced by incentives and intentions of an
emperor, such person should be killed.

In brief, such robotic retainers should be tamed by such rule and deployed around
an emperor. His theory seemed to be very favorable for an emperor, and an empire
Qin had finally achieved to synthesize all smaller countries as an entire China.

---

[1] https://en.wikipedia.org/wiki/Qin_dynasty.

In this respect, a theory of Han Fei probably resembled Machiavellianism in Western culture. In addition, Spinoza also mentioned the effectiveness of the king's rule of the state based on "fear and reward" in his article "*Tractatus Politicus*". Thus, despite the differences in historical backgrounds between Eastness and Westness, many philosophers argued for the desirability of a "reward and punishment" system of state governance.

2.2   Meeting Between Han Fei and the AI

Here, in this regard, the author specifically discusses the similarities between AI-based scoring services (e.g., Zhima Credit) and a theory of Han Fei.[2] Han Fei's theory discussed the ways of ruling the nation based on thorough a rule of *reward or punish*, but at the same time backed by a cynical and radical attitude towards the human existence. It could be said that this provided a theoretical foundation for the governance of the nation in both ancient and contemporary China. Furthermore, this idea likely extends to digitize the AI scoring mechanism, which may be a distant cause of China's early adoption of AI-based monitoring systems and their usage to govern the peoples. Certainly, because there are issues of the free rider such as *the tragedy of common*, monitoring and punishment system is still required in various situations despite contemporary society.

Digital transformation has been progressing rapidly in modern China, please remind the AI scoring service such as Zhima Credit (Shibuya, 2020a). Such implemented algorithms driven by the AI measures what is value in order. And a calculated score measured by the AI can examine a credit of someone, and it offers automatic services for anyone by the extent what the AI evaluated. For such goals, it sought to be aspirated by the strong intensives which leveraged by economic reasons or devastating risks. As it is likely Han Fei's theory, it can paraphrase reward or punish. In this point, scoring services driven by the AI will likewise approach an emperor who exclusively decides. Of course, AI based evaluation system such as deep learning mechanism always requires fairness and accuracy for everyone (Chouldechova & Roth, 2020).

2.3   Descending a Perfect World

When such social situation has established, as mention later chapter, western philosophy of Spinoza and Leibniz interlinked with Fei's regulating way to keep the morality among the citizens. Leibniz (2017) described such ideal situation in his book Monadology: "*Finally, under this perfect government no good action would be unrewarded and no bad one unpunished, and all should issue in the well-being of*

---

[2] https://en.wikipedia.org/wiki/Han_Fei.

the good, that is to say, of those who are not malcontents in this great state, but who trust in providence".

One of the misfortunes of the democratic nations around the world is that they do not have enough *reward or punish* for their citizens. Like an emperor in ancient China, it should evaluate and determine what is good and what is bad, based on government leaders' self-worth and self-intention, and to reward or punish them. If not so, inducing people to act on the basis of their value standard through specific policies does not work well enough, nor does specifying penalties ensure the compliance with them. If the AI-based scoring services are to take root in democracies of the Westness in the future, there will be a difference in the effectiveness of their governance between those linked to all administrative services and private property and that are not. The same was true for the COVID-19 pandemic. Governments came up with measures to guide citizens' daily actions, but few of them were effective.

Because the capitalism is closely linked to the democratic political system, which is based on the private rights of each individual, such system of the governance is strictly embedded in the individualism (Locke, 1998). Particularly, in Western societies, the legal system has been established with a view to minimizing state intervention and interference in the private rights of each independent an autonomous individual. In other words, viewing a rule of reward or punish, the value of the rulers cannot induce each citizen to obey the value of ruling leaders of the governance, rather they usually prioritize their individual preferences and the evaluation of each commodity in the marketplace.

On the other hand, self-organized and effective institutional design is also known. There is a comparative historical institutional analysis that suggests that the mutual sharing of reputation and trust information made commercial transactions possible even in ancient times when information and communication technologies were quite poor (Greif, 2006). In such socio-economic institutional system in the medieval age of western countries, many stakeholders must monitor each other to make sure that the agent does not betray the client or violate business transactions. Such historical information of transaction is considered important for reputation of each individual for trust. In particular, it effectively designs the "punishment" as sanction.

To be sure, in many modern countries, various incentives are provided to get markets and individuals to implement policy intentions, but the penalties for these themselves are often ineffective. Even if they behave in accord with the policy intentions, it would take a considerable amount of time.

2.4 The China's Hands Over the World

On the other hand, however, in a state-driven socioeconomic system like China's, even if the citizens are given capitalist incentives, each citizen cannot ignore the will of the central government leaders, because it is possible to put effective reward or punish controls into effect. Moreover, through the AI driven scoring services, each citizen from own birth to death must be living under strict and unavoidable control by the central government to be measured, monitored, and evaluated from the value of their own existence to market value. Moreover, in the nearly future, advanced medical technologies will make it impossible to escape such evaluation system before birth of each individual. It's also very interesting matter to see under what assessment the more than 1.4 billion citizens will be able to stay living with when robots are mass-produced to replace their labors.

Furthermore, it may not be just their citizens that the Chinese government puts under governance by the value standards. In addition to the Asian Infrastructure Investment Bank (AIIB), the Chinese government will also embark on the operation of electronic currency in the nearly future. With the creation of the digital CNY as a variety of cryptocurrencies, it is clear that the aim is to consolidate the control conducted by central China government, as well as foreign business partners, under their centralized control. Since cryptocurrencies rely on block-chain technology, its distributed ledgers can ensure high security and are difficult to tamper it, but since it is likely to be operated by the central government, there are concerns about whether each individual's private rights are protected and guaranteed.

2.5   Chinese Capitalism and AI Driven Valuation

It was said that the communism and socialist-planned economy had little personal working morale and little incentive to pursue economic profit, which led to the collapse of the Soviet Union and other former communist blocs. In this regard, China has been behaving more smartly. The Chinese government has established a legal system that will strengthen state control of data starting in September 2021. Furthermore, according to a report[3,4] by the China Institute of Information and Communication Technology, the scale of China's digital economy will be $5.35 trillion in 2020, which was about 40% of the scale of the USA. However, they aim to increase by 50% from the 2020 level by 2025 by promoting data capitalism.

The author thought that they figured out how to scientifically link merits among the individuals with the national benefits. It pushes capitalist policies to elicit the desire for profit and approval of the individual, and succeeds in leading to the development of the nation as a whole. These economic incentives match with the AI-based scoring services. In fact, AI scores have been successful in shifting toward such for-profit behaviors by rewarding individual efforts. While monitoring individuals online, it can also induce them in a direction that is economically convenient for the state, and it is an opportunity to develop China's version of capitalism into a digital space.

---

[3] http://www.caict.ac.cn/english/.

[4] http://www.caict.ac.cn/kxyj/qwfb/bps/202111/P020211119513519660276.pdf (in Chinese).

3 Rising Issues on Valuation by the AI

In order to conquer the COVID-19 pandemic, various behavioral changes are required, and some ideas are being implemented to avoid the spread of infection by constantly monitoring them. As noted before, if information such as GIS, behavioral history, trajectory of the dynamics, and mobility patterns of the citizens can be obtained, it is possible to calculate the risk of epidemic growth with a considerable degree of measure and trace the possibility of contact with infected persons (Shibuya, 2020a, 2020b). In addition, if it is analyzed in conjunction with daily health and genomics data, it is possible to build a health management model for each individual, which is certainly very useful (Coeckelbergh, 2010). However, the problem of the risk of them being monopolized by one state or one company has emerged.

The main problems can be summarized as follows: (1) management of sensitive data such as individual genomes, personal preferences, mobility log and other data, (2) analysis of such big data may reveal not only many useful findings but also results that are not intended by each individual, which may violate privacy, and (3) evaluation of social value through the AI driven services may make individuals blindly obey to specific social norms, and there is also a concern that the national government induce the value as they need.

Value induction, in particular, is fraught with major problems in the political process, voting behavior, and public opinion trends, and is likely to deny the foundations of a democratic political system. But certainly, with respect to purchasing behavior, if well guided, it may be possible to shift to environmentally friendly purchasing behavior (e.g., SDGs) or to bring about useful behavioral change even under extraordinary daily circumstances such as the COVID-19 pandemic. However, we have come to the point where we need to question whether or not it is really ethically and legally permissible for these trends to be monitored and analyzed by the central government (Cenci & Cawthorne, 2020).

For example, there were discussions about the Chinese government's crackdown on the people of the Uighur autonomous region, re-education, networked monitoring systems (Noorden & Castelvecchi, 2020). Particularly, the genome data and facial recognition data accumulated and analyzed by the AI driven mechanisms of residents directed by the Chinese central government were unethical and denounced as there was no guarantee that they had given their full consent (e.g., informed consent). Those facial data of the people can be automatically monitored to identify each individual in everywhere of China, and each genome data enables the central government to crucially examine the synthesized health and daily living data of them. The existence of each individual and his or her essential data will be fully taken over by the state.

4 Moral Science

On the other hands, sociology is often said to be the study on the self-organizing order (Luhman, 1984). In other words, it is an area of the researches that is interested in how social norms and legal systems generate the dynamism of human behavior and the entire society, and how the social order exists through them. Thus, it deals with various issues that connect individual, two or more interaction processes, groups, and nations.

In particular, values and attitudes are an integral part of connecting the individual to the society. As a philosopher Kant (2017) articulated, acting in a way that is consistent with the moral law of the society and the rule of conduct of each individual must be certainly considered as an aspect of sociology as well. At the same time, it will be necessary for each individual to act in a way that is consistent with those values in terms of compliance with the values from the legal system that is being worked on by society and the nations.

In the AI oriented society, sociology will have to consider the mechanical meritocracy that individuals are not only bound by such traditional studies on value systems, but are also exposed to automated monitoring and evaluation tools by the AI.

In historical background in sociology, the idea of a "moral science" ("morality and the science of morals": "*la morale et la science de moeurs*" in French) led and practiced by Comte de Saint-Simon and Durkheim in the eighteenth century during the industrial revolution may suggest a unique sociological perspective on scoring in the age of the AI. They described it as "*science*", oddly enough, but nowadays it is possible in some sense to attract and "control" people's behavior to a particular value. In that sense, it might be said that they have realized a part of their ideas. At the same time, they said that religion can also be a "religious science". It makes sense that it can be a science as well, if it converges citizens to a particular value, that is, the doctrine of religion. However, one could argue that the extent to which their thinking is consistent with modern AI scoring services and value assessment systems (i.e., AI based meritocracy).

Moral sociology ("*sociologie morale*") in France specifically also sought to scrutinize morality as a social phenomenon, which was a difference to discern from the ethics tradition. There should be opportunities to learn some kind of norm that facilitates the internalization of specific values, including initial training to enable individuals to participate in society and groups. Since the foundation of social identity assimilation and commitment is an essential factor in becoming a member of the society to which one belongs, morality should lead to such an attitude. Then, each culture, society or group has its own unique morals and norms, and scrutinizing these differences can identify specific characteristics of each culture and society (Albery et al., 2021). In other words, morality and ethics need to be viewed from both perspective between individuals and society, and the final destination that the AI scoring services will lead to specific values is to intentionally cultivate *better* morality of the citizens.

In addition, Saint-Simon also pointed out that the development of industry and science will be an essential and important factor for the nation in the future, stressing that engineers, doctors and scientists are far more valuable than ordinary citizens and rulers, and that they will determine the development and even the fate of the nation. That is to say, the survival of the nation depends on science and technology and the industrial economy, and then they were regarded as more top prioritized and valuable existences than other group of experts including rulers and ordinary citizens.

5 Measuring Compensation

According to sociologist (Simmel's, 1900) *"Philosophie des Geldes"* (*"Philosophy of money"*), his classical debate over "human value" had been investigated. An example was the question of how "compensation for the murder crime" should be determined. This is because, as the author (Shibuya, 2020a, 2022) already pointed out, evaluating the human existence with the dignity by economic indicators is inherently demeaning priceless dignity. However, in a court of law, they exactly dare to do so because they have to determine the amount of damages.

Thus, as discussed at this chapter, how is the human dignity assessed by the AI driven scoring service? How is the amount of compensation calculated? In many cases, the stipulated amount in law will be paid.

Is it justifiable to ask the other country to compensate for any loss of economic activity as a result of the COVID-19 pandemic? Even if one of the causes was the globalism that Westness has promoted, the call to blame China for the COVID-19 pandemic grows stronger day by day. As of the end of April 2020, governments, private companies and individuals in the USA, the UK, Italy, Germany, Australia, and other countries have filed lawsuits against the Chinese government seeking compensation for their significant damages caused by the COVID-19. It contributes to the entire international community's hostility to China over its responsibility. In response to them, Chinese government counters that the USA has never been held responsible for the spread of HIV. It also argues that it is too unfair and unjust.

As the Coase theorem states (Coase, 1960), when a company or government is held responsible for pollution or human-made disasters committed against the environment, it should be required transparency for the public to break information asymmetry among the stakeholders (Akerlof, 1970; Report of the NOAA Panel on Contingent Valuation (Jan. 11, 1993); Alm & Brown, 2020). However, criticisms claims that the view that the Chinese government's reluctance to disclose such information and its inadequate initial response were critical factors that led to the pandemic makes sense. The plaintiffs' countries have already suffered considerable deaths and economic damage, and it will be inevitable that this cause considerable conflicts or warfare among those countries.

6   AI Driven Valuation System and The Dignity of the Humanity

As the COVID-19 pandemic countermeasures have become a hot topic, it is also becoming possible to achieve more accurate predictions for drug discovery, pipelines for mass production, and quantification on pharmacological risks using the AI to learn existing clinical big data. But the COVID-19 pandemic still also posed a question on our dignity and medical data in terms of ethical concerns (Colombo, 2021).

There are tough issues in ethical context. From the perspective of "distribution of sacrifice", if each personal medical data is necessary for the COVID-19 prevention, should it publicly offer to medical solution (Harden, 2021)? Is it natural to sacrifice own data without any incentive? The question of how AI will analyze and manage the personal data of genetic and genomic information will become even more problematic in the future.

Similarly, as an ethical verification on AI, the AI-driven evaluation and assessment system is quite necessary to carefully examine whether it is based on the justice, fairness and equality or not (Cath, 2018; Gerlick & Liozu, 2020; Hagendorff, 2020; Mhasawade et al., 2021; Winfield & Jirotka, 2018). Let me introduce more concrete example. Some AI based facial recognition system is bad at classifying faces of Asian and African people, because such systems has not sufficiently learned their actual data enough yet. It is certain that the AI driven system should be required justice, fairness, and equality for everyone. However, in the valuation context by the AI, it should be focused on other facets, namely, it is the decisive relationships between genetic foundation and its development patterns. Likewise intellectual specialties or deficits (including somewhat latent diseases) may be clearly detected by the AI based diagnosis mechanisms.

For example, actually, there is an application named "*Face2Gene*[5]" which developed by the FDNA, and it is a startup venture company based in the USA and Israel. By simply taking a visual data of a person's face, the AI driven system can compare it to the image data as the face of a patient with a large volume of genetic diseases to diagnose the presence of the disease. On the other hand, although it now requires the support of a human medical doctor, it is expected to help detect hereditary diseases earlier, and some studies have actually identified infants with latent autism spectrum disorders with high accuracy. These advancements of the AI driven diagnosis system has been nearly approaching a kind of the Szondi's Test in clinical psychology (or merely, physiognomy), and such systems may perform more scientifically (Kosinski, 2021). Certainly, with a huge amount of machine learning data, it would be comparable. Namely, not only mental illness, but also the possibility of being able to mechanically may predict what kind of life-history that person lead if such person looks like (of course, as a biased prediction of the future).

---

[5] https://www.face2gene.com/.

7 Conclusion: From Genome Analysis to Human Control

Genes alone do not predetermine the behavior and adaptability of all life organisms. If all citizens have their cognition and behaviors determined only by genetic prescriptions, in such condition, it can be easy to ascertain the genetic tendencies and characteristics not only of individual members but of the entire population (Cyranoski, 2020). Nowadays, knowledge that utilizes population genetics and genetic statistics as well as measures of adaptability such as the comprehensive degree of adaptation of each individual could be put to good use (Papaioannou, 2009; Plomin & Deary, 2015). In addition, the understanding of SNPs (Single Nucleotide Polymorphism) and differences in motif structures is also an important factor for the analysis of the humankind.

In the future, the AI driven systems and simulation may make it possible to comprehensively analyze human behavior and social systems from the evidences based on genetic information. It means that sociology and ethics would have to drastically reshape their research styles. Based on the findings of normative ethics and behavioral economics, AI driven systems may become the foundation of a society where AI constantly executes and monitors necessary mathematical models that aim to control the social morality of citizens. Constant monitoring of people's behavior in the name of medical welfare, public safety and other reasons is already taken for granted in some countries. Such advanced technologies will form the basis of the AI driven assessment and control for each citizen. Will the humanity lastly accept a rule of *reward or punish* driven by the AI?

References

Akerlof, G. A. (1970). The market for "lemons": Quality uncertainty and the market mechanism. *The Quarterly Journal of Economics, 84*(3), 488–500.

Albery, I.P., et al. (2021). Differential identity components predict dimensions of problematic Facebook use. *Computers in Human Behavior Reports, 3*.https://doi.org/10.1016/j.chbr.2021.100057

Alm, K., & Brown, M. (2020). John Rawls' concept of the reasonable: A study of stakeholder action and reaction between British petroleum and the victims of the oil spill in the Gulf of Mexico. *Journal of Business Ethics*. https://doi.org/10.1007/s10551-020-04474-9

Bell, D. A. (2000) *East meets west: Human rights and democracy in East Asia*, Princeton University Press.

Bell, D. A. (2016). *The China model: Political meritocracy and the limits of democracy*. Princeton University Press.

Bentham, J. (1789). An introduction to the principles of morals and legislation. http://www.koeblergerhard.de/Fontes/BenthamJeremyMoralsandLegislation1789.pdf

Cath, C. (2018). Governing artificial intelligence: Ethical, legal and technical opportunities and challenges. *Philosophical Transactions Royal Society A (mathematical, Physical and Engineering Sciences), 376*, 20180080. https://doi.org/10.1098/rsta.2018.0080

Cenci, A., & Cawthorne, D. (2020). Refining value sensitive design: A (capability-based) procedural ethics approach to technological design for well-being. *Science and Engineering Ethics*. https://doi.org/10.1007/s11948-020-00223-3

Chouldechova, A., & Roth, A. (2020). A snapshot of the frontiers of fairness in machine learning. *Communications of the ACM, 63*(5), 82–89.

Coase, R. H. (1960). The problem of social cost. *Journal of Law and Economics, 3,* 1–44.

Coeckelbergh, M. (2010). Engineering good: How engineering metaphors help us to understand the moral life and change society. *Science and Engineering Ethics, 16,* 371–385.

Colombo, E. (2021). Human rights-inspired governmentality: COVID-19 through a human dignity perspective. *Critical Sociology.* https://doi.org/10.1177/0896920520971846

Cyranoski, D. (2020). China's massive effort to collect its people's DNA concerns scientists. *Nature.* https://doi.org/10.1038/d41586-020-01984-4

Gerlick, J. A., & Liozu, S. M. (2020). Ethical and legal considerations of artificial intelligence and algorithmic decision-making in personalized pricing. *Journal of Revenue and Pricing Management, 19,* 85–98.

Greif, A. (2006). *Institutions and the path to the modern economy: lessons from medieval trade.* Cambridge University Press.

Hagendorff, T. (2020). The ethics of AI ethics: An evaluation of guidelines. *Minds and Machines, 30,* 99–120.

Harden, K. P. (2021). *The genetic lottery: Why DNA matters for social equality.* Princeton University Press.

Inoki, T. (2014). *A history of economics.* Chuokhoron-shinsha. (in Japanese).

Kant, I. (2017). *Fundamental principles of the metaphysic of morals (English trans. ed.),* Createspace Independent Publication.

Kosinski, M. (2021) Facial recognition technology can expose political orientation from naturalistic facial images. *Scientific Reports, 11,* 100. https://doi.org/10.1038/s41598-020-79310-1

Leibniz, G. (2017). *Monadologie.* Hofenberg.

Locke, J. (1998). *Two treatises of government.* Cambridge University Press.

Luhman, N. (1984). *Social systems.* Stanford University Press.

Mhasawade, V., Zhao, Y., & Chunara, R. (2021). Machine learning and algorithmic fairness in public and population health. *Nature Machine Intelligence.* https://doi.org/10.1038/s42256-021-00373-4

Noorden, R., & Castelvecchi, D. (2020). Science publishers review ethics of research on Chinese minority groups. https://www.nature.com/articles/d41586-019-03775-y

Papaioannou, T. (2009). The impact of new life sciences innovation on political theories of justice. *Genomics, Society and Policy, 5*(2), 40–52.

Plomin, R., & Deary, I. (2015). Genetics and intelligence differences: Five special findings. *Molecular Psychiatry, 20,* 98–108. https://doi.org/10.1038/mp.2014.105

Shibuya, K. (2020a). *Digital transformation of identity in the age of artificial intelligence.* Springer.

Shibuya, K. (2020b). *Identity Health.* https://www.ncbi.nlm.nih.gov/pmc/articles/PMC7121317/;

Shibuya, K. (2020). *Digital transformation of identity in the age of artificial intelligence.* Springer.

Shibuya, K. (2022). An "artificial" concept as the opposites of human dignity. In T. Sikka (Ed.), *Science and technology studies and health praxis: Genetic science and new digital technologies.* Bristol University Press.

Simmel, G. (1900). *Philosophie des Geldes.* Nabu Press. (in Germany).

Winfield, A. F. T., & Jirotka, M. (2018). Ethical governance is essential to building trust in robotics and artificial intelligence systems. *Philosophical Transactions Royal Society A (mathematical, Physical and Engineering Sciences), 376,* 20180085. https://doi.org/10.1098/rsta.2018.0085

Five Virtues to Be Digitized Norms

1 A New Normal

At confronting with the COVID-19 pandemic, each government has been calling for a new rule of conduct for their citizens (Bonotti & Zech, 2021). This renewal of our lifestyle needs to be sustained in the long term as there is a similar risk of recurrence of infection, even if it becomes the periods of the After Corona.

In Japan, for example, the Ministry of Health, Labor and Welfare has established a *New Lifestyle* and requires each citizen to comply with this new way of normal life and ask for their cooperation. Certainly, a part of the workers such as medical doctors, health care workers, and civil servants are excluded, but the ordinary citizens are compelled to comply because they do not want to be infected.

1. Basic infectious disease control for every person
2. Basic lifestyle for daily living
3. Lifestyles for different aspects of daily life
4. A new way of working

Japanese's conformity pressure is similarly a "blind obedience to a value standard" that forces individuals to attune to somewhat "the social value" by mutually monitoring each other. It is not collective intelligence, rather many of them are usually prone to act in line with other's directions and without own deep contemplation. Lastly, when such majority trends of the mass opinion have been surrounded by their uncritical obedience among the citizens, their invisible atmosphere obeying to the pseudo consensus seems to be enforceable as the general will of all citizens. Social psychologically speaking, such atmosphere can be regarded as pluralistic ignorance or collective conformity. However, there is a very high risk of blind faith in a direction that is poorly endorsed by democratic deliberations and scientific evidences. In addition, criticism and hate speech against infected peoples and disregard against their privacy and human rights online, and a state in which everyone binds each other with self-restraint can be said to be a representative case of the collective oriented culture in Japan.

K. Shibuya, *The Rise of Artificial Intelligence and Big Data in Pandemic Society*, https://doi.org/10.1007/978-981-19-0950-4_6

In Tokyo capital city of Japan, a municipal governor announced their indexes[1] for preventing the COVID-19 to measure data per a week such as the number of new positives as infection, the unknown rate of contact history among new positives, the weekly increase in positives, the number of critically ill patients, the number of hospitalized patients, the rate of positive PCR (polymerase chain reaction) tests, and the number of consultations at the consultation desk. Every citizen in Tokyo had to be aware of those data during the pandemic, and many of them *spontaneously* kept socially moral actions. As their behavior patterns were also analyzed by spatiotemporal data using smart-phone, city congestion could be monitored by the government authority[2] and some companies.[3]

Otherwise, there are also countries like China, which aggressively use the AI driven system and big data analysis to monitor the behavior and health of the citizens at all times. This means unifying individual values and norms throughout the society and becoming a society with a common purpose of curbing the risk of infection. In this context, a legal system be not only will established to publicly monitor and punish such cases, but also each individual life will be incorporated into the automated system of the AI.

2   What Is an Individual's the Highest Value and Its Equivalent?

2.1   Cooperation with Each Other

A good nation can be formed by good citizens. Hence, what does a good citizen look like to a nation? This was a recurring problem in all times and places. In Western societies, such as Plato and Aristotle, it was synonymous with national system for governing and social morality by through recommendations of the cultivation of the goodness to the citizens. In ethics, it is called *civic virtue*.

However, the COVID-19 pandemic inevitably exposed the human nature, namely the nature of the humanity is egoistic and far from ethical, moral and altruistic (ÓhÉigeartaigh et al., 2020). At this pandemic, in some countries, the law stipulates that it is mandatory to wear own facial mask when going outside. A daily custom of wearing on mask was indeed accepted as a new common sense among the western citizens, but their battles to egoistically obtain masks were often observed all over the countries. Further, perhaps, many citizens felt compelled to be so hostile to the existence of the others around each citizen because they were unintentionally threatening by through the others who were already infected but unidentified to be infected. Such situations recalled us a fact, namely, *Leviathan* written by Hobbes

---

[1] https://stopcovid19.metro.tokyo.lg.jp/en/cards/monitoring-status-overview.

[2] https://corona.go.jp/.

[3] https://mobakumap.jp/.

(2009) represented an impressive phrase *"Bellum omnium contra omnes"* in Latin, and it crucially indicates our human nature. Similarly, Spinoza's statement that "the others are naturally enemies (*Homines ex natura hostes sunt*)" has been confirmed as correct (Spinoza, 2008).

In contrast, a series of the past studies on evolutionary psychology had asserted some theoretical assumptions that humans were able to adapt evolutionarily because they were able to cooperate better than other zoological species (Turner, 2001; Hooper et al., 2021). However, as the author already said (Shibuya, 2020a), such hypothesis is clearly false. There are any number of species that were well adapted and continue to survive, even in uncooperative zoological species. The humanity had survived from numerous infections since their emergence on the earth, but it cannot be explained by only their ability to cooperate. The factors that have allowed us to survive seemed to obtain language, representation, and the ability to regulate oneself as well as the ability to act strategically based on these factors, due to the accidental development of the neocortex within the brain. The fact that we could not keep self-regulation without externally enforcing cooperative behavior in times of the COVID-19 crisis was proof enough. Our need for morality and self-control in times of crisis is deeply dependent on such social structures and institutions.

2.2 Discerning a Meaning of Two Sacrifices

2.2.1 What Differences in Sacrifices?

Here, a concept of *sacrifice* that the author repeatedly noted can be divided into at least two categories. First type is a simple meaning, namely, to sacrifice either something or someone in the literal and ordinary sense of the word. It would often mean sacrificing something other rather than oneself for the sake of one's own survival (including selfish motives). Some of such cases cannot be usually excluded committing either crimes or suspicious actions.

On the contrary, next is *self-sacrifice*, which is the sacrifice of oneself to save the others (Ackeren & Archer, 2020), and this is associated with *altruism*. For example, *agape* [αγάπη (in Greeks)] means a philanthropic love by self-sacrifice in Christianity. And, further, this case subdivides two patterns either *voluntary* or *enforced* one. For example, the conscription system is generally the latter, but the voluntary one can be said to be the former. An old custom of *noblesse oblige* can be regarded as a latter pattern.

For the cultivation of ethics, the general public must be educated so that both *"Maxim"* of the individual and social morality coincide, as in the categorical imperative of Kant's theory of *duty* (i.e., deontology), without sacrificing others (Kant, 2017). A shared morality within a society would ultimately be one that reaches a *"Reich der Zwecke"* (i.e., Kingdom of Ends). However, can the maxim or duty in Kant's theory of duty be straightforwardly described as self-sacrifice? The difference between self-sacrifice without duty and self-sacrifice with duty may not only be due to the presence or absence of external factors such as legal regulations or customs. Self-sacrifice based on individual volition (Schwartz, 1970) may also be accompanied by an intention to fulfill some kind of duty (such as a religious act) in the name of a mission or a cause in line with some idealized value. In this context, Moore also (1903) told *"Self-sacrifice may be a real duty; just as the sacrifice of any single good, whether affecting ourselves or others, may be necessary in order to obtain a better total result"* in his ethical discussion.

Taking the COVID-19 pandemic as an example, the case that relies on selfishness corresponds to people who acted irresponsibly, infected many people around them, and forced them to make sacrifices including the death. The latter, on the other hand, would be medical doctors, nurses, and other professionals who were engaged in medical work and died in the line of duty. Whether they died because of their "role" as doctors (i.e., duty-based self-sacrifice) or they saved others and the nation through noble self-sacrifice or philanthropy as represented by agape (i.e., non-duty-based self-sacrifice), this mentality can be distinguished as an intrinsic rather than extrinsic motivation.

In addition, the difference between morality and ethics as described by Hegel (*"Grundlinien der Philosophie des Rechts"*) seems to be the difference between individual' maxim and social norms. The more social institutions such as family, community, and state are required, the more normative the society as a whole need to be in order to put certain restrictions on social evils as one of the emergent properties. This is the difference between self-sacrifice = self-control (i.e., morality according to Hegel), which involves the will to discipline oneself, and constraints that discipline society = sacrifice carried by society as a whole (i.e., social ethical aspects according to Hegel). As a legally enforceable social norm, it forcibly guides or adjusts the behavior and thought of each individual in a certain direction as needed. From the author's point of view, this means that the system or political function forces each individual to make sacrifices, or that each individual assumes the sacrifices. Actually, Hegel also clearly said, *"Sacrifice for the sake of the individuality of the state is the substantive relation of all the citizens, and is, thus, a universal duty. It is ideality on one of its sides, and stands in contrast to the reality of particular subsistence. Hence it itself becomes a specific relation, and to it is dedicated a class of its own, the class whose virtue is bravery"*.

Indeed, how we think about the obligations and institutional constraints that the state imposes on its citizens is a problem. For example, Rawls (1971) raised the issue of "civic disobedience" (conscientious disobedience) from his perspective of justice and fairness. In American society, it seems to have been particularly discussed in the context of military service. In the COVID-19 pandemic, the same was true for hesitations to vaccination and other disobedience patterns among the citizens. However, maximizing the utility of each citizen's freedom and individualism can nullify the utility of society as a whole. Such risks will become to be worse than a kind of free ride. In fact, some countries adopted a policy that it is still good if everyone is infected but the mortality rate is low. However, they sacrificed the essence of these obligations and institutions, as a result, they spread infectious diseases to other nations and the world at large, and imposed sacrifices on other nations and their citizens.

In the emergency, especially, true freedom will be more recognized. Restraining own behaviors definitively means a patterns of the self-sacrifice. Namely, each difference of self-restraints during the pandemic could be observed in each nation. In the EU, the USA and other countries, any national government leaders had to strictly conduct national policies against preventing epidemic proliferation and regulate their citizens by a set of the imperatives with expensive *punishments*. It means that they enforced the citizens to obey their rules, and then it required such attitude among the citizens as enforced self-sacrifices. In that situation, each individual shall be sacrificed more than the society and nation as whole, even if their regime was based on democratic and liberal institution. On the contrary, Japan government had roughly designed own a set of public health policies and requested their citizens to behave in accord with announced rules on self-restraints (no penalties during 2020, but penalties were established in February 2021). Rather, before and during the pandemic, many of the Japanese citizens *spontaneously* behaved keeping the total conformity in the society, and strict government requirements might be regularly unnecessary for them. Further, China government had required their citizens to be obeyed strictly their rules as well as monitored and traced them by the AI driven systems. Then, which country more liberal and democratic is? Therefore, self-sacrifice among the citizens should be paid more attention to one of the democratic and institutional fundamentals.

2.2.2 How to Give An Incentive for Self-sacrifices?

Normally, ordinary human has fundamentally instincts of both *approach* and *avoidance*. Each is linked to identifiable emotions with the former being interconnected with the positive emotions of curiosity, joy and ease, meanwhile the latter with the negative emotions of disgust, anger and pity, for example. The former brings the mutual distance between two individuals closer, while the latter drives them far away. The former is called fraternity, meanwhile the latter can be called prejudice. These primitive instincts are deeply relation to own self-survival motivation. These definitions are similar to those made by Spinoza's *Ethica* (2008). And scientific investigations which supported by brain science revealed the actions of oxytocin and serotonin components for fraternity among the people (Franks & Turner, 2012). From the perspective of the sacrifice, the self-sacrifice based on agape, the act of actively transcending self-preservation to serve others and dedicate one's life to them by own death is not comprehensible without assuming the higher level value or supreme humanity based on agape.

On the other hands, there is also another insight on self-sacrifice. As a matter of fact, war and morality are diametrically opposed and very similar at the same time. According to *"The Methods of the Sima*[4]*"* as one of the ancient books on military

---

[4] https://en.wikipedia.org/wiki/The_Methods_of_the_Sima.

strategy in China, this book said *"People will die for love. Dying to defeat the object of your anger. Dying for authority. Dying for the justice. Dying to make a profit. War is about educating soldiers, promising them rewards so that they are willing to die, and educating them on reason so that they can die for justice"*. It exhibited a fact that people live in order by setting the highest value for "something", and it reversely suggests that they are willing to battle, die and self-sacrifice for such "something" in the society as whole. Is it honor, merit, justice, love for someone, or something else? Thus, such thing means to be equivalent to own life. Ultimately, such incentives to abort own life can be considered as similar idea to "reward or punishment" as mentioned earlier chapter (Willer, 2009).

In the western culture, similarly Walzer (2006) explicitly mentioned the sacrifices of war. And he described a motivation for warfare in accord with above essences of the Sima book. *"The cause of many militant organizations is to establish a state; the cause for which states organize armies is to defend their own existence and the common life and individual lives of their citizens. It is a fact of our moral history that many individuals are willing to risk their lives for these causes. This is not easy to understand. If states exist primarily to defend life, then how can life be sacrificed to defend the state? That is the question posed most clearly in the political theory of Thomas Hobbes, to which he offered no satisfactory answer"*. As a natural consequence, Walzer also said, it is too common and essential to be required the young citizens *voluntary* engagement in military collective actions. On the contrary, according to criticisms by Foucault's social theory (e.g., criticism on governing social system like *Panopticon*), modern power rather actively intervenes in people's lives, and further it tries to manage and direct the general public in accord with the government purposes whether ordinary daily context or not. He called such characteristic of the normative power as *"biopower"*.

In ordinary context, certainly, many of the citizens are also needed to prepare affordable way for motivating them. Even when taken from the aspect of any *role* in sociology, without the rewarding mechanism of *"reward or punish"* mentioned earlier chapter, such actions would not normally occur or be left inadequate. As a phrase *"Quid pro quo"* noted, many cases require something value for obtaining something, and it usually calls as a fair exchange and an exchange something with equal value. The AI scoring service suggested that it can guide people in a significant direction with somewhat any incentives based on each individual preferences and values. Further, from this perspective, it is possible to generate "self-sacrifice" with "compensation" and at the same time, it can also be applied to *labor*.

Therefore, if the author inquiries more deeply into the perspective of sacrifice, it can discover the answer to sustain *"reward or punishment"*, how to realize the formation of ethics and morality in each level of individual and the society, and how to incite "self-sacrifice" and altruism, and how not to induce each individual to rife with selfishness.

2.2.3   How to Think About Indirect Reciprocity?

There is a novel entitled *"Pay-It-Forward"* (Hyde, C.R.). In this fiction, a little boy initially helps three unknown persons, but instead of receiving something valuable in return, he instructed them to do helping similarly three strangers. And then, persons who was helped by them will be requested helping the others. The idea was that if they repeated the process, the world would be gradually changing.

This corresponds to what is called *indirect reciprocity* in the interdisciplinary research (Nowak & Sigmund, 2005; Shibuya, 2012). In short, it means being willing to make some kind of "self-sacrifice" to serve a third party or the society. Indirect reciprocityIndirect reciprocity, after all, means that if you serve someone, but there is no direct return, and you may not be able to obtain any benefits in the future. Undoubtedly, such reciprocity is a goodness in ethical context. It would certainly be wonderful if the society were full of such reciprocity.

But a fiction depicted an ideal. It is not likely to be easy to achieve, as it involves the private rights of the individual. However, the essence of donations, charity, contributions, volunteerism, and non-profit organizations are to provide free labor, money and goods to third parties (Wilson, 2000; Wang & Graddy, 2008). It is the act of trying to realize their ideals all over the world. And, more recently, similar activities online have taken on an aspect of indirect reciprocity.

Therefore, if it considers such actions as self-sacrifice, there is no incentive to lead anyone to self-sacrifice. Then, we face a question why we help the others. In some cases, people risk their lives to help others rather than donating money or goods (Ackeren & Archer, 2018).

On the other hands, ordinary people do not only seek valuable goods, but they also desire to obtain the "honor". By volunteering or donating money, some of them are making history of the fact that they have contributed to the society, and they will be making own presence known. Some people are motivated to preserve own existential value for the future. Indirect reciprocity also improves the stability and credibility of local communities, as cultural studies and anthropology researches (e.g., Mauss *"Essai sur le don"*), so certainly such practices are likely to have more than a certain effect (Kawachi & Berkman, 2003; Bowles, 2012).

Regarding above concerns, many societies and cultures have articulated the provisions that facilitate individual adaptation and encourage the formation of common values throughout the society through some form of religious values and other constraints.

3   A System as Culture

3.1   Embedding in Social Norms

Morality and norms are situated in the social context and cultural diversity. Next, we need to think about the three layers of the culture. Namely, these are representation, identity, and institution. Firstly, representation shows a socially shared cognitive base, a social reality shared among its members, and history, language and lore as shared knowledge. Second, identity indicates the definition of each individual self-existence, which is combined as part of the self through social groups and relationships. And thirdly, institution, which defines the laws, norms, customs, and value systems (e.g., it determines what is justice) that are the basis for ordering, forming, and maintaining the system.

Triandis (1995) said a dichotomy such as *individualism* and *collectivism* among the global cultures. Here, individualism can be referral to the majority which is usually reached by deliberated discussions among the stakeholders, whereas collectivism can be often observed a consensual motivation to be harmonized predetermination by the governing leader.

3.2   Labor as Self-Sacrifice or Not in Cultural Backgrounds

Probably, it is the labor issues that makes the most difference between Eastern and Western values. In Japan, particularly "self-sacrifice" closely tied to one's "living existence" that dedicates to working, such labor is unconditionally embedded in common prerequisite for their society. That includes almost everything in working situations such as childcare and employment. Self-sacrifice is implicitly required and culturally situated to maximize loyalty to a company, organization, or country (Hosono et al., 2020). Irrelevant to explicit consent between them, such norm is shared in implicit common senses. They are forced to be workaholic, but their productive work has not been achieved higher than other OECD countries. For example, using an index of the GDP per a worker,[5] there is no large difference between Japan (42.8 US dollars, at 2018) and France (45.1 US dollars, at 2018), regardless their annual average working time have larger differences (Japan: around 300 h, France: around 160 h at 2014). Unpaid overtime in Japan has been accused by external professionals, but those traditional customs cannot be changed. Namely, they ought to be self-sacrificed in irrelevant to their wills and intrinsic motivation. Many of Japanese may feel to commit self-sacrificing work that means virtue for them.

On the contrary, a common view on labor in the West (e.g., the EU, the USA) seems not to be ordinarily associated with self-sacrifice. They are only based on the concept of the role and contract. The idea that those with the right skills, experience and level

---

[5] https://data.oecd.org/economy.htm#profile-Productivity.

of education required for the individual job will take on a role. Each employee is only expected to fill that role at a designated time in line with an agreement and a contract. Although some exceptions such as medical doctors, policemen, soldiers, and presidents have the specific risk of being died in the line of duty. This is with reasonable agreement (Fia & Sacconi, 2019).

China, on the other hand, although they have the same oriental value standards as Japan, seems to commit rather Western view of labor. They don't think labor as something that involves self-sacrifice.

There is no doubt that communism was intended to prevent the uneven of the wealth through the inhibiting of private and property rights. The reason why this was hard to establish was because it forced each individual to make self-sacrifices and considerations for balancing own desires for self-interest. The lack of such balances was probably one of the reasons for the failure of communist's system. On the other hand, capitalism's success is based on the minimization of forced self-sacrifice and the maximization of each individual's desires for self-interest as much as possible. The aim of capitalistic system was to build such socio-economic system. While the former is a social system based on the state control of each individual (e.g., despotism, tyranny and monarchy), the latter is clearly a social system based on the autonomy of each independent individual. On the contrary, it can be said that Japan is a society where maximization of both self-sacrifice and self-interest is implicitly required.

4   Five Virtues

4.1   Looking Chinese System on the Morality

Sociologist Weber articulated and theorized his religious speculations (*"Gesammelte Aufsätze zur Religionssoziologie"*). Confucianism in the ancient China had been picked up by his eagerly intentions, and he compared it with cultural differences such as Western Christianity, Buddhism in Asia and Hinduism in India. However, he imagined such differences by investigating the literature resources, and then he could not analyze actual data by own social survey, at least. However, cultural differences in doctrine and religious attitudes could be compared along with sociological and anthropological considerations.

On the other hands, Rawls (2001) enumerated the six goodness on the justice as fairness. For example, he noted the idea of rational goodness, basic goodness, political ones and other several goodness for social relations. This goodness is usually required in any social and political fairness situations, and certainly his political philosophy has been succeeded from the western cultural heritages of deliberations on the goodness and social justice.

On the contrary, the same was true of Eastern culture, where Confucius ideas emphasized the morality of human relations from the nation to the family based on the order of the paternalism. Exactly, Confucianism in the ancient China told the *five virtues* for ordinary citizens, and it was historically spread around the east Asia. Common background of the Asian culture could be founded from social morality defined by Confucianism.

Five virtues in descending order respectively consists of *(1) benevolent tenderness for the others, (2) social justice, (3) rules of conduct in various situations such as daily life, social institution and ceremony, (4) wisdom,* and *(5) trustworthy attitude with each other.* These facets of the morality might depict a concrete scene of the daily living among the citizens of the ancient society. Of course, sharing these attitudes could sustain sound human relationships, harden their social background, and grip the handle of governing collective citizens by the emperor and kings. In brief, such paradigm of the morality indicated the collectivism based morality rather than individual independence, because each element of the five virtues was clearly underlying in the moral standard for daily living weaved with the nexus of human relations in the community. If ancient China society prioritized each individualistic value and pursued personal moral standard, they would arrange their virtues such as independence from the others, freedom from the authority, engagement for their democratic regime and other necessary values. In such ways, morality in each society can exactly reflect their hardcore of the value doctrine and its living styles among the citizens. Further, such behavior patterns inscribed in their dogma might also suggest the most adaptive styles for their survivals in the serious ancient environment and civilizations.

First of all, benevolent tenderness for the others apparently implies mental generous capacity for the uncertainty and heterogeneity at encountering unknown persons who came from outside. Cultural adaptation based on this virtue might coin in temporal oscillation for assimilation of unknown knowledge as well as developing more innovative chances derived from the heterogeneous combinations. Secondly, social justice would purse achieving larger value standard and prioritizing more important purposes of the entire society rather than personal goals. Third, rules of conduct in various situations would certainty denote to describe predetermined procedural strictness for the social order. Fourth, sagacious virtue could cultivate their learning incentives and be adopted in more intellectual and insightful manners among the citizens. And fifth, the trust could facilitate to interconnect with each other, and such harden relationships might override the constraints of time and space. And ancient people convinced that the trust should be rooted in all of the human relationships.

4.2   Five Virtues at the COVID-19 Pandemic

Let me exemplify observed matters at the COVID-19 pandemic in term of the five virtues. First of all, it must be no blame against the infected person, and further it must be tolerance for working medical doctors and medical professionals at the front of the hospitals. However, there was an excessive hatred responses and discrimination against those peoples. Aversion to living in the neighborhood itself has been reported to medical doctors and health care workers alike. In other words, such claimed person was only overly asserting own risk of infection.

Secondly, while it might be unavoidable from the point of view of social justice to denounce those who did not understand the epidemic prevention in the society as a whole, it was not permitted to slander others unconditionally. Much more, there was no legal background for such violence.

Third, the infected person must take care not to transmit the disease to the others. And the most important is to behave honest and not to deceive own infection fact. It was also too natural for the non-infected people to keep their distance from each other and immediately follow the instructions of the administration and medical doctors. Everyone had to refrain from a business-first approach that maximized own profits.

Fourth, counterfeit data on pharmacology, misleading of medical information, and hate speech about the COVID-19 pandemic were disseminated online. Namely, it was fake news (Escolà-Gascón, 2021). It was up to the general public to correctly discern necessary truths and select only those discourses that were endorsed by both scientific evidence and verification. However, flamed emotion and anxiety easily invoked much hatred against the infected persons and protest against the government.

And fifth, the trust would have been especially reaffirmed during the COVID-19 pandemic. It was not to mention how much trust was required for the government leaders, the administration and the medical profession. Exactly, government leaders had to led national citizens their safety, and they immediately must command appropriate executive directions.

Above all, all of them shall be integrally led to social norms among the citizen, especially for the purpose of conquering crisis of the pandemic. First point is to prevent *moral hazard* among the citizens. This case can be often underlying in *information asymmetry* between clients and medical doctors (e.g., *adverse selection*) (Akerlof, 1970). If the side of client does not appear any symptom of the COVID-19 yet, but when he is already infected, infectious risk to the other citizens will increase. If the side of client can be enough aware of some symptoms of the COVID-19, but when he will not drive to the hospital or eagerly move free in the city, similarly infectious risk will be accelerated by his thoughtless egoism. Further, the citizen who purchased any insurances may behave more irregular and undesirable actions against the prevention control by the government, because he is supported from the insurance if he gets infected (Erev et al., 2020). Otherwise, it's not easy to test all the concerned and uninfected citizens. PCR tests and other necessary medical checks should be certainly applied for many patients and those who desires to be examined, but such policy will easily reach the limit of the medical capacities such as checking machines, instruments, medical tools, beds and medical staffs. Namely, citizens' morality must be also required in emergency cases.

Secondly, the WHO announced their program on donation and cooperation for global prevention of the pandemic and promotion of medical care (*ACT (The Access to COVID-19 Tools) Accelerator*[6]). As of June, 2020, they said that the contribution was $3.4 billion with a shortfall of $27.9 billion. The urgently needed amount was $13.7

---

[6] https://www.who.int/initiatives/act-accelerator.

billion. However, who undertakes such self-sacrifices without any returns? Equally, every citizen shall be supported own burden for our total survival, not selfish. But the most significant is how to work such program effectively.

4.3   Five Virtues in the Digitized World

Dogmatic ruling styles to manage and tame the ordinary people have been observed anywhere and anytime. Religious and normative directives definitively initialized each belief system of the people. Those invisible mechanisms for the governing peoples could shape an outline of the lawful paradigms in each society. Each content of the lessons for ordinary citizens could cultivate both individual and social values. The demands of the five virtues had been naturally assimilated into the depth of mentality among ordinary citizens, and it offered to sustain governing imperial regime by the emperor and kings as well as paternalism in each home and rural communities.

In Japan, hereditary generals in the Tokugawa (Edo) era (AD 1603–1867) had been recommending a theoretical school derived from Confucianism (i.e., neo-Confucianism), because the dogmatic imperatives of this school could provide a still bases for governing the ordinary citizens and other samurais in hierarchical order. Similarly, ruling government leaders in the Meiji Era (AD 1867–1912) legitimated their imperial constitution to harden foundations of their modern industrial nation directly authorizing by the supremacy of an emperor.

Social norms usually gave the invisible mental barrier among ordinary citizens, and it tranquilized bottom-up disturbances conducted by the uprising leaders and confusions for interpretation on theoretical understandings. As it was established at once, ordinary citizens often obeyed traditional customs conducted by the preferable manners of the canonical government leaders, descending imperatives from the ruling government to ordinary citizens could exclusively command to suppress those social uncertainties.

In digitized society, as well now known, such conditions can be replaced by the AI driven monitoring mechanisms equipped by the ruling government. It cannot deny that such ancient governance by the old customs and traditional institutions was still similar to our AI driven technologies based on big-data such as monitoring for ordinary citizens, automatic detection against criminal conducts, and other panopticon-like mechanisms.

On the contrary, individualistic culture and society could not relatively do such ways. Western nations whether ancient age or not would prioritize democratic decision making and its principles for the majority. Such bottom-up driven incentives among ordinary citizens might be understood as a dynamism to integrate each living desire, but it often delays to reach a total consensus among them. Predefined laws and institutions legitimatized by the ruling government also permitted them to behave own autonomy as a maximized freedom from the authority. In the present, digitized tools usually gain each private information and privacy rich data among the users, but ordinary citizens confront with a trade-off between usefulness of smart-devices and freedom from the entire authority.

4.3.1 Benevolent Tenderness for the Others

Since the beginning of time, in human history, coexistence with others in the same environment has been a sign of mutual survival in the struggle of natural selection. Anyone who kills or injures another is a mutual enemy to each of us. The most adaptable people must have finally learned the significance of cooperation between different people to achieve and maintain social order after the devastation of national wealth and heavy casualties in repeated large-scale wars. Game theory has revealed the nature of strategic situations, which can usually be defined as cooperation or competition (Axelrod, 1997; Cederman, 2005). Needless to say, the most important point in human history is that cooperative strategies may be better in terms of survival than competitive strategies.

Cooperation is based first of all on the benevolent and generous attitude of its citizens. A community populated by such persons would keep social uncertainty and unrest low.

However, excessive nationalism, extremists, and hacktivists frequently appear in digital communities. Such words and actions by citizens often further evoke populism and xenophobia. This can be exemplified by hate speech, fake news, and propaganda against certain races and individuals (Shibuya, 2020a, 2021a).

4.3.2 Social Justice

Whether it is online or not, democratic decision making in politics merely means procedural justice to legitimate imposing the sacrifice. For example, there is a way called deliberative discussion (Johnson, 2008). Johnson picked up a Canadian case on nuclear power plant in rural community. This discussion style involves all of the many experts and stakeholders including citizens to thoroughly discuss and debate each other on a particular issue until their opinions reach a consensus among them. Such method is certainly recommended as representative of the democratic decision-making.

But frequently, no one among these stakeholders whose sacrifices are imposed on them is even participating in the discussion. If such persons to be sacrificed don't participate, it will almost certainly sacrifice them. When such persons participate, the conclusion will almost reach "defer" or "bear equally with everyone". Of course, any citizens of the future generation cannot participate in such discussion, and they will be merely forced to obey. The Fukushima's case of nuclear power plant accidents in Japan, the most sacrificed peoples are rather future generation (Shibuya, 2017a, 2020b). In short, what most citizens assume is fairness is based only on their own current generation and the situation around them, and does not fully take into account future generations and the global environment as a whole. Therefore, there is no guarantee that it will be able to find a solution that essentially takes all sacrifices into account simply by making decisions based on democratic deliberation and limited rationality.

Certainly, civic engagement and social movement online can be often synchronized by the extent of larger interconnected participants who are taking with own smart-phone (Shibuya, 2017b, 2020a). Such social momentum of massive opinions in any social networking site steeply converges to the consensus and majority whether pro or con among the online users. Either decoupling or clustering with other like-minded partners online can be self-organized by each desire of individual who hopes to stays in the position for supporting social justice.

Social justice in daily life is often contextualized in the atmosphere of opinion dynamics among the citizens whether it is online or not. The majority consisted of larger size of existing opinion groups usually governs the handle of what justice is in such situation. It seems that social justice can be defined comparatively by such pseudo democratic proving process. It means the majority can stand on the side of justice, whereas the minority is enforced to occupy the injustice status as a loser.

Online, it is also possible to recruit large numbers of anonymous volunteers and patrons to organize and run volunteer and public charities to help the poor (Shibuya, 2012). Social movements and online debates on local political issues can easily bring together people with shared interests. There is a similar theory in Eastness, and it is clearly documented in the Chinese books. For example, Confucius said "*Virtue is not left to stand alone. He who practices it will have neighbors*". Moreover, "*The Book of Changes*" described "*if someone is honesty, he gives hands for the neighborhood, and then they share any merits among them*". In these books, ancient Chinese took close look around the neighborhoods and community to keep harmony.

4.3.3 Rules of Conduct in Various Situations

Digital society unconditionally requires somewhat rules of conducts for keeping social harmony and avoiding chaotic situations. It often calls aesthetics and manners to be better human relationships such as media literacy, *Right to be Forgotten* and a *netiquette*" which is scripted in GDPR (General Data Protection Regulation) regulations (Slokenberga et al., 2021).

In the pandemic of the COVID-19, the significance that wearing a mask has become common practice in the EU and other countries as part of corona countermeasures has not only improved their manners, but has also helped to improve their lifestyle habits. The wearing of masks was already taken for granted in Asian countries, and the WHO and other institutions have finally had to acknowledge the effectiveness of mask wearing. This is an explicit indication of "the concern for the others" rather than self-centered standpoint. It was one of the reasons why the cultural values of Asian countries are said to have surpassed Western value standards in this pandemic. Quite simply, the emphasis on individual values meant that the government did not provide a uniform standard, but simply left them self-governing. But lastly, it enforced mobility restrictions and urban lockdown, and further it seemed to be a self-denial against individual liberty and even the very foundations of the Western system of the national governance (Mill, 1989).

4.3.4   Wisdom

Information and knowledge-based economy, without saying, has been governed by somewhat value as much as people desired and expected. It means that data accuracy and quality which are guaranteed in specific manner. But on the contrary, ordinary citizens as well as professionals must be regularly ready to be insightful for all data. Evidence and scientific based examinations cannot be also ignored in regular daily contexts.

First, it should deal with fake-news (Viviani & Pasi, 2017). What is the matter on fake-news? As the author mentioned before (Shibuya, 2020a, 2021a), social media has become to one of the opinion manipulation tools for the citizens as well as external malicious intentions (Aral & Eckles, 2019). Indeed, the level of education, morality, and well-educated scientific attitudes of each country's citizens will be clearly visible in the event of an emergency. For example, during the pandemic, there was a news that several deaths due to accidental ingestion of disinfectant had been reported in the USA. It was believed to have been triggered by the president's statement at a press conference, and there had been a number of cases where some citizens, who trusted this statement, had intentionally swallowed it out of fear of epidemic infection. However, this was clearly fake news, and it was very far from scientific judgment. Not only were measures to prevent panic, hoaxes, and prejudice based on hoarding and selfish intentions, but there was also a greater need to promote scientific understanding among the citizens.

4.3.5   Trust

It roughly presumes that digitized interconnected world has been given by mutual trust and understanding among the global citizens. However, our actuality does not match such ideal condition. Decoupling each opinion and ideology arises serious inevitable conflicts whether online or not. In the digitized global world, it must be definitively dedicated by every citizen and government leaders for mutual trust. International cooperation and collaboration have been required in the various contexts such as resolution for the COVID-19 pandemic, systemic risks in financial market, global warming and other catastrophic situations.

At the pandemic, Fukuyama said[7] "This proves that for a nation to survive, it must first have experts, fair people who are committed to the common good, and leaders who listen to them and ultimately make decisions. As far as our president is concerned, he has been saying for the last two months that the pandemic has nothing to do with us". As well known, the USA suffered the world's worst case of infections

---

[7] https://headlines.yahoo.co.jp/article?a=20200425-00000005-courrier-int&p=3 (in Japanese).

and deaths during the COVID-19 pandemic, despite the fact that they had the best infection control research centers and resources in the world. Fiscal year 2020 was also the USA presidential election period, and attention was focused on the impact of this pandemic.

As the advent of social media, legitimation crisis had been often observed in global Habermas, 1976). Democratic movements and revolutions can be easily occurred by collective dynamics of ordinary citizens. The political crisis during the COVID-19 pandemic was certainly borne out by the example of the USA as well. For example, the approval rating between the Mayor of NY and the President of the USA was entirely in the favor of the former. It was the year of the USA presidential election, and his opponent seemed to be better than the incumbent. The reason for this was not only the delay in responding to the COVID-19 pandemic, but also the adverse effects of state and federal cuts in public health and medical budgets, lacking on scientific evidence, that had led to the increase in the number of American civilian casualties.

The same was true of other countries with respect to these political crises. For example, support rate for the prime minister in Japan became extremely too low, even by the global standards.[8] The reasons for this were that it took too long to decide on specific policies against the epidemics, and that the content of the specific policies was not enough to provide relief to needy citizens.

5   Conclusion

Regarding above discussions, every citizen must be self-sacrificed obeying social norm and working themselves as whole in the emergency, whereas what sacrifices the ruling governor should be undertaken? Similarly, they also must govern the citizens in compliance with the five virtues and something like this. Such mutual symmetry based on the common virtues will be institutionalized in the social stability (Meulen, 2016). In brief, any government leaders have to keep own trusteeship supporting from the citizens.

References

Ackeren, M., & Archer, A. (2018). Self-sacrifice and moral philosophy. *International Journal of Philosophical Studies, 26*(3), 301–307.
Ackeren, M., & Archer, A. (2020). *Sacrifice and moral philosophy*. Routledge.
Akerlof, G. A. (1970). The market for "lemons": Quality uncertainty and the market mechanism. *The Quarterly Journal of Economics, 84*(3), 488–500.

---

[8] Online survey conducted by Singapore's Blackbox Research and France's Toluna: https://bla ckbox.com.sg/everyone/2020/05/06/most-countries-covid-19-responses-rated-poorly-by-own-cit izens-in-first-of-its-kind-global-survey.

Aral, S., & Eckles, D. (2019). Protecting elections from social media manipulation. *Science, 365*(6456), 858–861.

Axelrod, R. (1997). *The complexity of cooperation.* Princeton University Press.

Bonotti, M., & Zech, S. T. (2021). *Recovering civility during COVID-19.* Palgrave Macmillan.

Bowles, S. (2012). *The new economics of inequality and redistribution.* Cambridge University Press.

Cederman, L. S. (2005). Computational models of social forms: Advancing generative process theory. *American Journal of Sociology, 110*(4), 864–893.

Cui, H., & Kertész, J. (2021). Attention dynamics on the Chinese social media Sina Weibo during the COVID-19 pandemic. *EPJ Data Science, 10*, 8. https://doi.org/10.1140/epjds/s13688-021-00263-0

Erev, I., Plonsky, O., & Roth, Y. (2020). Complacency, panic, and the value of gentle rule enforcement in addressing pandemics. *Nature Human Behaviour.* https://www.nature.com/articles/s41562-020-00939-z

Escolà-Gascón, A. (2021). New techniques to measure lie detection using COVID-19 fake news and the multivariable multiaxial suggestibility inventory-2 (MMSI-2). *Computers in Human Behavior Reports, 3.*https://doi.org/10.1016/j.chbr.2020.100049

Fia, M., & Sacconi, L. (2019). Justice and corporate governance: New insights from Rawlsian social contract and Sen's capabilities approach. *Journal of Business Ethics, 160*, 937–960.

Franks, D. D., & Turner, J. H. (2012). *Handbook of neurosociology.* Springer.

Habermas, J. (1976). *Legitimation crisis.* Heinemann.

Hobbes, T. (2009). *Leviathan.* Oxford University Press.

Hooper, P. L., Kaplan, H. S., & Jaeggi, A. V. (2021). Gains to cooperation drive the evolution of egalitarianism. *Nature Human Behavior.* https://doi.org/10.1038/s41562-021-01059-y

Hosono, A., Page, J., & Shimada, G. (2020). *Workers, managers, productivity: Kaizen in developing countries.* Palgrave Macmillan.

Johnson, G. F. (2008). *Deliberative democracy for the future: The case of nuclear waste management in Canada.* University of Toronto Press.

Kant, I. (2017) *Fundamental principles of the metaphysic of morals (English trans. ed.).* Createspace Independent Publication.

Kawachi, I., & Berkman, L. F. (Eds.). (2003). *Neighborhoods and health.* Oxford University Press.

Mill, J. S. (1989). *On liberty.* Cambridge University Press.

Meulen, R. (2016). Solidarity, justice, and recognition of the other. *Theoretical Medicine and Bioethics, 37*, 517–529.

Moore, G. E. (1903). *Principia Ethica.* Cambridge University Press.

Nowak, M. A., & Sigmund, K. (2005). Evolution of indirect reciprocity. *Nature, 43*(7), 1291–1298.

ÓhÉigeartaigh, S. S., Whittlestone, J., Liu, Y., Zeng, Y., & Liu, Z. (2020). Overcoming barriers to cross-cultural cooperation in AI ethics and governance. *Philosophy & Technology.* https://doi.org/10.1007/s13347-020-00402-x

Rawls, J. (1971). *A theory of justice.* Harvard University Press.

Rawls, J. (2001). *Justice as fairness a restatement.* Harvard University Press.

Schwartz, S. H. (1970). Elicitation of moral obligation and self-sacrificing behavior: An experimental study of volunteering to be a bone marrow donor. *Journal of Personality and Social Psychology, 15*(4), 283–329.

Shibuya, K. (2012). A study on participatory support networking by voluntary citizens-the lessons from the Tohoku earthquake disaster. *Oukan, 6*(2), 79–86. (in Japanese).

Shibuya, K. (2017a). An exploring study on networked market disruption and resilience. *KAKENHI Report.* (in Japanese).

Shibuya, K. (2017b). *Bridging between cyber politics and collective dynamics of social movement.* In M. Khosrow-Pour (Ed.), *Encyclopedia of information science and technology* (4th ed., pp. 3538–3548). IGI Global.

Shibuya, K. (2020a). *Digital transformation of identity in the age of artificial intelligence.* Springer.

Shibuya, K. (2020b). *Identity health.* https://www.ncbi.nlm.nih.gov/pmc/articles/PMC7121317/; Shibuya, K. (2020). *Digital transformation of identity in the age of artificial intelligence.* Springer.

Shibuya, K. (2021a). Breaking fake news and verifying truth. In M. Khosrow-Pour (Ed.), *Encyclopedia of information science and technology* (5th, ed., pp. 1469–1480). IGI Global.

Shibuya, K. (2021b). A risk management on demographic mobility of evacuees in disaster. In M. Khosrow-Pour (Ed.), *Encyclopedia of information science and technology* (5th ed., pp. 1612–1622). IGI Global.

Slokenberga, S., Tzortzatou, O., & Reichel, J. (2021). *GDPR and biobanking individual rights, public interest and research regulation across Europe.* Springer.

Spinoza, B. (2008). *Ethica, ordine geometrico demonstrata,* BiblioLife (English translated edition).

Triandis, H. S. (1995). *Individualism and collectivism.* Westview Press.

Turner, J. H. (Ed.). (2001). *Handbook of sociological theory.* Springer.

Viviani, P., & Pasi, G. (2017). Credibility in social media: Opinions, news, and health information—A survey. *Wires Data Mining and Knowledge Discovery, 7.*https://doi.org/10.1002/widm.1209

Walzer, W. (2006). *Just and unjust wars: A moral argument with historical illustrations* (4th ed.). Basic Books.

Wang, L., & Graddy, E. (2008). Social capital, volunteering, and charitable giving. *Voluntas: International Journal of Voluntary and Nonprofit Organizations, 19*(1), 23–42.

Willer, R. (2009). Groups reward individual sacrifice: The status solution to the collective action problem. *American Sociological Review, 74*(1), 23–43.

Wilson, J. (2000). Volunteering. *Annual Review of Sociology, 26,* 215–240.

Synchronizing Everything to the Digitized World

1 Cyberspace to be Synchronized

1.1 Cyberspace

Nowadays, sociological discussion extends to *cyberspace*, which is a place of online communication via the Internet. Moreover, with the realization of the ubiquitous society and the effective use of the AI technology and big data, the boundary between private and public sphere has been becoming even blurrier (e.g., metaverse) (Shibuya, 2020a). If we think about it in this way, what is it that the humanity is ultimately aiming for? The distinction between the self and the others naturally requires social morality and institutions, but if only an entity in the cyber-world dominates the identification between the self and the others, these differences will become irrelevant. So, where will the AI and its related technologies lead the humanity (Rahwan et al., 2019)?

The open questions that the author wonders in general are the following issues.

1. Is the ubiquity of the AI and its realization of the autonomous governance for the humanity desirable? Does this have something to coordinate with the view of the nature and values rooted in East Asian culture, or with a pantheistic perspective?
2. How is it possible to design a value standard and evaluation system that is fair and unlikely to cause envy and dissatisfaction?
3. Can it steer our citizens towards international cooperation and permanent peace? Is it possible in the age of the AI to induce and educate citizens to a *public opinion* with global common values?

First, in these contexts, while Spinoza's pantheistic ethics and Eastern thought had been well understood in early days, the author (Shibuya, 2020a, 2022) pointed out the relevance of the AI's ubiquity to our cyber-physical world. In fact, the Japanese government's Sixth Basic Plan[1] for Science, Technology and Innovation includes an item to *"To realize a society in which people are free from the constraints of their bodies, brains, space and time by 2050"*. Then, talking about these technologies and future society is rather actual thought than the realm of science fiction (Bostrom, 2013, 2016; Von Braun et al., 2021).

The idea of the ubiquity of AI may seem strange. But, according to Elhacham et al. (2020), *"anthropogenic mass"* is estimated to have exceeded the mass of naturally occurring matter. Human civilization has already filled the earth with artificial materials for its own convenience. It has been proposed that the era in which we live be called the *Anthropocene.*[2] And if we continue to replace the earth, which is a symbol of the natural world, with artificial materials, the earth will eventually become an artificial planet. Lastly, the AI driven governing system may control all of them as well as the humanity.

Secondly, a part of the fairness as justice, which proposed by Rawls (1971, 2001) should be achieved in the AI driven surveillance system. The AI driven system requires to judge safety, equality, justice, and fairness for everyone, engineering has to be equipped such ethical concepts (Cath, 2018; Winfield & Jirotka, 2018; Shibuya, 2020a). Certainly, if possible, all contradiction and other social issues cannot be settled in those mechanisms, but at least, it might step forward going to the equivalent and fairness society. When AI control during the pandemic becomes even more sophisticated and far beyond the level of the human perception, will such digitized systems converge into a single ideology across libertarian, conservative and authoritarian differences? But someone does not have tolerances against such AI driven surveillance and monitoring world.

And thirdly, sociologist Etzioni categorized four patterns of the society. Namely, the ability to build consensus and control as the two basic components of social induction. Depending on the combination of these elements, the following four social types were suggested.

(1) Passive societies: those with low levels of control and consensus building. Many developing countries.

(2) Over-managed: a society with a high level of control, but with a low level of consensus building. A society that has not developed the consensus-building capacity, and it is close to a totalitarian state.

(3) Drifting: the ability to build consensus is advanced, but the control is not as strong as it should be. A society in which capacity is not fully operational. It is close to a capitalist, democratic society.

(4) Active: A society in which control and consensus-building abilities are balanced and heightened. It is nearly a post-modern society (i.e., future society).

---

[1] https://www8.cao.go.jp/cstp/stmain/mspaper3.pdf.

[2] https://www.nature.com/news/polopoly_fs/1.17085!/menu/main/topColumns/topLeftColumn/pdf/519144a.pdf.

This category is certainly reasonable if it considers that we are already living in an era where the vectors of a society are categorized by the extent to which it actively depends on the AI related services and big-data oriented economy.

1.2 Decoupling Worlds

In the international situation where Eastness and Westness are observed to stay in mutual conflicts, the digital transforming innovation promoted by China embodies the differences between political systems such as democracy, state control, the problems of liberalism and digitized monitoring systems driven by autonomous AI (Falco et al., 2021).

These differences were also evident in the COVID-19 pandemic. This is the difference between Western civilized societies, which allow maximum individual value, private rights (such as freedom of mobility and private ownership) and other freedom, while Eastern civilized societies, which coordinate with the interests of the individual and prioritize the optimization of the whole society. The former was a complete blunder in countering the emerging corona virus, resulting in countless casualties, while China also successfully applied its emerging digital technology to bring the infection process under near complete control.

These are not limited to the differences between civilized society and human history, but will become even more pronounced as digital transformation progresses. In other words, Eastern civilized societies (represented by China) will rise to prominence thanks to the benefits of digital transformation, while Western civilized societies may converge to the Eastern value standards that digital transformation represents, and gradually decline and self-decay (Root, 2020). If such a process is certain, it can expect a growing phase of international politics in which Eastern values and Western values will become severe conflict rather than international cooperation.

1.3 Warning of Mill's "On Liberty"

The debate about a cultural difference between Europe and China possibly goes back to before the nineteenth century, at least. According to Mill (1989)'s book "On Liberty"(it was originally published at 1859), he clearly noted that the mode of control of public opinion in modern Europe was merely an unorganized form of doing what the Chinese were organizing themselves to do through their educational and political systems. Therefore, as he noted, if the individuality of the European could not resist these institutions and assert itself, finally Europe would become a second China, even though it might boast a noble ancestry and Christianity.

He further expressed what the factors that had saved Europe from such a fate until now were. Why did European civilizations and citizens continue to progress instead of stagnate? It was not because Europeans were superior to the others. Even if they were excellent, it was a result, not a cause. It was because Europeans had been diverse in their characters and cultures. Event at the Mill's age (19 century), however, those advantages had already begun to be lost in significant part. Mill had already warned that Europe was beginning to move towards the Chinese ideal of making all its citizens uniform.

2   AI Driven Cyberspace to Monitor the Humanity

2.1   *Moral Standardized and Monitored by the AI*

Philosophically speaking, in general, the *God* stands for the "ideal image" that each individual holds in equal measure. If it imagines a "personified concept" that codifies a "moral code", it will be shared and adhered to equally throughout the society from the rule of conduct as a source of the highest good to the bases of each individual. Since such ideal image is an abstraction of the *value system*, the individual makes an effort to approach that ideal by internalizing it, and thus realizes the unified substance as the ideal existence beyond the real self. As discussed in the previous chapters, since ancient times, people desired to become ethical and moral beings, and it could be practionalized within the beauty, the truth and the goodness. In ancient times, when writing and communication were not sufficient, cultures and nations that succeeded in orienting their constituents and citizens through such *concepts* of legal systems, customs, and social norms that underlie groups, societies, and nations would have prospered. As a result, the concept that they held equally was dogmatized as a *religion* and transmitted it to the world transcending national and cultural boundaries.

Recently, Whitehouse et al. (2019) analyzed historical big-data, and they published an intrigued work on emergent possibility of *the moralizing God* in the world history. When population size in larger community exceeded somewhat degree of criteria, such God had appeared in each history of human civilization. Their estimations based on big data could facilitate the globally accumulated demographic evidences to understand the facts. And their results meant that each civilization did not require the God to be initially organized, but rather growing process of civilization consequentially engendered *the moralizing God* for ruling own members in community. If such assumptions were true, it was possible that social identification committed in own community and group memberships had been steadily initialized for their morality in such process.

This kind of big data analysis provides an answer to the question, "Why do people want a god?" The above study implies that people needed God for utilitarian reasons to enable them to lead the entire community. As mentioned in earlier chapters, people often desire reward and dislike punishment. However, in other words, it is not necessary for God to be the one who guarantees the reward. Whether it is a real government or somewhat AI services, it is not irrational for us to live according to its will and value standards as long as we are sure to receive satisfactory compensation and rewards for our good deeds and labor.

Similarly, after the pandemic, the question will emerge whether the government repealed or proceed the law that permitted automatic surveillance of the citizens. It can handle big data at high volume and speed, and the AI driven services are able to monitor human behavior and thinking. In this context, the question is whether the AI driven mechanism will guide humans by monitoring their adherence to moral codes and institutions. Even if it is a behavioral economic tool like a *"nudge"* (Thaler, 2018; Thaler et al., 2012), if it is guided by AI (or the central government that commands AI what to do) (Jesse & Jannach, 2021), isn't that the equivalent of AI manipulating people from behind? A growing number of countries around the world including China are using AI driven mechanisms with such intentions of the government. It is clear that they are trying to preside over state control with "the standard of the values" and present a national norm to guide their citizens and force them to abide by it. However, while various legal systems are naturally enforced in each country, excessive monitoring by AI driven system has been identified as a problem. Such system also applies to the deterrence and prevention of the crimes, and further in emergency situations such as the COVID-19 pandemic, it could be effective for the isolation to avoid infection from infected to other sound persons for the purpose of public health.

2.2 Spinoza and His Discussions

Here, the author redraws an auxiliary line from Spinoza's pantheism and his theory in order to see the future in this age of digital transformation (Shibuya, 2020a). Principally, Spinoza was a philosopher who lived at a time when modern science had reached a plateau (e.g., the mid-1600s AD). He was often said to be a successor of Descartes' philosophy, and his generation inherited intellectual assets of the previous ages because they lived at a time when many scientific and technological advances were remarkable. In short, it was around the time that *scientists* began to awaken to the fact that it could understand the laws of the universes and the natural world through the power of science not theological faith. For example, Newton, Leibniz, and others. As a result of their pioneering work, calculus, mathematical methods, and scientific experimentation, which had become available to the humanity were making great strides, and intellectual advances were beginning to blossom.

Even at that time, with his main book *Ethica*, Spinoza explored the truth with his enthusiasm for the God (e.g., *Amor Dei Intellectualis*). He sought to elucidate the singularity of his philosophical inquiry and the nature of the God as the substance in a geometric logic space. However, there were many people who denounced him at the time. Many of them were said to have preached pantheistic ideas that equated with the universal nature. There was also the decisive break with the synagogue of excommunication, an idea that was considered heretical from the very beginning. The points made at the time were generally correct at the time, but from a modern perspective, it is now hard to understand whether there was any significance in denouncing it.

Mullins (2016) looked back at the past discussions between the *Panentheism* and *Pantheism*. Critically, he stated that there were a lot of extreme interpretive fallacies. In brief, panentheism can be said as "all things $\in$ the God", whereas pantheism can be said as "all things $\leqq$ the God". In the former case, the God encompasses everything in the world (even all things like the world and the natural world) and is considered as an external reality, while in the latter case, the God's will and its aspects (or transformation) fill all the natural world, and therefore the God is real. Both are common in their affirmation of the reality of the God.

In addition, Leibniz (2017) who had originated modern mathematics also investigated *monad* as a kind of the universe, which holds all possibility of the nature, and it led us to consider the God through a representation of the monad. His theory was not just an ordinary dualism, and there were those who could be related to Spinoza as well as East Asian thought on the nature. In this concern, Leibniz and Spinoza discussed such topics each other, because they were living at the same time, and their conclusion seemed to be different to understand for the nature of the God with each other. The author thought that one of their main differences was that Spinoza stressed on the world without contradiction derived from the God and its deductive logic on the nature and the humanity, whereas Leibniz conceptualized a simple substance (i.e., monad) which holds all possibility to be the perfect world. Rather, Leibniz's monad seems to be realized as one of the multiverses, fractal as self-similarity, quantum physics and the possible worlds in modal logic (Brown & Chiek, 2017). In addition, in Leibniz's case, his correspondence with missionaries sent to the East Asia led him to contemplate the plurality and diversity of values and cultures. In China, even without the Western concept of the God, the legitimacy of governance was maintained by a concept such as "heaven".

Further, a philosopher Husserl, in his book "*Cartesianische Meditationen*" (2012), connected Leibniz's concept of monads with the issues in phenomenological inter-subjectivity in the daily world. The world serves an existential linkage for both self and the others to approach a conceptual proximity to the monad. However, in terms of the simultaneous coexistence of mutual subjectivity between self (ego) and the other (alter ego) with a sense of reality that involves a mutual corporeality, it is different from the space on the extended cyberspace.

2.3   The Ubiquity of the AI

In those contexts, today, we are witnessing innovations from new technologies such as AI, big data, ubiquitous devices, high-speed wireless communication (e.g., 5G: 5th Generation wireless communication), IoT (internet of things), smart cities, automated driving technology, and quantum computing. In addition, medical technology has been advancing remarkably including regenerative medicine and medical diagnosis technology using the AI (Von Braun et al., 2021). There is a high probability that the so-called *Singularity* will be reached soon, and the fear that AI will transcend even the level of human intellectual understanding is beyond the scope of the science fiction world and will become a reality. The relationship between the science fictional world and philosophy may be a source of old and new ideas.

In the first place, a term *ubiquitous* is derived from the Latin word for *"the omnipresence of the God"* (i.e., *ubique*) (Weiser, 1993). There's also another expression *pervasive*, which strips away the theological nuances, but in reality, it's just the same thing. Thus, it seems that the contact point between the philosophies that Spinoza preached is surprisingly close to each other even in the AI age.

In the ubiquitous digitization world, Spinoza's theory has high affinity with the view of the nature in East Asian thought (including polytheism and animism). If it assumes that the central AI is a "God" as the substance, and its distributed agents of a kind of such AI ubiquitously show a part of the God (it means here the AI's ubiquity). Such AI-driven system will be possibly observed anywhere on today's interconnected network such as XaaS (everything as a service), and it automatically progress many different kinds of big data and begin to adjust and optimize social systems as a whole including human behaviors. Additionally, through wireless communication such as 5G, VR (virtual reality) and tele-existence technology, every people will be interconnected online to each other and their brains, transcending physicality and sensing reality and converging into a singleton (Giannotti et al., 2012; Agar, 2013; Zeng et al., 2021). If such a society is realized, people may have no choice but to know about the world through the information presented by the central AI. If that happens, it brings us closer to a real world that can only be understood through *"Sub Specie Aeternitatis"*, and it is the only recognition that can recognize the God, as Spinoza preached. If such situation will be achieved, it shows a loss of identity of each individual and the disappearance of both the *raisons d'etre* of state governance and the necessity of democratic political system.

Similarly, Leibniz also definitively described such status in his book *Monadology: "Finally, under this perfect government no good action would be unrewarded and no bad one unpunished, and all should issue in the well-being of the good, that is to say, of those who are not malcontents in this great state, but who trust in providence"*.

The result of digital transformation may lastly invoke the loss of boundaries between the self and the others, and proceed sharing of synthesized personal experiences on both a sense of own reality of the world and actuality, and consequently it becomes the excessive attunement to each other. The optimization and mutual coordination of these features controlled by the central AI will indicate the unification of the individual identity with the one world provided by the central AI, and will mean nothing less than the abandonment of the identity of each individual. The world governed by the AI, where all social phenomena are perfectly aligned with the values and norms among all individuals and the morals of the society is nothing but one of the *worlds* as envisioned by the central AI.

2.4   Relations Between East Asia Culture and the Ubiquity of the AI

Does the digitized monitoring society driven by the AI demonstrate a possible world of the AI systematized world? Given that the similarity between Spinoza's pantheistic ideology and Chinese thought piled in the Eastern cultural background is "the heaven" and "the natural world" itself, is the ubiquity of the AI equalizing to Spinoza's ideology? Further, it seems that this conceptual relation nearly equals to a ruling ideology which conducted by the top of the governing members of China using the AI and big-data in pervasive interconnected network. To what extent can it say that there is any relevance to China's ambitious goals of Pan-China strategy? Do the ubiquitous society and somewhat "the divine nature" perspective of China have anything in common?

Habermas (1991) told structural transformation of the *public sphere*. It refers to a group or sphere that lies between the private family or market economy and the public state apparatuses such as parliaments, bureaucracies, and armies. And it means to the area in which social life is carried out on a daily basis. At the same time, it can be considered as an open area for debate among citizens and can be the bases of public opinion formation. After the advent of internet era, his notations of public sphere enlarged into the cyberspace and online interconnected world, and those who participated with each motivation and purpose can be assembly associated in the pervasive and virtual contexts. For instance, larger social movement and political demonstration by massive citizens whether it organizes online or not could be contextualized in such background.

However, particularly in the digitized age of China, collective dynamics of the citizens taking with own smart-phone can be embedded in the ubiquitous monitoring environment driven by the automatic system of the AI, which is developed and deployed by the central government. Namely, it means that even though their liberty of the private right and freedom of expression is apparently permitted by the legal authority, but their sequences of all behaviors within a tractable range of monitoring system cannot be inevitably escaped from the global web of the government. These conditions are certainly quite far from the ideal of democracy, and it finally approaches a situation among all citizens synchronized with the ideology induced by the top of governing members in China. Reversely saying, each citizen has this common ideology and behaves in line with the imperative from the leader, and all of the social phenomena participated by the massive citizens might be harmonically assorted by the ruling reasons of the governments.

It means that the possibility of an emergent public sphere would be diminished by the interaction among heterogeneous members in communication networks. In other words, the private sphere (a sphere of private life) and the sphere of the nation (a system of social institutions in the nation) have become fused and ambiguous, and the public sphere, which is supposed to lie in the middle of these spheres has also disappeared. The nation does not necessarily internalize only the common cultural backbones. But, assuming an absolute view from the top of the oppressive and ruling governance, they unnaturally deny that there are multiple and diverse cultural systems, opinions, and historical backgrounds within the same nation abandoning internal and spontaneous evolution. It can be paraphrased so that they are too afraid to reach a situation of self-decay and loss of control their ruling governance caused by democratic diversification and overexposed individualism among the citizens.

Moreover, as discuss later chapter, the manipulation of information and election results through social media has made it possible to try to improve the soundness of proactive governance. If any leader aspires to stabilize national governance through long-term regimes, it will be nothing short of moving closer to a science fictional system of the governance that only deifies the AI driven central system and obeys its commands. In fact, in tyrannical states like Russia and China, that such feature is becoming more prominent.

In early 2021, USA President Biden[3] said that there would be two major categories of countries in the contemporary world, *"those that consider autocracy to be the best"* and *"those that are democratic"*, with China and Russia in mind. However, this discourse has lost sight for the future. In other words, the issue is no longer a simple "tyrannical system vs. democratic system" debate, but a difference between "a system of state governance with a control system that makes maximum use of AI and Big Data technologies vs. traditional democratic system (which is vulnerable to attacks by AI and Big Data technologies, as it details later chapter)". The world of AI and big data technology will inevitably lead to the ubiquity of AI as described above. We are now facing an era in which the governance and political systems will be transformed based on such a social information infrastructure, and it seems that we have little sense of crisis if we are too caught up in the old schemes of understanding.

2.5 Changing to Digital Memory in Cyberspace

What happens if it accelerates the melting of the boundaries between self and others using metaverse, VR, and BMI (Brain-machine interface) as well as AI technologies possible (Shibuya, 2022)? For example, *social consciousness*, which was originally conceptualized by an ancient sociologist Durkheim may be also embodied in the cyberspace (Saha et al., 2017). By through digital transformation and ubiquitous interconnected communications such as VR and tele-existence among the multiple members, our sensible and physical capabilities will converge to compacting a shared memory and sensibility with the others (Yuste et al., 2017).

However, a concept on social consciousness that defined by Durkheim had been extrapolating into sociological meaning. It might be reluctant to be acquired an idea from consciousness of psychology within the human brain, because sociology had to stand independently on own firm discipline, and rather he definitively characterized it as socially sharing common knowledge and representation among the members. It did not mean sharing image and knowledge by interconnecting with each individual,

---

[3] https://youtu.be/4tyUK2uk_D8.

and it just theorizes a fact that our knowledge commonly shares with other group members.

Further, regarding such concerns, a recent theory which named as social representation theory (Deaux & Phiogene, 1999; Sperber, 1996; Wegner et al., 1991) has been inherited from such Durkheim's theory. Particularly, Wegner and his colleagues (1991) argued the "Transactive memory", and it depicts a functional part of social representation. It can be considered as a socially sharing common knowledge among those who are rooted in own community. This was one of the "*externalizations of own memory within brain*" that went beyond the limits of an individual's lifespan, and at the same time resulted in the cultivation of an indigenous culture and identity through the sharing and transmission of mutual knowledge among community members. Further those mediated and memorized styles could support them being adapted in the ancient environment. Their committing *group mind* could be initialized and cultivated by social identification context, and such patterns would be also defined as racial or national identity (Shibuya, 2020a).

In the age of big-data, such conceptual idea will be eagerly realized as inter-shared electronic data in the cyberspace, and it probably means not only the life histories of each individual at the global level, but also a part of the world history formed by their interconnections. For example, after the advent of social media, ordinary citizens face own vulnerability against information manipulation by external intentions of the government leaders, hacktivists, antagonists and others. Fake-news and counterfactual information about social events can be misconducted to be inflamed situations online (Kirkpatrick, 2020; Shibuya, 2021).

2.6   The General Will and the Ubiquity of AI

In a democracy, the freedom of each citizen to express his or her opinion and political participation through own will are important elements, and it is necessary to consider whether the unification to the entire society that the ubiquity of AI can bring about should be regarded as the same as the general will or a different form of "something".

In "*The Social Contract*" (2009), Rousseau wrote about the relationship between the state and the citizen: "*How can we find a form of union that can protect and preserve the identity and property of each member with all the power of the communities? In this union, each person must be able to remain united with all others and yet remain free as before, subject only to himself*". According to Rousseau's argument, all citizens embody the answer to the above question by thinking that each member has a form of "*social contract*" in which he or she assigns all rights to the community and receives a corresponding value (e.g., rights and duties) from the community. The argument that we are thus approaching the general will in the community as the citizenry as a whole, certainly has a certain persuasive power in the debate on legislation and freedom.

Then, the current situation has been pseudo- or indirectly realizing the digitization of the general will. And if the current stage of discussion in the interconnected cyberspace, where people maintain their individual identities, progresses to a more molten form of the ubiquity of AI, it will be more like "unified will" than the general will. It will be a society in which people are synchronized with the "world" constructed by the ubiquity of AI. In this respect, the relationship between the state and citizens will clearly take a different form.

In the event of COVID-19 pandemic, various disasters, and wars, according to the social contract theory, the democratic state will take the place of each citizen and make arrangements to deal with the crisis, and each citizen will cooperate as necessary. And the same is true for despotic states, and will be true when the ubiquity of AI arrives. In this respect, tyrannical states and AI driven state governance may be able to respond quickly to crises, or even prevent them from occurring, because it is easier to reach the unified will without the time-consuming coordination of opinions and social preferences that democracies necessarily require.

2.7 What If Maximizing AI Driven System of Surveillance and Rewarding

The author serves an example. One of the ideals envisioned by the distribution of sacrifice is the minimization of the self-sacrifice that all people must undertake if possible. Ultimately, what if maximizing the AI driven system of surveillance and rewarding for the ordinary citizens? And if quantum computing will be completely realized in the future, as a pattern of the optimization problems, can the sacrifice involved complex causal relationships and these coefficients in the real world be detailed and resolved?

The past deliberation on socioeconomic fundamentals can be recalled for me. For example, Hayek (a Nobel Prize winner in economics) and Mises made criticisms about *economic calculation problem* based on complete central governance conducted by the nation and complete information such as all individuals and other necessary variables in the society, but in nearly future, it seems that such similar society has a possibility to be realized when big-data of all social phenomena as necessary as the AI driven mechanisms can be accumulated and analyzed within such AI driven world. Rather, during the cold war, western liberal and democratic society laid weight on autonomous and decentralized socio-economic systems (Arthur et al., 2020), which were based on personal preferences and values in the global. Their decentralized models would be relied on the simple assumption that cannot calculate all of the values and solve any contradiction by the centralized system.

However, recent advancement of the computer technologies will be probably possible to achieve a dream of the past theories. In these days of big data, the collection and analysis of large amounts of diverse data can reveal unexpected causal relationships and complicated patterns are being discovered. For example, Phillips and

Rozworski (2019) argued the fact that online companies represented by Wal-Mart and Amazon have been able to operate a much larger centrally planned economic system than the planned economy promoted by the former Soviet Union (now Russia), it is now possible for entire nations to similarly achieve a socioeconomic system. In fact, in the near future, when we can freely utilize quantum computing and other advanced technologies, many problems will become computable and operational in many fields. Whatever complex systems of the human behavior may be, it can find the clues that can resolve the sacrifices it create how promising is the potential? To some extent, it will require the AI driven system to control all ordinary people's behavior in advance and in everyday life.

It will further strengthen the surveillance system as seen during the COVID-19 pandemic, and based on the daily activity data, algorithms for simultaneous optimization and adjustment of economic, traffic, CO_2 emission and energy demand forecast will monitor the global trends. People's lifestyles will be guided by the daily scenarios that AI predetermine. If people are reliably rewarded for complying with norms, there will be no social unrest and no crime. If sacrifices are minimized, there will indeed be peace. In fact, in Japan, Toyota has begun to actually build a smart city.[4] They are building cities from the ground up that utilize advanced technologies such as AI, big data technology, and automated driving systems, and analyzing the big data of daily life as people actually live there. In this way, humanity will gain control over everything from urban planning to socio-economic systems and individual lifelogs in general.

Some people believe that the AI surveillance systems infringe on freedom. But is this really true? Freedom is indeed the core of democracy. Individual liberty is guaranteed to the maximum extent that it is not an obstacle to the society and public purposes (Mill, 1989). This idea is related to Kant's moral law. The idea is that individuals should be free to engage in actively own economic activities and participate in the governance of the country. Here, this pattern of freedom names "F1" for further discussion. On the other hand, a nation based on the central governance is characterized by the assumption of freedom guaranteed within the context of total control. The state proactively assumes the central role in the economy and politics, and citizens have freedom to the extent that they are allowed to do so within those legal systems. This pattern of freedom names "F2".

Importantly, many people implicitly assumes that the degree of freedom F1 and its utility of each individual is better than F2 among the citizens (F1 > F2). However, can it really judge that F1 is superior to F2? In an emergency situation, both freedom and its utility may be practically equalized (F1 = F2). When examined comprehensively in detail, F2 may be actually a better system, and those citizens are not only freer but also have less of a burden of responsibility and moral obligation. It is practically impossible to impose the same burdens, duties and other sacrifices on the same "citizens", because their abilities, intelligence, education level, health conditions, social background are significantly different (e.g., Rawls' theory of fairness). The idea that it is preferable to share controlled social conditions across society and

---

[4] https://youtu.be/jh7FHx8M3G0.

minimize those disparities is not wrong in itself. Even in the welfare state system, there is a sense that they are dedicated to improving their social welfare and health care systems (e.g., Sweden, Cuba[5]), and then it may be similar to F2.

Further, as seen in the COVID-19 pandemic and smart cities (Ross et al., 2020), China is looking to enhance its advanced surveillance systems (Góis et al., 2019). If this comes to fruition, the lifestyles of all people will be guided and lived according to personalized scenarios.

If recognizing that the safety, security, economic benefits and other factors gained by the citizens are more important than the self-sacrifice they must undertake, and their approval degree are higher and more certain than in Western democratic societies, and then its validity is unquestionable. It means to be lower risks, crisis, and sacrifices than profits. Still now, such story is only within a science fiction or sociology of the future design, but this isn't perfectly fictional, because the AI technology, smart cities, big-data, and related technologies are making progress in such direction.

Living within such a system will lead to a variety of risks, crises and sacrifices that are minimized. Clearly, there are greater benefits to be enjoyed by each individual. On the other hand, one might argue that the *"freedom"* as democracy or *libertarianism* (Sandel, 2009), which Western society has been adhering to is more valuable. But, it should be recalled that the COVID-19 pandemic has caused massive deaths in Europe and the USA. Many countries have stood on the sidelines measuring the policies of both the economy and the health care system. And yet, at a sacrifice too enormous to be the *"freedom"*, they still lead the way in their values. How significant was it?

2.8 Beyond the True Liberalism

The ultimate in liberalism and individualism can be concisely rephrased as a status of the maximization of one's own free will, emancipation of physical and social constraints (including threat of epidemic diseases, warfare and other troublesome), and independence from subordination to the others. But, only place where it could be achieved would be on a desert island or in cyberspace. Thus, especially in the cyberspace of the future, the emancipation of humanity from all physical and social constraints led by advanced science and technology will meet the demands of those who desire to sustain true individualism and liberalism. However, this means, after all, that each individual has no choice but to live in a virtual "simulation world" where they live in each scenario. For example, it becomes a computational process in which only necessary information is extracted from each individual and a software entity based on the data is executed on the computer. As noted earlier in this chapter, this may be a future that Japan's science and technology programs will promote to realize (Shibuya, 2022).

---

[5] https://covid19cubadata.github.io/#cuba.

If such future world cannot be accepted, we must recognize the actuality of our world. Claiming freedom without self-sacrifice is just an abandonment of *self-responsibility*, which means *irresponsibility* and *anomie*. Strict adherence to social norms and other institutional designs is quite needed for coexistence with the others.

In the fight against the corona pandemic, there had also been conflicts over the freedom of vaccination in various countries. These cases clearly ignored the coexisting relationship between oneself and others, and it must balance self-sacrifice with social utility in such situation. The fact that some people insisted on irresponsibility in the name of freedom was a serious problem for the entire humanity, regardless of whether they were democratic or not.

3 Conclusion: Decoupling or Fusing Each Other

The COVID-19 pandemic served a discussion on freedom of the individual and social order. The digitized surveillance system should be minimized self-sacrifice of each individual, equalized each utility of them and maximize total safety of the whole society whether in the crisis or not. In the future, will it be realized that the AI-driven society coordinates and guides the citizens who are interconnected inside a vast cyberspace, transcending the boundaries between the self and others (Shibuya, 2022)? Such cyberspace beyond all differences of values and forms ultimately become to be free of any contradiction and conflict. Individual differences no longer matter and converge in interrelationships. Thus, the author wonders how to change the differences between Western society, which continued to seek the dignity of the autonomous individual and Eastern society, which continued to seek the adaptation of the whole. Will they continue to interfere with each other within a single cyberspace or lead to the same depth?

References

Agar, N. (2013). *Truly human enhancement: A philosophical defense of limits*. MIT Press.

Arthur, W. B., Beinhocker, E. D., & Stanger, A. (Eds.). (2020). *Complexity economics: Proceedings of the Santa Fe Institute's 2019 fall symposium (dialogues of the applied complexity network)*. SFI Press.

Bostrom, N. (2013). Existential risk prevention as global priority. *Global Policy, 4*(1), 15–31.

Bostrom, N. (2016). *Superintelligence: Paths, dangers, strategies*. Oxford University Press.

Brown, G., & Chiek, Y. (2017). *Leibniz on compossibility and possible worlds (the new synthese historical library book 75)*. Springer.

Cath, C. (2018). Governing artificial intelligence: Ethical, legal and technical opportunities and challenges. *Philosophical Transactions Royal Society A (mathematical, Physical and Engineering Sciences), 376*, 20180080. https://doi.org/10.1098/rsta.2018.0080

Deaux, K., & Phiogene, G. (1999). *Representations of the social*. Blackwell.

Elhacham, E., et al. (2020). Global human-made mass exceeds all living biomass. *Nature, 588*, 442–444. https://www.nature.com/articles/s41586-020-3010-5

Falco, G., et al. (2021). Governing AI safety through independent audits. *Nature Machine Intelligence, 3*, 566–571.

Giannotti, F., et al. (2012). A planetary nervous system for social mining and collective awareness. *The European Physical Journal Special Topics, 214*(1), 49–75.

Góis, A. R., et al. (2019). Reward and punishment in climate change dilemmas. *Scientific Reports, 9*, 16193. https://www.nature.com/articles/s41598-019-52524-8

Habermas, J. (1991). *The structural transformation of the public sphere.* The MIT Press.

Husserl, E. (2012). *Cartesianische Meditationen.* Meiner Felix Verlag GmbH.

Jesse, M., Jannach, D. (2021). Digital nudging with recommender systems: Survey and future directions. *Computers in Human Behavior Reports, 3*.https://doi.org/10.1016/j.chbr.2020.100052

Kirkpatrick, K. (2020). Deceiving the masses on social media. *Communications of the ACM, 63*(5), 33–35.

Leibniz (2017). *Monadologie.* Hofenberg.

Mill, J. S. (1989). *On liberty.* Cambridge University Press.

Mullins, R. T. (2016). The difficulty with demarcating panentheism. *Sophia, 55*(3), 325–346.

Phillips, L., & Rozworski, M. (2019). *The Peoples Republic of Walmart: How the biggest corporations are laying the foundations for socialism.* Verso Books.

Rahwan, L., et al. (2019). Machine behaviour. *Nature, 568*, 477–486.

Rawls, J. (1971). *A theory of justice.* Harvard University Press.

Rawls, J. (2001). *Justice as fairness a restatement.* Harvard University Press.

Root, H. L. (2020). *Network origins of the global economy. East vs. west in a complex systems perspective.* Cambridge University Press.

Ross, A., Banerjee, S., & Chowdhury, A. (2020). Security in smart cities: A brief review of digital forensic schemes for biometric data. *Pattern Recognition Letters.* https://doi.org/10.1016/j.patrec.2020.07.009

Rousseau, J. J. (2009). *Of the social contract, or principles of political right.* Regnery.

Saha, S., et al. (2017) *Handbook of research on applied cybernetics and systems science (Advances in Computational Intelligence and Robotics).* Information Science Reference.

Sandel, M. J. (2009). *Justice: What's the right thing to do?* Penguin.

Shibuya, K. (2020a). *Digital transformation of identity in the age of artificial intelligence.* Springer.

Shibuya, K. (2021). Breaking fake news and verifying truth. In M. Khosrow-Pour (Ed.), *Encyclopedia of information science and technology* (5th ed., pp.1469–1480). IGI Global.

Shibuya, K. (2022). An "artificial" concept as the opposites of human dignity. In T. Sikka (ed.), *Science and technology studies and health praxis: Genetic science and new digital technologies.* Bristol University Press

Sperber, D. (1996). *Explaining culture: A naturalistic approach.* Blackwell.

Spinoza, B. (2008). *Ethica, ordine geometrico demonstrata,* BiblioLife (English translated edition).

Thaler, R. H., Sunstein, C. R., & Balz, J. P. (2012). Choice architecture. *The Behavioral Foundations of Public Policy.* https://doi.org/10.2139/ssrn.1583509

Thaler, R. H. (2018). From cashews to nudges: The evolution of behavioral economics. *The American Economic Review, 108*, 1265–1287. https://doi.org/10.1257/aer.108.6.1265

Yuste, R., et al. (2017). Four ethical priorities for neurotechnologies and AI. *Nature, 551*, 159–163.

Von Braun, J., et al. (2021). *Robotics, AI, and humanity.* Springer.

Wegner, D. M., Erber, R., & Raymond, P. (1991). Transactive memory in close relationship. *Journal of Personality and Social Psychology, 61*(6), 923–929.

Weiser, M. (1993). Some computer science issues in ubiquitous computing. *Communications of the ACM, 36*, 75–84.

Whitehouse, H., et al. (2019). Complex societies precede moralizing gods throughout world history. *Nature, 568*, 226–229.

Winfield, A. F. T., & Jirotka, M. (2018). Ethical governance is essential to building trust in robotics and artificial intelligence systems. *Philosophical Transactions Royal Society A (mathematical, Physical and Engineering Sciences), 376*, 20180085. https://doi.org/10.1098/rsta.2018.0085

Zeng, Y., Sun, K., & Lu, E. (2021). Declaration on the ethics of brain–computer interfaces and augment intelligence. *AI and Ethics, 1*, 209–211.

A Living Way in the Digitized World

1 Science of Careers and Life Style in the AI Driven Age

Principally, in today's AI driven society, issues surrounding careers and lifestyles are already in the realm of *"science"*. For example, there has been a technological innovation called "HR (Human Resources) technologies", which utilizes big data and AI technology in human resource management (Vrontis et al., 2021). In addition to optimal personnel selection and team formation, job performance, behavioral patterns in the workplace, and psychological satisfaction can be measured and analyzed as needed. A number of product case studies such as the Business Microscope,[1] and a longitudinal study in China (Dong et al., 2014), which examined the impact of mentors and social networks on the career success of new employees based on social capital theory and mentoring theory have begun to be reported.

On the contrary, the COVID-19 pandemic also exposed the absurdities that existed in society, such as socioeconomic inequality and imbalance of opportunities, and reminded us of how people were left in a state of uneven sacrifice in their way of life. However, denying such structural inequality or criticizing it alone will not save anyone in the end. Regardless of whether exploitation and inequality are imposed as sacrifices, unless each citizen continue to work on self-improvement, each of them will not be able to obtain a higher position or employment opportunities by choice from the others.

How should people live in the world based on such socioeconomic inequality? It continues to be a difficult question for each individual in the age of digital transformation (Acemoglu et al., 2021). It is not enough to complain about the injustice of systems and institutions. In some cases, those who had improved their chances of survival and honed their skills even under the worst conditions of exploitation and sacrifice have seized the success opportunity in the devastation brought by the pandemic. Conversely, some of those who thrived in capitalist globalism experienced catastrophic bankruptcy in the Corona pandemic. The difference may be not only in

---

[1] http://www.jfma.or.jp/award/05/pdf/paneldata07.pdf.

their understanding of the causal relationships between events in their lives, but also in the way they lived and thought, and the possibility that they misjudged important factors other than cash flow.

It can be said that online networking through digital technology rather expanded the diversity of entrepreneurship, ways of life and the freedom of life choices. Digitized online networking made them possible to seize such opportunities to improve achievements and skills of each individual by reversing restrictions such as the COVID-19 pandemic and environmental issues. For example, there are ways to learn online such as Coursea, and even get a degree. It could be said that the range of life planning improved. Such situation can be related to the traditional customs of networking as human relationships, and to the Chinese classics about the usefulness of acquiring acquaintances. These matters will be discussed in this chapter.

2   The Nature and Limit of the "Capitalism With Democracy

2.1   Sacrifice Over the Privilege of Freedom and Democracy

In daily life, the economy and labor issues are one of the hot topics where the "distribution of sacrifice" really comes to the fore. In economics, it is often defined as some kind of resource allocation problem, but it is undisputed that not only the optimization and profit maximization of economic capital and human resources, but also various economic utility and social preference issues are still important (Arthur et al., 2020).

In particular, the issue of correcting the imbalance of wealth and inequality of redistribution often become a point of discussion (Bowles, 2012; Sen, 1973). To what extent are the developed countries and the rich obligated to save the poor? In other words, who should bear what sacrifices so that they are evenly distributed across the whole of the society (Hauser et al., 2014; López-Roldán & Fachelli, 2021; Putterman, 2014)? Even before the Corona pandemic, according to the World Bank,[2] there are still more than 1.4 billion people living on less than $1 a day (e.g., *International Poverty Line*). The additional "sacrifices of the poverty", as indicated by the *human development index*, includes life expectancy at birth, adult literacy rates, gross enrollment rates, and GDP per capita. It is often called an issue on global distributive justice in the ethics and politics (Stefan & Bernhard, 2020).

Modern capitalism also developed at the sacrificed many of lower-wage labor forces (Ellerman, 2010; Williams, 1994). In the first place, it pointed out that slaves were an absolutely essential factor in the establishment of ancient democracy. According to the social philosopher Rousseau, in his book on the Social Contract (2009), these problems were clearly defined. "*...in a civil society in an unfortunate*

---

[2] https://www.worldbank.org/ja/news/feature/2014/01/08/open-data-poverty.

situation, it is not possible to preserve one's freedom without enslaving others, and it is not possible for a citizen to be completely free without the extreme subjugation of slaves. Modern people like us do not own slaves, but you yourselves are slaves. We sell our own freedom to buy the freedom of our slaves".

Therefore, modern capitalism with democracy had been probably traced its same route, and it has been exploiting uncountable lower-wage workers and peoples in global (Held et al., 2000; Simon & Glynn, 2020). The people leading such values are similar to those of the "civic class" of the ancient Greece times. They are more well-educated and wealthier than other peoples. On the other hand, those who are not have come to be summarily identified in many countries as *"the people left behind"*. Coincidentally, as *"Black Lives Matter"* (BLM) movements in the USA (Fukuyama, 2018) raised at June, 2020 again, those structural prejudice and justification became also salient due to the aftermath of the COVID-19 pandemic.

The questions of why nations are divided into poor nations and rich nations, and why nations develop or decline, have been asked many times throughout history. In "The Protestant Ethic and the Spirit of Capitalism" (2002), Weber, who explored the sociology of religion, argued that not only the education of human resources but also the ethic of workers backed by a high religious backbone contributes to the development of the state. However, Acemoglu and Robinson (2012) clearly rejected Weber's theory in their book ("Why Nations Fail"). However, Becker et al. (2009) raised the alternative hypothesis that while there was indeed an impact of Protestantism on economic development, the channel was the accumulation of human capital through education.

On the contrary, from the author's perspective, "real freedom" and "real democracy", after all, are the privilege that only the wealthy and the wealthy nations can hold. The reason why developing countries remain poor is that many of their citizens are under the delusion that they will soon be able to attain the same or higher economic fruits and status as their counterparts without bearing the burden of self-sacrifice. Such nation will never develop and become stable, not only socio-economically, but also institutionally and culturally. The result is a free and democratic citizenry with disposable time after improving their education level and achieving economic development. The productive surplus resulting from the accumulation effect is the invisible national wealth, which leads to a cultural level of living and forms the basis for a high quality of life. Naturally, it will take a long time and several generations at least (OECD calls inter-generational social mobility "social elevator"[3]). The most important is to educate them cultivating their independent minds to launch own business by small incentives in line with the design of institutions and systems that provide the living basis such as micro-credit and social business proposed by Yunus (2007). Whether it is social morality or the legal system that guides them, it is a common task for any democratic or tyrannical state to functionally engage people in various kinds of labor and services as appropriate self-sacrifice, and to return the benefits of the nation as a whole to its citizens. This is a common task for both democracies and

---

[3] https://www.oecd.org/social/broken-elevator-how-to-promote-social-mobility-9789264301085-en.htm.

tyrannies. The essence of this task remains the same regardless of the progress of online digital technology.

Therefore, the gap between the rich and the poor can be concluded as either a process of history or an inevitable consequence of history. Even if it is a systemic problem based on prejudice and discrimination, as long as the reason for not being able to escape the negative cycle is due to the privileges of those who have already earned their positions (i.e., establishment), it will not easily change and the sacrifices will only be directed to the poor.

Such situation cannot be remedied simply by viewing it as a matter of fairness (Rawls, 1971). What is sacrificed and whom is sacrificed has nothing to do with whether it is fair or unfair. Such a situation is inevitable, a mere actuality in the world. If someone is willing to change the reality according to the subjectivity of hope, the only way to do so is for each individual to grasp the current situation and facts objectively and strategically, learn on own initiative, improve own abilities and skills, and increase own chances of survival.

Living itself means a risk. In fact, it has been reported that the COVID-19 pandemic impeded labor mobility in Asia (ADB Institute, 2021), and there are many middle class and highly educated workers whose livelihoods were threatened by unexpected bankruptcies and layoffs due to the economic downturn. However, this may have been an opportunity to distinguish between those who have been constantly refining their skills and others who have not. As digital transformation progresses, such a situation is undoubtedly accelerating.

Capitalism justifies an institution to utilize freedom and democracy as disposable income. Therefore, the poor, as well as those who do not continue to develop their skills and abilities, will find it more difficult to survive in the digital society.

2.2   A Question

What is labor? In the eighteenth century, when Adam Smith lived, the following was already defined. "Labour, therefore, it appears evidently, is the only universal, as well as the only accurate, measure of value, or the only standard by which we can compare the values of different commodities, at all times, and at all places"(Smith, 2012). However, labor is about the quality of the results from working hours and its performances. Furthermore, why is it that the unit price of labor is different for the same labor in developed and developing countries? The causes and reasons for this are too complex.

Regarding this issue, when the author was a young student, the author once asked an economist a question. When *lower*-wage labor forces become abhorrent at home, each nation accepts large numbers of foreign workers and immigrants (Baldassarri & Abascal, 2020). So, developing countries will develop their economies based on such demands, but their labor costs will also rise. Then, the developed countries will shift to buy other *lower*-wage labor from the different countries. The process will repeat itself. In this point, the author further asked next question, from where does modern capitalism get labor as necessary in the end? An economist became to be silent and seemed to be troubled. In other words, the economics driving globalism (Autor, 2018) seemed to be a system of sorting out poorer people and countries rather than solving the gap between the rich and the poor (Root, 2020). The author found the nature of "capitalism with democracy" and its limits (Rennwald, 2020; Schumpeter, 2008).

There's also a lot of economic debate, as the Washington Consensus (Williamson, 1989) was already criticized, this posed question is still now one of the hard problems[4] to be solved in economics. In other words, modern economics is indeed justified in both sacrificing low-wage labor and its price difference as a growth force for each national and company development. It is a theoretical base that approve the globalism (Piketty,[5] 2014).

Of course, there are diverse reasons to fluctuate values for the goods, services and products in global (e.g., Pareto insufficiency, market failure (market imperfection), social and political instability). Interestingly, in fiscal and monetary policy, the trade-off between inflation and the output rate is called the "*sacrifice ratio*" in economics (Durham, 2001). It can be calculated by taking the cost of lost production and dividing it by the ratio alter in inflation.

On the contrary, standing on a perspective of the author, it critically thinks about labor itself of the workers. As allocation of fairness in economics said, an amount of surplus subtracted from the lower-wage workers can shift to higher skillful workers and talented managers, and then it means that such differentiation can be better to achieve more rational, fair, and effective allocation of capitals among the workers (Weinar & Koppenfels, 2020). And, essentially, after conveniently subtracting "working hours" out of "each worker's lifespan", it neglects workers' dignity as the human. This is a problem caused by devaluating the life as far as it goes with other measurable and comparable value. In short, it seems that capitalism just allocate the sacrifices to workers in developing countries for the purpose of intentionally expropriating their surplus. Moreover, a company based on the capitalism takes up jobs from the workers in homeland, and it sacrifices them.

In this point, many economists maybe say that it's good for our country because it's a clear sign that our economy is more competitive than other countries. But it also reveals the fact of imposing sacrifices on other countries. In other words, where the sacrifices are imposed is as essential in economic phenomena as similar as a state of war. Such logic similarly exhibits that living of the people is rooted in an imposition of sacrifices among them. Therefore, any workers should ask themselves how they should live in such environment.

In the first place, democracy has never resulted in a flat society. In legal, it is just correct that everyone is equally. But for example,[6] it is known that in Japanese

---

[4] The Japan Institute for Labour Policy and Training (in Japanese): https://db.jil.go.jp/db/seika/zenbun/E2000011231_ZEN.htm#01000000.

[5] World Inequality Database: https://wid.world/#Database, https://wir2018.wid.world/files/download/wir2018-full-report-english.pdf.

[6] https://www.u-tokyo.ac.jp/focus/en/features/f_00082.html.

education, there is a higher linear positive correlation with parental wealth and children's education (e.g., a pattern of intergenerational correlation between parents' social status and education level of their children). This is not a result of competition, but rather a situation that is not fair between generations and cannot even be called competition. In addition, while frequently promoting individual liberty, the current economic system forces citizens to sacrifice themselves or work for low wages, leaving the society in a state of desperate economic disparity, which was not corrected by democracy. "No alternative situation" cannot be said to be freedom, and as Rawls said (1971), educational opportunities often stay a condition of the inequality among peoples.

Actually, the reality is that we have already seen many cases of fix-termed, part-time and temporary work in the many countries. Another important point is the loss of the ideal for them. In the past, the unfounded fiction that communism and socialism would be replaced capitalism was widespread. But now, such fictions have been completely rejected. Even in the USA, economic disparities continue to be so severe that young people are screaming "new socialism" to improve the situation. In some cases, even if the Scandinavian welfare policies are adopted, but the more such countries need a simple labor force in the field of welfare and other essential working services.

2.3   Factors of Disparity

Certainly, there are many factors that contribute to the growing socioeconomic disparities resulting from the complexity of the economic system, such as the distribution of labor, statistical variability and dependency, and the accumulation of wealth.

First, there is the issue of the ratio between the labor and capital shares. The ratio of value added to wages and salaries received by workers out of the value added produced, which is also called the distribution rate. If the amount of value added is Y and the total amount of wages and salaries is W, the distribution rate becomes W/Y, and (Y-W)/Y is the distribution rate of capital. In the USA economy, for the past 30 years, the labor share has been falling, but this is not due to an increase in the capital share (Barkai, 2020). When the distribution of capital is calculated based on the cost of capital, such as the real interest rate, the distribution of capital has fallen rather more than the distribution of labor. Since the decline in the labor share affects not only income but also the standard of living, it can be inferred that the real lives of American citizens have fluctuated accordingly.

Second, there are statistical variations and cumulative effects. One fact that should be of concern is the possibility that the cumulative gap between the rich and the poor will gradually emerge and widen, even though initially there is not a large economic gap between people. This can be seen from the fact that economic phenomena such as uneven distribution of wealth can be reproduced in simulations by randomly repeating economic transactions among arbitrary people using physics and mathematical models (Boghosian, 2014). This is called the "yard sale model" and is a

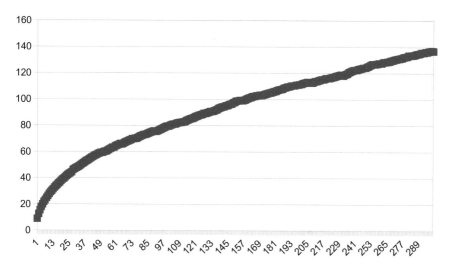

Fig. 1 An example of variance of disparity in the simulating society

model that provides an explanation and prediction of how economic disparity can be generated. Initially, a hypothetical society consisting of N agents (i.e., it means citizens) is assumed to all have the same amount of wealth. Then, it sets up the rule for transferring wealth. First, a pair that consists of two agents is chosen at random. This pair then transfer a specified percentage of wealth (an amount that falls within a certain ratio of the amount of wealth of the poor side) between two agents. For example, if all agents start with wealth of 100 yen each, at some point, 50 yen will be transferred from agent A to agent B. As a result, agent A will have 50 yen in wealth and agent B will have 150 yen in wealth (while the others still remain at 100 yen). In this way, each time, a very small fixed amount of wealth is continuously transferred between two randomly selected agents. As this process is iterated, an economic disparity is created in a society consisting of N agents. Figure 1 shows the variance value of the wealth of the entire society at each turn. (Fig. 1).

Since only one pair transfers wealth at each turn, it seems that the society as a whole apparently maintains an *equilibrium*, but rather the disparity gradually widens. Of course, actual economies and markets are not determined by simple factors alone. However, this model suggests that economic disparity can be easily generated by even a trivial rule. These factors are, after all, statistical variability, dependence (i.e., randomness), and the effect of accumulation of wealth. Such simulations of the emergence of economic imbalances can be reduced to an understanding of physical phenomena and statistical fluctuations based on human behavior, and may provide clues for correcting the causes of economic imbalances and inequities through appropriate taxation and institutional design.

Such statistical variability, dependency, and cumulative effect can make a huge difference in the lives of individuals. Not only capital such as cash flow, but also "invisible capital" such as genetic capital (i.e., genetic inheritance, intelligence, physical attractiveness) (Hill et al., 2019; Yamagata et al., 2013) and social capital (i.e., human relationship) (Kawachi & Berkman, 2003) held by each individual can be important, especially in a digital society (Shibuya, 2020a). These factors can define subtle differences in each person's talents, physical attractiveness, early life encounters and foundation of human relationships. Even if it seems that there is only a small difference, the accumulation of daily efforts and results in the course of growth suggests that a large disparity will inevitably arise. The disparity will be born inevitably, rather than accidentally, and will determine each person's life.

The debate over reducing inequality is vigorous in the global. However, it also suggests that this is inherently difficult. Taxes can be imposed to correct economic disparity, but it is extremely difficult to impose taxes or other "constraints" (fairness as Rawls (1971) explains, and sacrifice as the author claims) on genetic inheritance and social relations. Genetic inheritance (e.g., inheritance of qualities between parents and children through marriage) and the construction and inheritance of human relationships (e.g., wealth family relationships, second-generation legislators) are clearly based on individual *"preferences"*, and are not determined by completely random factors alone, but by apparently arbitrary preferences that determine the differential distribution of such useful resources. Moreover, if disparities appear due to statistical variation and cumulative effects even if there are no differences at first, it is inevitable that trivial differences are even more likely to appear.

In addition, in research on the sociology of education and mathematical sociology, the question of how parental education, occupation, and job position affect children is an issue of social stratification and social mobility (Boudon, 1973, 1974; Miller & Bogal, 1979). For example, as mentioned earlier, Japanese society has been becoming more and more stratified. Depending on the parents' class of origin and educational background (e.g., university or higher), they can have considerable influence not only on the intellectual qualities of their children, but also on their living environment and even on the development of their lives (Bourdieu, 1984). Therefore, in a democracy based on fair competition, the existence of such disparities is considered problematic because it is tantamount to self-denial. However, correcting such disparities is not possible for genetic qualities. Public educational opportunities, scholarships, and other programs may correct some of the problems, but they are not effective in all cases. Particularly when issues such as origin, race and immigration are involved, many discrimination and stereotyping problems arise, not only with regard to educational opportunities, but also with regard to employment and promotion opportunities (Jost & Major, 2001).

Thus, it is pivotal to accumulate results by successfully controlling necessary factors that are determined at an inevitable level, while also keeping in mind the contingent risk factors that we experience in life and the numerous crises that accompany them. In particular, as discussed later, one's own career development and career choices (and those of others) affect how one is positioned in the network of human relationships and what useful resources (e.g., jobs, positions, honors) one has the potential to acquire. In a crisis such as the COVID-19 pandemic, there is a concern that the degree of damage that may be caused and what kind of response strategy is possible may be determined by the superiority or inferiority of socioeconomic status.

2.4 Basic-Income

There were already discussions such as minimum wage to eliminate various economic disparities and inequities in workers' wages (Autor et al., 2016). On the contrary, as China demonstrates, if their *"capitalism with state control"* will be recognized as a prominent design, adopting such system using the AI and big-data may increase. If their AI and big data driven socio-economic system may become a model case for the future of basic income[7] and flatting working condition of the workers, it will be a reference for the pioneering experiments, and further some hopes that such system has the potentials to take the after-corona economy to another dimension.

Actually, UNDP (2020) proposed Temporary Basic Income (TBI) in developing countries during the pandemic. Due to the pandemic prevention, poor citizens and needy peoples should be treated by the basic income. They said *"As the rate of new COVID-19 cases accelerates across the developing world, it exposes the potentially devastating costs of job losses and income reversals. Unconditional emergency cash transfers can mitigate the worst immediate effects of the COVID-19 crisis on poor and near-poor households that do not currently have access to social assistance or insurance protection. This paper provides estimates for a Temporary Basic Income (TBI), a minimum guaranteed income above the poverty line, for vulnerable people in 132 developing countries"*. When such policies will be activated, this is one of the advanced social experiments including budget and institutional design to see whether it can be used to reduce economic disparities between the citizens or not.

2.5 Living Ways in the Democracy

Modern democracy is based on the autonomous free will of the individual. The distinction between self and others and the independent will of the individual are seen as the core of identity, and through the expression of individual opinions and attitudes, the individual is expected to participate in society. However, in many cases, in order to carry this out in society, it involves competitive situations and various sacrifices. Then, let me take a deeper look at the economic and political aspects.

First, the economic aspect of modern democracy is based on the basic premise that each individual must take the initiative in the practice of labor production and achieve own individual goals through self-sacrifice. In parallel, capitalism is a system in which the practice of labor can be entrusted to others through capital (including debts and credits). It is a contractual relationship of production-management through the exchange of capital and labor, and it also takes for granted the investment of capital and economic practices in general through incentives. Naturally, the rule is to assume the risks and crises such as bankruptcy and uncollectibility of receivables.

---

[7] Pandemic speeds largest test yet of universal basic income: https://www.nature.com/articles/d41586-020-01993-3.

Secondly, the political aspect of modern democracy is representative democracy, also known as indirect democracy. Individuals elect their representatives and presidents through fair elections. However, from a different perspective, this is just an act of voting on who they want to be their "master". The irony of Rousseau's statement that indirect democracy would only lead to a new form of slavery is correct in this respect.

In other words, modern democracy is a system that claims to respect autonomous free will to the maximum extent possible, but takes it for granted that there will be a division between those who choose (or are forced) to be "subordinate" in both economic and political aspects, and those who can take (or inherit) a "governing" position. On the other hand, while tyrannies are criticized for their lack of freedom, it does not correct their own inequality and hierarchical structure in democracies. Both systems have one thing in common: both system require large numbers of citizens to make sacrifices. Regardless of the differences between both systems, in the end, they are just different means to achieve a hierarchical society.

It is difficult to say that there is proper guidance on these points, at least in the field of education in Japan. There is no career education or entrepreneurship education from an early age. In addition, human network is effective for job change and employment opportunities. However, the quality of an individual's life is clearly different depending on whether he or she is in a governing or subordinate position in the economic and political meanings. What is needed and what must be sacrificed in order to be accepted as one who should occupy a position of leadership or expertise? It is often not clear in advance. Not only merit and achievement, but also luck and origins are sometimes important. Self-evaluation alone does not determine one's value, because evaluation is also dependent on the preferences and values of the others.

As mentioned earlier chapters, with the development of AI and big data related technologies, human bosses may be replaced by AI. Will citizens, who simply accept democracy, accept management and evaluation by AI if it is fair?

3   Changing the World

3.1   Natural Environment

There are some reports that the global environment has improved because economic activity of the global citizens has ceased or stagnated during the pandemic (Allam, 2020). If it continues year after year, one might even think that environmental problems would be considerably alleviated. There has been a significant reduction in carbon dioxide emissions in urban areas around the world as well as a reduction in energy consumption and traffic congestion in urban areas (Negi et al., 2020). It could also be said that these were demonstrative experiments in the interconnections between environmental issues and the human activities. Therefore, it seems that it can find clues as to which the key factors can be controlled and how much they can be used to reduce the environmental impact without curbing economic activities (i.e., imposing sacrifices on the natural environment).

To be sure, there is a need for systematic improvement in the interlinking of environmental problems and emerging infectious diseases, as well as secondary threats to humanity such as poverty and the economy. Moreover, even before the COVID-19 pandemic, we need to recognize that the situation is more urgent. According to IPBES (The Intergovernmental Science-Policy Platform on Biodiversity and Ecosystem Services, 2020), their reports warned *"Pandemics represent an existential threat to the health and welfare of people across our planet. ... Without preventative strategies, pandemics will emerge more often, spread more rapidly, kill more people, and affect the global economy with more devastating impact than ever before. Current pandemic strategies rely on responding to diseases after their emergence with public health measures and technological solutions, in particular the rapid design and distribution of new vaccines and therapeutics. However, COVID-19 demonstrates that this is a slow and uncertain path, and as the global population waits for vaccines to become available, the human costs are mounting, in lives lost, sickness endured, economic collapse, and lost livelihoods. ... The recent exponential rise in consumption and trade, driven by demand in developed countries and emerging economies, as well as by demographic pressure, has led to a series of emerging diseases that originate mainly in biodiverse developing countries, driven by global consumption patterns. Pandemics such as COVID-19 underscore both the interconnectedness of the world community and the rising threat posed by global inequality to the health, wellbeing and security of all people. Mortality and morbidity due to COVID-19 may ultimately be higher in developing countries, due to economic constraints affecting healthcare access. However, large-scale pandemics can also drastically affect developed countries that depend on globalized economies, as COVID-19's impact on the United States of America and many European countries is currently demonstrating"*. Thus, humanity is forced to look for ways to survive not only infectious diseases but also environmental problems.

3.2 Necessary Services

Many countries such as Japan, the USA, Europe and others are beginning to resume own economic activity. However, it will not be the same everyday life. After the pandemic, expected social changes due to technological innovation begin to advance drastically. ICT based innovation and business through remote work, working from home, and online collaboration further progress the development of the international R&D and business (Shibuya, 2006; Dwivedi et al., 2020; Lee & Han, 2021).

When shifting from an economic system determined by the *human mobility flow* (Bonaccorsi et al., 2020) to an economy determined by *data and information flow*, it is a facilitated digital transformation in the all level of the social systems. In such era, collective dynamics of human behavior can be monitored by the automatic AI driven systems, and their minimum necessary mobility will only be required. The realization of an energy-saving society through automated driving systems and smart cities will indicate clearly the AI and big data use cases (Chong et al., 2020). In this context, maximizing that the economic system will be also consolidating demand and business opportunities by attracting massive people to the larger cities becomes less meaningful. By connecting demand online, customer data and various types of information can be aggregated, analyzed and predicted, and then it will be able to provide any services at the best price when customers need them. In addition, the services that specialize in local demands (e.g., delivery, medical administration), the consolidation of production sites (e.g., domestic production of foodstuffs, medical supplies, and daily necessities for each country) will be also expected to advance. Digital interconnected supply chains using IoT and 3D printers can invoke further innovative solutions in rural fabrication.

Therefore, the ways people work will naturally have to change as well (Soltanifar et al., 2021). Any workers have to confront with their dramatic change of working environment, and brush up own professionalism beyond such new spatial and necessary constraints.

3.3 Living Ways After the Pandemic

The pandemics revealed further social problems such as the high number of collective immigrants, illegal undocumented workers,[8] homeless people, the high population density in urban areas, the constant poverty problem, the lack of public health and medical services for these people. These matters can lead to clusters among the poor in urban areas, which can lead to further contagion of the epidemics. Moreover, in terms of the risk of natural disasters, the severity of urban disasters in megalopolis has been increasing every year.

Further, it could be said that the risks associated with the coexistence of heterogeneous ethnicity within the same nation became salient for everyone in a different way. The USA and EU countries have often actively accepted innumerable immigrants and refugees (and undocumented workers), saying loudly that heterogeneity is the source of innovation and vitality. But how well have the entities granted them public health services, educational opportunities, and opportunities to improve their social status? In the USA, blacks and Hispanics in a white-dominated society are relegated to a relatively low position and are unlikely to make it into the mainstream. And in the EU, as the 2015 immigration crisis also highlighted, immigration policy did not always become better situations (Forman & Wu, 2016). Those peoples were clearly aimed at passing through the gates of the EU by the Schengen agreement.

---

[8] Census director dodges legislators' questions about Trump memo on undocumented residents: https://www.sciencemag.org/news/2020/07/census-director-dodges-legislators-questions-about-trump-memo-undocumented-residents.

Because each ratified member country of the EU shall permit citizens to travel and move across these countries (Yuval-Davis et al., 2017).

Actually, calls for inhibition and regulation of immigrants' policy were also made in the EU member states during the pandemic. At the ceasing of the pandemic, leading countries such as France and Germany have clear intentions to open the gates of border for free mobility, but some of member states still claims such policy because of their prevention of further epidemic reasons. On the other hands, the UK after Brexit has own independent policy for immigration and mobility of the citizens.

On the contrary, the countries that send out immigrants have naturally experienced a slowdown in economic development because of the outflow of useful human resources to foreign countries (Powell, 2015; Kahanec & Zimmermann, 2016; Bauboeck, 2019). Globalism namely stands for the logic of capitalism, which has shut out countries that cannot provide high salaries and high social status for the peoples as well as companies.

Certainly, assimilation and diversity of cultural backgrounds and multi-purposes in each company, organization, and nations should be equivalently considered (Aoki, 2010), and those adoptions ought to be collaborative and inclusive memberships unless there are any exceptionable reasons. However, higher educated workers with higher skills cannot be firstly treated as equivalent as uneducated workers from a viewpoint of human resource management in any countries (Autor, 2019). Innovative and intensive R&D has been achieved by only those who earned own higher careers and status beyond the borders and barriers in global. It is possibly that low-wage workers can only occupy in the bottom of the pyramid even if they can immigrate to the other countries.

Further, if remote work and online collaboration move forward, and if online broadcasting of lectures becomes mainstream at the world universities (Mishra et al., 2020), the physical mobility of labors and students will become merely a risk including such heterogeneous relationships within the same nation (Brader et al., 2008; Buzby, 2016). The presence of heterogeneous people in the same environment increases uncertainty in the society and enhances the likelihood of negative prejudice and discrimination. This pandemic gave us a glimpse of the invariance of human nature and the foreshadowing of future changes in society due to the progress of digital transformation.

Therefore, as social distance means moderating physical distance among people and resources, this pandemic may have overturned the essence and value of immigrating to a big country, not to mention living in a big city, and it can be probably more stable to live in a moderate size city. And the citizens eagerly stay in one's own country and should be better to enjoy a moderate life, while taking advantage of ICT based remote work.

3.4   A Necessary Person in the Age of the After Corona

Digital transformation through AI and big data will drive industrial innovation, but the "After Corona" world will be far more advanced than we expected. As consequentially, the *useless class*, which named by Israeli historian Harari may increase in the age of After Corona. Those who are stayed in such social class will be impossible to find any jobs and workplaces, because the AI driven working machines and intellectual robots can be replaced in simple labor markets. He predicts that such hierarchy of the social classes will be brought by a series of technological progress of the AI, but rather the consequences of the COVID-19 pandemic will boost such fictional story faster than he expects.

It was certain that this pandemic facilitated every labor type to divide between possible jobs to do remote-working or not. The former is categorized such as remote lecture of higher education, remote medical services, remote professional consulting, ICT based design and others, whereas the latter is conditionally grounded on physical working such as home delivery services, agriculture and builders to develop something onsite. Namely, the more advanced working styles proceed, the less simple jobs will be replaced by the AI and robots. Meanwhile a part of the cleverer workers will occupy upper social classes, a large number of the physical workers will force to be stayed on the bottom of the hierarchy.

What is expected of the people in living in a new working environment? While someone must live in physical and essential working environments based on the old tradition, many workers are encouraged to rebuild a new adaptive way of own working life. In the digitized era, any workers should be adapted in distributed workplaces and decentralized networking. Probably, their working styles drastically alter remote work and interconnected collaborations. Such heterogeneous backgrounds such as race, gender, idea, and opinions facilitate to coexist and cooperate with each other. Their identification, mental attributions and commitments to the employed institution will be reshaped by digital transformation.

In brief, there are some necessary points in these matters.

1. *Human Relationships*: how many people are there who believe in and appreciate you, whether online or offline, and interconnected evidence will also clarify your own reputation if such people are more competent and outstanding than others.
2. *Skills and Knowledge*: The extent to which you have the skills and knowledge required even in the new After Corona world is important.
3. *Career*: The question is how much of a career you can make of yourself in an After Corona world.
4. *Internationality*: The ability to work across the national borders is also an important factor. If online telecommuting becomes mainstream, ICT jobs will be competing with workers in emerging economies such as China and India for jobs. It is also inevitable that higher skills and talents will be required for the AI and big data-related work.

4 A Comparative Cultural Study of Identity and Way of Life

4.1 Identification to Commit in the Nation

The EU and the USA have long been aware of the need to reconsider the nature of the Chinese people in the context of global trends. For example, recently, Perez-Garcia (2021) discusses the history of China and their cultural uniqueness by examining the historical facts of Spain and the Qing Empire. As mentioned before, although there have been repeated instances of confrontation between Western civilization (i.e., Westness) and Eastern civilization (i.e., Eastness), or interaction through cross-cultural contact, identity and its cultural attitudes are the essential factors of the conflict.

The Westness totally represents democratic political system, liberalism and market centered economics, whereas the Eastness prioritizes social harmony and group-centrism. In other words, the Westness culture requires each individual to be autonomous and independent one, and more identifiable existence among the others. After that, the collectives consisted of each individual can cooperate and collaborate with heterogeneous such members. On the contrary, the Eastness culture is apt to synthesize the assemble as each individual, which can be imperatively ordered and hierarchically assorted by the central authority, is totally about to be coordinating their actions.

An example of people living between Eastness and Westness is the continuing upheaval in Hong Kong. There was a mood of stagnation in the civic movement due to the Corona pandemic, and during that time, a number of Hong Kong citizens had chosen to move abroad. The UK government is also pursuing a policy of accepting Hong Kong citizens, which deepened the conflict with the Chinese central government. As in the case of Hong Kong, the focus is on where the citizens feel their own identity. In fact, in 2020, according to the Hong Kong Public Opinion Research Institute,[9] more than 80% of young people said that they are Hong Kong people.

The differentiation of identity on cultural background certainly seems to be a part of the nature of national identity at the root of national governance as seen by Hegel himself. In Fukuyama's theory (2018) and Eurasia Group's report (2018), there are many who see identity as the essence of the social foundations. Especially,Hegel (2018) said *"The collective national psyche and the divine entity are the product of a sphere of the forms of the national psyche, which now encompasses the whole of the natural and ethical world. These forms also stand on the basis of one person's rulling authority. It is better to say so rather than it is under its control"* (*"Phänomenologie des Geistes"*).

Based on Hegel's view that people's desire for recognition of the others is the driving force behind the changes in the human history, Fukuyama based his *"identity politics"* on this perspective (2018). He said *"The terms identity and identity politics*

---

[9] https://pori.hk/.

are of fairly recent provenance, the former having been popularized by the psychologist Erik Erikson during the 1950s, and the latter coming into view only in the cultural politics of the 1980s and '90s. Identity has a wide number of meanings today, in some cases referring simply to social categories or roles, in others to basic information about oneself (as in "my identity was stolen"). Used in this fashion, identities have always existed. Identity grows, in the first place, out of a distinction between one's true inner self and an outer world of social rules and norms that does not adequately recognize that inner self's worth or dignity. Individuals throughout human history have found themselves at odds with their societies".

If their thought is valid, how should we think about the life of Fan Li,[10] the famous vizier of ancient China? He has been praised for his living way in the history of ancient China. He assisted the King of Yue, Goujian, in destroying the State of Wu, and then left the State of Yue. However, this did not mean that he betrayed the Yue Kingdom, but rather that he defected to another country (the State of Qi) because of the high possibility that the suspicious king would purge him, and he sensed his intentions in advance. On the other hand, the king of the State of Qi admired Fan's talent and not only accepted him generously, but also tried to make him the grand chancellor (i.e., the prime minister and the president may be equal in a modern state). In this context, his identity that did not committed in the specific nation might not be measured by the index of Fukuyama (and Hegelian philosophy). But his intention was still calmly living in rural village, because he would like to avoid confronting with conflict situations at actual politics.

4.2   Selection

To choose a way of own life clearly means to sacrifice other choices. Choosing one way of life at the sacrifices of those options is seen as the most worth. In other words, it reflects the values of each person. Living process is about narrowing the alternatives of the possibilities of how to live, and sacrificing any potentials and efforts. And it can be said that our life is a series of selections that you make and others make of you. These largely are conformed as each preference and value standard of each individual. Alternatively, the consequences of that choice may also be called prejudice and discrimination, but there is nothing one can do about it. Such selection may not be made by a person, but by an AI decision based on big data analysis. And the bosses and coworkers may be robots in nearly future.

A meaning of *selection* in natural selection biologically occurs on a species basis with each zoological species undergoing repeated kin selection and adapting to their natural environment. However, evolutionary theorists made the terrible mistake of the argument by extending it to the civilization and social structure of the entire humanity. Evolutionists (Turner, 2001) were merely believers who, in conjunction with Marxism (Popper, 1963), were eagerly assume a historical model of the depth of

---

[10] https://en.wikipedia.org/wiki/Fan_Li.

social structures and their variable processes. They denounced the poverty invoked by the capitalism, but they had failed to sustain their political system or economy at the same time. If poverty is the result of evolutionary factors, then human history could be rephrased as a reproduction process in which losers produce losers.

Certainly, even though whether it is based on Marxism or not, contemporary economics has own severe problems. Capitalism was nothing more than capitalist-optimal economy, and globalism is a networked system associated with rich countries. It depends on both networking effects and a principle on the *"rich get richer"*, and then it is nothing more than the fact of concentration of the wealth to few people and selective exclusion of the rest of the peoples.

The COVID-19 pandemic, similarly, highlights the disparity between rich and poor, and shows that the accumulation of the capitals and the quality of health and well-being in each nation not only affect the economic wealth of the citizens but also their life expectancy (Laurencin & McClinton, 2020; Shadmi et al., 2020). Such consequences were too depended on a series of accumulation in the human history.

4.3 Networking as Social Selection

A network is, in the end, an association cluster mediated by human relationships that provide benefits and achieve common goals. Peoples' choices also determine which the others connect with or not (Kadushin, 2002; Robins et al., 2001). Each individual's life choices are dependent on others is generally correct. Furthermore, similar people connect with each other and eventually become a huge cluster, self-organizing into a power structure based on strong relationships. This dynamism is a motivating factor in the creation of history. It is not only the power to change one's life and working environment, but also the power to acquire and rise in the ranks through the hierarchy of power structures. Even if we live a life that is not bound by the position, we cannot be free from the interpersonal relationships that exchange various resources and services to each other. Whether online or offline, as long as our social life is rooted in the actuality, it is the place where we stand in our relationships that matters.

In sociology, there are two types of social status that can be distinguished (Linton, 1936). Further, Parsons extended this Linton's model to distinguish between *ascription* and *achievement* as one of the conforming frames (Parsons & Shils, 1951). First is an ascription. Ascribed status cannot be replaced by individual abilities, efforts, and personal features such as age, gender, race, ethnicity, family background, and others. It is a social status perfectly defined by the origin. It is also referred to as attributive status. And secondly, it can consider an achievement, and it denotes achieved status. A position that is allocated based on the performance of each individual, a position that each individual acquires through own ability and effort. It is also referred to as an obtaining position by each performance after the birth of each individual.

The essence of the achievement oriented society is how to recruit others and get them to appreciate you. As described in Granovetter's famous book *"Getting A Job"*

(1995), it is self-evident that the influence of weak ties is great, but whether or not one is accepted as a member of the network of human relationships will test one's ability to adapt as a social being and the significance of one's existence.

Even if it is not about the power structure and the hiring of human resources, we need to take another look at the fact that in order to connect people and become part of the circle of the numerable peoples, everyone must have the appropriate fame, personality, talent and skills to be a part of it.

Moreover, the principle that capitalist societies uphold is the allocation of economic resources based on performance evaluation (i.e., meritocracy) and the practice of *"from each according to his ability, to each according to his contribution"* (Bell, 2016). However, in the case of a state where AI-based evaluation systems and surveillance systems are being strengthened as mentioned earlier chapters, there is a concern that these principles may become too much, leading to what Rawls (1971) calls *"a callous meritocratic society"*.

4.4 Social Decoupling as Emerging Social Stratification

The immigration crisis of 2015 was a clear manifestation of this negative aspect of the Western society. And, even during the COVID-19 pandemic, socioeconomic activity through lower-wage workforce such as agriculture and welfare services were to stagnate because of the suspension of immigration and entry of immigrants and refugees to prevent more infection. And yet, there is no significant recognition of their human rights, and identity conflicts over racial ethnicity and mental barrier separate them. Even though it is democratic society in the Western countries, it has been keenly decoupling between them. In other words, even if they profess freedom, equality, and benevolence, they are just talking only about *"ordinary citizens who share a same national identity"*, and they unconscionably exclude both immigrants and refugees. Certainly, on the surface, the dignity for human rights may seem to be mentioned, but the mental divide between them has not been bridged yet.

Likewise, over the past decade, Chinese society has achieved tremendous economic development by accepting large numbers of peasants from rural areas into urban and industrial production areas. However, the differences in social class (or social stratification) continue to persist, as it can be clarified from the fact that there are two types of the rural or urban registers. In addition, the *"Hundred Talents Program"* has lured back many researchers and engineers with outstanding achievements from foreign countries with high salaries, but the large gap between those who live in poverty at homeland has been widening. On the other hand, in Japan, the majority of opinions are still negative about the acceptance of immigrants and refugees, and even though Japan has accepted a substantial foreign labor force, their livelihood base is far from stable. In addition, the income gap between the bottom and the top in economic status has been beginning to widen, even for those with the same Japanese nationality. At the time of the COVID-19 pandemic, firstly those who faced job crisis in both countries were those who were situated at the bottom of the pyramid.

Actually, standing on a sociological point of view, the issue of social stratification and social mobility has been observed in the COVID-19 cases. In fact, in the USA during the time of the COVID-19 pandemic, those who belonged in the bottom of such social strata were primarily the victims, and these were clearly blacks and Hispanics rather than whites. Moreover, the COVID-19 pandemic has revealed that the social stratification which divided among the citizens also indicated very important in determining whether they are eligible for health insurance and social security benefits.

4.5 Knowing Yourself: A Study on Expansion from Sociometer Theory

Psychologically, there is a view or hypothesis that the "self" is an adaptive product. In other words, the environment to which people have adapted is not the natural environment, but the social environment. Then, from the point of view of evolutionary adaptation, it is easy to assume that exclusion from such kinship groups means, in the end, a crisis of one's own survival in such a social environment (Bowles & Gintis, 2011).

Therefore, our perception of the self may have evolved to be capable of reasoning to which we belong (including complex and multilayered relationships), which is the hypothesis of the *sociometer* proposed by Leary, and is the research and theoretical core of the self-process from an adaptive perspective (Leary et al., 1995; Leary & Baumeister, 2000). In other words, its main argument is that it is advantageous to be able to guess the psychological state of the others in order to interact with them smoothly, and that it is able to recognize one's own psychological state accurately in order to do such manners. They argued that people should have needed to detect how accepted (or excluded) they were by the group to which they belonged, and that self-esteem is a built-in gauge (i.e., a sociometer) to detect such situations. It can be said that this study focused on the adaptive function of both self and other evaluation, which had been the focus of traditional self-study.

In an actual experimental social psychological study, Rudich and Vallacher (1999) showed that high and low self-esteem influences the choice of the person with whom the subject interacts, and considered the consistency with sociometer theory. Among them, it has been suggested that people with high self-esteem may aspirate to higher status and other-selection to belong to or interact with more prestigious positions.

4.6　Human Relationship During a Period of the Three Kingdoms in Ancient China

In cultural context, these have also important implications for the way in which self and the others relate to their fundamental nature. For example, a dense human relationship as the brothers-in-law in the Three Kingdoms[11] in ancient China is interesting from this point of view as well. In the first place, a history of the Three Kingdoms did not consist only of individual biographies and chronological accounts. For example, Liu Bei, Guan Yu and Zhang Fei's harden their relationship. It was a human relation based culture, a relational self-concept that constitutes the society through interpersonal relationships, and that it has become an Asian cultural backdrop (Triandis, 1995).

In the ancient China, reputation among the peoples indicated each birth of an influential person or a prestigious family, literally meant fame or aspiration, or it was often an evaluation of academic fields or talents. Whether or not one was an eminent person (acknowledged or not) was of great significance for own adaptation and promotion. In other words, it was known that being recognized as personage or being certified by the authority could sometimes significantly improve the adaptability of the community (e.g., promotion, formation of factions by the same family, social class or clans).

In particular, in the Three Kingdoms period (180 BC–280 AD) to the reunification of the Sima and Jin dynasties (265–420 AD), it formed the networking of human relationships and influence among the intellectuals and the ruling class, which sometimes determined the fate of the nation. It was well known that Xun Yu recommended many superior personnel to Cao Cao through his connections with influential people, and contributed to Cao Cao's success. Also known are Zhuge Liang's academic masters, such as Sima Hui and the Xiangyang community of Huang Cheng-yin.

On the other hand, there was a system for the recruitment of central officials known as Hsiao-lien, who were promoted from the provinces including Cao Cao[12] who would later become a King of Wei (one of the Three Kingdoms). Of course, the nominees were often excellent people who were convenient for the local powers or the offspring of prestigious families, and thus contributed to the maintenance and improvement of the power structure. Incidentally, in the reign of Cao Cao's son, it was said to have been the establishment of a system by *the Nine-Rank System*, which resulted in the strengthening of the power and foundation of the eminent and powerful clans (e.g., Sima clan), leading to the later decay of the Wei dynasty (i.e., the establishment of local officials in charge of recommendation).

Thus, at least, prior to the *Chinese imperial examinations* as a civil servant examination (598 AD–1905 AD), such ruling system in the ancient China could be said to have encouraged the selection of people who were as convenient as possible for those in power to form and maintained hierarchical clusters of dominance-subordination

---

[11] https://en.wikipedia.org/wiki/Three_Kingdoms.

[12] https://en.wikipedia.org/wiki/Cao_Cao.

relationships. This is analogous to the current *"Hundred Talents Program"* mentioned above, and it really means that the discovery and promotion of excellent human resources is fundamentally dependent on human relationships.

5 Conclusion

COVID-19 inspired the reform of our working environment, but it still broadly divided into intellectual forms of labor that are independent of time and space, and physical forms of labor in the field. In this chapter, national belonging was also discussed, matching the contemporary issues of each worker and individual identity with examples from Chinese history. Each person's choice is an opportunity to change his or her way of life and ultimately to drastically alter the course of one's life. Certainly, not everyone in the same environment can share the same hope, and some people in despair may leave their homeland for other countries. Connecting with the human network of outstanding people will not only lead to the acquisition of status and honor, but also to a stable socioeconomic status and working environment, in other words, it will increase the possibility of what can be achieved by reconnecting with human relationships. The digital society offers its citizens the potential to expand their economic opportunities to rise in social status, but at the same time, the dynamics of such globalization have the potential to increase economic disparities among people. The question of what sacrifices each individual should make in his or her own life will become a question of personal choice.

References

Acemoglu, D., Autor, D., Hazell, J., & Restrepo, P. (2021). AI and jobs: Evidence from online vacancies. *NBER Working Paper No. 28257*. https://economics.mit.edu/files/21056

Acemoglu, D., & Robinson, J. A. (2012). *Why nations fail*, Currency.

ADB Institute. (2021). *Labor migration in Asia: Impacts of the COVID-19 crisis and the post-pandemic future*. https://www.adb.org/sites/default/files/publication/690751/adbi-book-labor-migration-asia-impacts-covid-19-crisis-post-pandemic-future.pdf#page=69

Allam, Z. (2020). *Surveying the covid-19 pandemic and its implications urban health*. Elsevier.

Aoki, M. (2010). *Corporations in evolving diversity: Cognition, governance, and institutions*. Oxford University Press.

Arthur, W. B., Beinhocker, E. D., & Stanger, A. (Eds.) (2020). *Complexity economics: Proceedings of the Santa Fe Institute's 2019 fall symposium (Dialogues of the Applied Complexity Network)*, SFI Press.

Autor, D. (2018). Trade and labor markets: Lessons from China's rise. *IZA World of Labor, 2018*, 431.

Autor, D. (2019). Work of the past, work of the future. *American Economic Association: Papers and Proceeding, 109*(5), 1–32.

Autor, D., Manning, A., Smith, C. L. (2016). The contribution of the minimum wage to U.S. Wage inequality over three decades: A reassessment. *American Economic Journal: Applied Economics, 8*(1), 58–99.

Baldassarri, D., & Abascal, M. (2020). Diversity and prosocial behavior. *Science, 369*(6508), 1183–1187.

Barkai, S. (2020). Declining labor and capital shares. *The Journal of Finance, LXXXV*(5), 2421–2463.

Bartik, A. W., et al. (2020). The impact of COVID-19 on small business outcomes and expectations. *PNAS*. https://doi.org/10.1073/pnas.2006991117

Bauboeck, R. (2019). *Debating European citizenship.* Springer.

Becker, S. O., & Woessmann, L. (2009). Was weber wrong? A human capital theory of protestant economic history. *The Quarterly Journal of Economics, 124*(2), 531–596. https://doi.org/10.1162/qjec.2009.124.2.531

Bell, D. A. (2016). *The China model: Political Meritocracy and the limits of democracy*, Princeton University Press.

Boghosian, B. (2014). Kinetics of wealth and the Pareto law. *Physical Review E, 89*(4–1), 042804. https://doi.org/10.1103/PhysRevE.89.042804

Bonaccorsi, G., et al. (2020). Economic and social consequences of human mobility restrictions under COVID-19. *PNAS, 117*(27), 15530–15535. https://doi.org/10.1073/pnas.2007658117

Boudon, R. (1973). *Mathematical structures of social mobility.* Elsevier.

Boudon, R. (1974). *Education, opportunity and social inequality: Changing prospects in western society*, Wiley

Bourdieu, P. (1984). *Distinction: A social critique of the judgment of taste.* Harvard University Press.

Bowles, S., & Gintis, H. (2011). *A cooperative species: Human reciprocity and its evolution*, Princeton University Press.

Bowles, S. (2012). *The new economics of inequality and redistribution.* Cambridge University Press.

Brader, T., Valentino, N. A., & Suhay, E. (2008). What triggers public opposition to immigration? *Anxiety, Group Cues, and Immigration Threat, American Journal of Political Science, 52*(4), 959–978.

Buzby, A. (2016). Locking the borders: Exclusion in the theory and practice of immigration in America. *International Migration Review, 52*(1), 273–298.

Chong, S. K., et al. (2020). Economic outcomes predicted by diversity in cities. *EPJ Data Science, 9*(17). https://doi.org/10.1140/epjds/s13688-020-00234-x

Dong, Y.-J., Wei, J., & Li, M.-M. (2014). The influence of mentor network to new employees' career success. In *International Conference on Management Science & Engineering 21th Annual Conference Proceedings* (pp. 1064–1069). https://doi.org/10.1109/ICMSE.2014.6930346

Durham, J. B. (2001). Sacrifice ratios and monetary policy credibility: Do smaller budget deficits, inflation-indexed debt, and inflation targets lower disinflation costs? https://www.federalreserve.gov/pubs/feds/2001/200147/200147pap.pdf

Dwivedi, Y. K., et al. (2020). Impact of COVID-19 pandemic on information management research and practice: Transforming education, work and life. *International Journal of Information Management.* https://doi.org/10.1016/j.ijinfomgt.2020.102211

Ellerman, D. (2010). Inalienable rights: A litmus test for liberal theories of justice. *Law and Philosophy, 29*, 571–599.

Forman, R. T. T., & Wu, J. (2016). Where to put the next billion people. *Nature, 537*, 608–611.

Fukuyama, F. (1992). *The end of history and the last man*, Free Press.

Fukuyama, F. (2018). *Identity: Contemporary identity politics and the struggle for recognition*, Profile Books.

Granovetter, M. (1995). *Getting a job: A study of contacts and careers* (2nd ed.). University of Chicago Press.

Hauser, O. P., et al. (2014). Cooperating with the future. *Nature, 511*, 220–223.

Hegel, G. W. F. (2018). *Phänomenologie des Geistes (The Phenomenology of Spirit) (English Translated Edition)*, Cambridge University Press.

Held, D., M cGrew,A., Goldblatt, D., Perraton, J. (2000). Global transformations: Politics, economics and culture. In C. Pierson & S. Tormey (Eds.), *Politics at the edge: Political studies association yearbook series*, Palgrave Macmillan. https://doi.org/10.1057/9780333981689_2

Hill, W. D., et al. (2019). Genome-wide analysis identifies molecular systems and 149 genetic loci associated with income. *Nature Communications, 10, Article number:574.* ,https://www.nature.com/articles/s41467-019-13585-5.pdf

IPBES. (2020). *IPBES workshop on biodiversity and pandemics.* https://ipbes.net/sites/default/files/2020-12/IPBES%20Workshop%20on%20Biodiversity%20and%20Pandemics%20Report_0.pdf

Jost, J. T., & Major, B. (2001). *The psychology of legitimacy emerging perspectives on ideology, justice, and intergroup relations*, Cambridge University Press.

Kadushin, C. (2002). The motivational foundation of social networks. *Social Networks, 24*(1), 77–91.

Kahanec, M., & Zimmermann, K. F. (2016). *Labor migration, EU enlargement, and the great recession*, Springer.

Kawachi, I., & Berkman, L. F. (Eds.). (2003). *Neighborhoods and health*, Oxford University Press.

Laurencin, C. T., & McClinton, A. (2020). The COVID-19 pandemic: A call to action to identify and address racial and ethnic disparities. *Journal of Racial and Ethnic Health Disparities, 7*, 398–402.

Leary, M. R., Tambor, E. S., Terdal, S. K., & Downs, D. L. (1995). Self-esteem as an interpersonal monitor: The sociometer hypothesis. *Journal of Personality and Social Psychology, 68*(3), 518–530.

Leary, M. R., & Baumeister, R. F. (2000). The nature and function of self-esteem: Sociometer theory. In M. P. Zanna (Ed.), *Advances in Experimental Social Psychology, 32,* Academic press.

Lee, J., & Han, S. H. (Eds.). (2021). *The future of service post-COVID-19 pandemic* (Vol. 1), Springer.

Linton, R. (1936). *The study of man.* Appleton-Century Crofts.

López-Roldán, P., & Fachelli, S. (2021). *Towards a comparative analysis of social inequalities between Europe and Latin America.* Springer.

Miller, S. I., & Bogal, R. B. (1979). Theoretical social mobility models and congruence with comparative findings: An analysis of Raymond Boudon's works. *Quality and Quantity, 13*, 289–306.

Mishra, L., Gupta, T., & Shree, A. (2020). Online teaching-learning in higher education during lockdown period of COVID-19 pandemic. *International Journal of Educational Research Open, 1.* https://doi.org/10.1016/j.ijedro.2020.100012

Negi, A., et al. (2020). *Sustainability standards and global governance experiences of emerging economies.* Springer.

Orrenius, P. M., & Zavodny, M. (2016). The economics of U.S. immigration policy. *Journal of Policy Analysis and Management, 31*(4), 948–956.

Parsons, T., & Shils, E. A. (Eds.). (1951). *Toward a general theory of action*, Harvard University Press.

Perez-Garcia, M. (2021). *Global history with Chinese characteristics: Autocratic states along the silk road in the decline of the Spanish and Qing Empires 1680–1796.* Palgrave Macmillan.

Piketty, T. (2014). *Capital in the twenty-first century.* Harvard University Press.

Popper, K. (1963). *Conjectures and refutations.* Routledge.

Powell, B. (2015). *The economics of immigration.* Oxford University Press.

Putterman, L. (2014). A caring majority secures the future. *Nature, 511*, 165–166.

Rawls, J. (1971). *A theory of justice.* Harvard University Press.

Rawls, J. (2001). *Justice as fairness a restatement.* Harvard University Press.

Rennwald, L. (2020). *Social democratic parties and the working class.* Palgrave Macmillan.

Robins, G., Elliott, P., & Pattison, P. (2001). Network models for social selection processes. *Social Networks, 23*(1), 1–30.

Root, H. L. (2020). *Network origins of the global economy. East vs. west in a complex systems perspective*, Cambridge University Press.

Rousseau, J. J. (2009). *Of the social contract, or principles of political right*, Regnery.

Sen, A. (1973). *On economic inequality*. Oxford University Press.

Shadmi, E., et al. (2020). Health equity and COVID-19: Global perspectives. *International Journal for Equity in Health, 19*, Article number: 104.

Shibuya, K. (2006). Collaboration and pervasiveness: Enhancing collaborative learning based on ubiquitous computational services, including as Chapter 15. In M. Lytras, & A. Naeve (Eds.), *Intelligent learning infrastructures for knowledge intensive organizations: A semantic web perspective* (pp. 369–390). IDEA.

Shibuya, K. (2020). *Digital transformation of identity in the age of artificial intelligence*. Springer.

Shibuya, K. (2020b). *Identity health*. In K. Shibuya (Ed.), *Digital transformation of identity in the age of artificial intelligence*, Springer. https://www.ncbi.nlm.nih.gov/pmc/articles/PMC7121317/

Schumpeter, J. A. (2008). *Capitalism, socialism, and democracy* (3rd edn.). Harper Perennial Modern Classics.

Simon, M., & Glynn, R. (Eds.). (2020). *Exploiting people for profit: Trafficking in human beings*, Palgrave Macmillan.

Smith, A. (2012). *An inquiry into the nature and causes of the wealth of nations*, University of Chicago Press.

Soltanifar, M., Hughes, M., & Göcke, L. (2021). *Digital entrepreneurship*. Springer.

Stefan, T., & Bernhard, K. (Eds.). (2020). *Need-based distributive justice: An interdisciplinary perspective*, Springer.

Triandis, H. S. (1995). *Individualism and collectivism*, Westview Press.

Turner, J. H. (Ed.). (2001). *Handbook of sociological theory,* Springer.

UNDP. (2020). Temporary basic income (TBI), https://www.undp.org/content/dam/undp/library/km-qap/Temporary%20Basic%20Income-V4.pdf

Vrontis, D., et al. (2021). Artificial intelligence, robotics, advanced technologies and human resource management: A systematic review. *The International Journal of Human Resource Management.* https://doi.org/10.1080/09585192.2020.1871398

Yamagata, S., Nakamuro, M., & Inui, T. (2013). Inequality of opportunity in Japan: A behavioral genetic approach. *RIETI Discussion Paper Series, 13-R-097*. https://www.rieti.go.jp/jp/publications/dp/13e097.pdf

Yunus, M. (2007). *Creating a world without poverty*, Public Affairs.

Yuval-Davis, N., Wemyss, G., & Cassidy, K. (2017). Everyday bordering, belonging and the reorientation of British immigration legislation. *Sociology, 52*(2), 228–244.

Weinar, A., & Koppenfels, A. K. (2020). *Highly-skilled migration: Between settlement and mobility*, Springer.

Weber, M. (2002). *The protestant ethic and the spirit of capitalism*, Penguin Classics.

Williams, E. (1994). *Capitalism and slavery*, The University of North Carolina Press.

Williamson, J. (1989). What Washington means by policy reform. In J. Williamson (Ed.), *Latin American readjustment: How much has happened*, Peterson Institute for International Economics. https://www.piie.com/commentary/speeches-papers/what-washington-means-policy-reform

International Affairs Against Crisis of the After Corona

A New World in Motion

1 After Corona: Reshaping the New World Order

1.1 Global Shift of the After Corona

The COVID-19 pandemic played a role as a catalyst to reorganize our global situations (Shibuya, 2021). People's lifestyles, economic systems, balance of power in global tension, and digital transformation will be reshaped in various ways.

As of 2021, an economic recovery is underway, and some industries are lagging behind. On the contrary, digital transformation will boost our next industrial and social progress, some economists pessimistically forecast that all of the global economic system will not be revived as the same one before the pandemic. Of course, international system for global cooperation will be also reconsidered among the nations.

Globalism before the pandemic was a constructive way of the world domination by the Western societies through capitalist economy that they originated. In contrast, the *OBOR (One Belt, One Road Initiative)* that China has been promoting is nothing less than a new structure of control to replace the global capitalism that Western societies had constructed. Through the aggregation of the global wealth to China through interweaving with each nation, namely, the Chinese Government is going to achieve their domination to complete an economic system that makes them extremely dependent on China. In this way, their influential power over the world has been exponentially increasing since the last decades. If such situation progress further, will the Western-dominated economic system be turned over to a Chinese-dominated system, or will the two antagonistic forces divide the world? With the use of AI and big data technology, in the near future, it is clear that the world cannot ignore entire influence of China. For the history of the humanity which has been driven by Western societies, is it a matter of inevitable rise and fall, or is it merely the nature of humanity which had repeatedly been divided and consolidated?

For these concerns, we have to gain insight into the future from historical trends. Certainly, we must have a discerning eye for those dynamics of the history. Indeed,

© The Author(s), under exclusive license to Springer Nature Singapore Pte Ltd. 2022
K. Shibuya, *The Rise of Artificial Intelligence and Big Data in Pandemic Society*,
https://doi.org/10.1007/978-981-19-0950-4_9

"*History repeats itself*" (Claudius). Namely, the humanity's ability to respond to the crises is expected to increase and the cost is to be naturally less. Because, each time, the humanity learned from its mistakes, identified the causes, and took measures to prepare for future crisis and risk. The decision making of the politicians will be also possibly precise and prompt. However, as the humanity has progressed, the situation has also become more complex, and the bureaucratic organization and decision-making system itself has become more difficult to see each crisis as the whole picture. This is because either risky events don't play out as we have learned from the history, or the response is often delayed because of crises that are always inexperienced by the peoples involved in the decision-making process. In such a situation, the severity of the crisis becomes precisely proportional to the severity of the sacrifice. Thus, it is still necessary to learn from the history and take sufficient preparations in advance assuming the inevitable serious cases.

1.2   Reforming Regimes Among the Nations?

Now it is time that a balance of power between *Westness* and *Eastness* should be investigated by multiple perspectives again. Such a twilight of Western society was already foreshadowed (Ringmar, 2005). The global expansion of the market economy on a global scale, namely globalization continued to bring about major social change (Inoki, 2014). Compared to the beginning and end of the twentieth century, the leading position among all nations in the world economy was replaced. As Inoki said, after the Great Depression in the UK at the end of the nineteenth century, the world's reserve currency was the UK pound sterling until the First World War brought the gold standard to an end. Even after the gold standard was abandoned and the end of the Second World War, the UK retained some of its international currency handle. However, after two world wars, the pound as a position of the international currency gradually declined, and lastly the US dollar took over its the major leading position. Still then, the post-war world economy was the era of Pax Americana, and it also meant the end of Pax Britannica. Towards the end of the twentieth century, the economic power of the USA, Germany, and Japan also declined relatively, and the economic threats of Asia (e.g., China, India) and Brazil began to be recognized. Similarly, the breakthroughs of Russia and Vietnam in the former socialist bloc were notable.

This global shift in hegemony is also becoming a reality (Bell, 2000, 2016). The WTO, the World Bank, the IMF, and other global coordinating bodies cannot ignore China's economic influence, and the standards that drive the world have come to be influenced by China's vision (You & Bu, 2020). What's more, China is likely to start managing not only the existing CNY (Chinese Yuan), but also its own cryptocurrency, increasing the likelihood that the reserve currency for the entire world will shift again. In other words, the postwar world structure, which was driven only by the USA, seems to have shifted to a world structure of two-horse carriages that is even more difficult to control. Will the "One Belt, One Road" initiative led by the Chinese government itself go so far as to surpass existing international political and economic frames such as the G7 and G8 and bloc economies such as the EU? In fact, the proximity of the "One Belt, One Road" initiative to Central-Eastern European countries is seen as a threat by EU countries (Jakimów, 2019). With regard to the gradual extension of China's soft power influence to Central-Eastern European countries, there is room to delve into the issue from the perspective of *"Securitization"* and *"Desecuritization"*, a concept of international security studies by the Copenhagen School, not only from an economic but also from a political and military perspective.

On the contrary, other emerging economies such as India and Africa are likely to move forward new reforming from their own perspectives. In the midst of such process, the government of Japan have to come up with a clear vision of what kind of strategy for Japan future, and the world's third-largest country as an aging population and a declining birthrate must be pursuing its further development (Committee on the History of Japan's Trade and Industry Policy RIETI, 2020).

Of course, the structural change of industries and balance of power in technology progress among the nations have also changed. In general, the development of science and technology of Western civilization was biased and interpreted to be superior to Eastern civilization. However, Needham (1956) argued that Chinese civilization was superior to Western civilization from 200 BC to 1450 AD at least. There was much criticism of his claim, but given the history of China, it was certainly a high level of civilization. However, Western civilization later reversed the gap with the development of the modern science from the seventeenth century onward. And now in the twenty-first century, China has been trying to lead the world through the use of the AI and big data again. In other words, if we look at the entire history of the human civilization, it can see how science and technology have been shaking up. Then, how will AI and big data transform the world's politics and economy, forms of governance, and the foundations of people's lives, including global standards in the twenty-first century?

1.3   The Four Horsemen

The United Nations Secretary-General's greeting[1] at the beginning of 2020 was summarized as following contents: "The Humanity now belongs to unprecedented crises, including environmental problems and disasters. These four horsemen…can jeopardize every aspect of our shared future". Coincidentally, just as he said, as "the fourth rider" (i.e., pale rider: this symbol stands for the death and the epidemic disease) of "the Four Horsemen of the Apocalypse" described, it was the COVID-19 pandemic that suddenly attacked our humanity.

Exactly, during 2020, such symbol actualized a global pandemic caused by the COVID-19. It further called a situation to raise an anti-globalism as a catalyst, and embodied as an incarnation of the global conflict. These symbols implied that we

---

[1] https://news.un.org/en/story/2020/01/1055791.

must confront with such epidemic virus as well as ourselves. At March 2020, French president also announced his policy as "We are at the War (against the threat of the COVID-19)" for their citizens. But this phrase was too ambiguous to interpret literally. All virus cannot be consecutively alive in the outer environment of the living host. Here, such host means ourselves. Then, it means that we are mutual enemy with each other, and such war means also to fight with ourselves.

Regarding those actualities, such virus of the pandemic seemed to successfully infect and invade into the regime of democratic institution and liberal ideologies among West side countries (i.e., the EU, the UK and the USA). As democratic and liberalism are relying on the maximization of freedom among the individuals, their activities (e.g., mobility, private rights, socioeconomic pursing) have serious potentials to spread such virus across the borders as consequentially. Cynically saying, for conducting such available ways to prevent the pandemic, they had to restrict their open mobility across the EU nations and democratic freedom. Thus, it can be paraphrased as a battle with Westness value by themselves. Namely, they could not help triggering against their precious Westness in the COVID-19 pandemic, and they lastly aborted their highest value.

Therefore, as someone said, *hegemony in global will turnover from the Westness to the Eastness?* It means a power shift by reshaping from open and liberal world structure to directive and top-down structure. On the contrary, here, Eastness nuance means majorly only China, and other nations such as Japan and South Korea should not be included into this side. But their cultural and historical backgrounds are commonly shared by their citizens.

2   Morgenthau's Prospects

The cold war as balance of power between west-side and east-side nations had already gone to the past. Such paradigm piles now in the underground of our subliminal memory. Those who experienced the cold war age became one of the witnesses for the tremendous global changes at that time. Exactly, the author was also a witness watching both collapse of the Berlin wall and Soviet Union at the end of twentieth century. It was just 30 years ago.

During the cold war, Morgenthau gazed into the reality. He was one of the foremost and distinguished scholars of political studies in international issues. When the author read through his book entitled as "*Politics among Nations: the Struggle for Power and Peace*" (1978), the author felt a surprising insight for the international order and balance of power in our contemporary world. Although he definitively noted that world politics can harder predict the future trends, his realism can still provide very clear perspective for understanding on international affairs in current digital society. As he was a realist, he sharply pointed out many issues, for example.

1.   Intercommunication using media will boost 'World Opinions' globally.
2.   Balance of Power and global tensions beyond the Cold War.

3. China's Presence in the twenty-first century (Department of Commentary People's Daily, 2020)

During the age of the cold war, he probably anticipated our future of world conditions. When ordinary citizens harden their attitudes (e.g., world opinion) for nationalism and exclusionism by globally intermediated communication tools, with enthusiasm, it may be inevitably to next world war in the future. His noted "future" is our living present age, and political propaganda in his era can be regarded as fake news in our time. His prediction has already clearly stated the negative aspects of social media before internet technology came into existence. And for such reasons, there also seems to reconsider opinion dynamics, contagion of fake-news, decoupling of the mutual trust, and balance of power. Many of the experts were worried about that those dynamics have the serious potential to trigger the Third World War. However, the author guess that all-out war has already invoked by ordinary citizens taking with a smart-phone. Venture to say, social media became one of the cyber weapons (e.g., sharp power).

In addition to the structures and dynamics among the nations, citizens are synchronizing with each other across the borders of the world, and transnational networking has come to influence global trends. In the globally interconnecting world, as articulated earlier chapters, each individual also sought to be synchronized with each other by reacting from the others' influential opinion, collective trends online, and false consensus based on the social psychological mentality. The phenomenon on *identity lost* in global online community has been emerged, each individual flattered with social trends to integrate with those who interconnected through the cyberspace (Shibuya, 2020a). Then, such digitizing trends nearly reach and converge the somewhat autocratic collectivism than autonomous individualism. One of such social phenomena stands for populism, nationalism and exclusionism in global (Lorenz & Anders, 2021).

Further, he predicted that China presence will be heighten by their population and technological advancements. All nations in the East Asia have been entangled with their progress, and their explicit influences cannot be neglected. In such context, the author worries about that the contemporary world is now confronting another balance of power between Westness and Eastness. Therefore, nevertheless such situation shows negative potentials, it should be deepen understanding the emerging possibility of perpetual peace in the global.

3 China Risks

At 2020, present USA government has published own strategic report that entitled *"United states strategic approach to the people's republic of China"* (USA (the White House), 2020). This report intends to counterpart against China risk and its emerging growth, and such attitudes of the USA clearly told competitive rather than cooperative relationships. Additionally, DOD (department of defense) also published their warning report on recent development of the China's military forces (DOD, 2020). Furthermore, during the presidential campaign as of August 2020, Esper who was a Secretary of Defense of the USA announced *"The Pentagon Is Prepared for China"* and *"PLA (China's People's Liberation Army) modernization is a trend the world must study and prepare for –much like the U.S. and the West studied and addressed the Soviet armed forces in the twentieth century"*. As Morgenthau predicted, the time has come for Western society to once again learn deeply about China, their values, cultural and historical background as well as advanced technological and military trends.

As we known, new technology war between the USA and China has already provoked serious tensions in the AI and big-data fields (Wu, 2020). Eurasia Group (2018) reported 'top 10 risks' in the world, its first place was "China" and third place was *"Global tech cold war"*. The Global axis between the USA and China will be distinctively amplified by their chicken race participating with their alliance's member countries. Actually, after the collapse of Berlin Wall, democracy's progress has achieved global economic success, but in recent, democracy faces its recession. While USA President Trump called for egocentrism and populism, China and Russia established panopticon regimes through AI and big data technology to strengthen state control.

Besides, representative extraordinary events and unexpected crises often reveal those implicit common senses among the differences. In particular, it is the difference in the specific response to the crisis that the political function should play. For example, Bremmer (2016) said before this global crisis of COVID-19, *"The United States can lead a more ambitious multinational effort to invest in local public health systems and improve coordination when health crises occur. By mobilizing the U.S. Centers for Disease Control and Prevention more effectively, the United States can provide doctors and other health care workers, technology, medicine, beds, volunteers, logistical support, and other resources needed to help manage the risk of pandemic—and the longer-term investment needed to develop vaccines. It can also respond much more quickly and comprehensively than Washington responded to any of these previous crises"*.

However, as everyone had known, actual crisis caused by the COVID-19 demonstrated a tragedy of the USA. Namely, his expectation was too optimistic. The COVID-19 exposed both the threats by explosive infections among American in their homeland and their vulnerability to maintain public health. In fact, this fear came true, and the USA became the worst in the world.

On the contrary, many citizens of East Asia countries could adapt in the imperatives ordered by the government or spontaneous prevention behaviors by them for the public health. The COVID-19 certainly triggered first domino-break in China, but further cascading collapses of socioeconomic systemic risks, medical services, and other necessary supply chains could be concluded that liberal democratic side nations enlarged those catastrophes in global. Although liberalism and democratic regimes based on the individualism could progressively motivate economic success and intensives of capitalism among citizens in global, but their risk managements against extreme emergency cannot always assure their accurate destinations. In this point, global politics have to confront with irritating matters on the international order (Ferrara, 2019).

4   Conclusion

Therefore, the world politics becomes controversial issues to discern the true directions beyond the gap between Before Corona and After Corona. Global power shift will be steeply provoked by consequences of this pandemic. Unless the USA takes urgent responses, the USA loses their international leaderships, whereas China steeply gains their comparative presence for global balance of power. Antagonistic disputes between democratic individualism and imperative collectivism are still clearly coupling with each value based on their national ideology. It appears that the world will once again begin to proceed a history of the global decoupling. Global conflicts are a recurring chronic disease because there is still no panacea to satisfy our existential identity. The craze for the nationalism is indeed an infectious disease to which we are not immune yet.

In the succeeding chapters, the author continues to discuss global political issues in the After Corona by following viewpoints. Namely, digitized shift of regime and hegemony such as cyber-warfare (Chapter 10), simulations on the balance of power among the stakeholders (Chapter 11) and finally, a comprehensive discussion on international affairs in the After Corona (Chapter 12).

References

Bell, D. A. (2000). *East Meets West: Human rights and democracy in East Asia*. Princeton University Press.

Bell, D. A. (2016). *The China model: Political meritocracy and the limits of democracy*. Princeton University Press.

Bremmer, I. (2016). *Superpower: Three choices for America's role in the World*. Penguin.

Committee on the History of Japan's Trade and Industry Policy RIETI. (2020). *Dynamics of Japan's trade and industrial policy in the post rapid growth era (1980–2000)*. Springer.

Department of Commentary People's Daily. (2020). *Narrating China's governance: Stories in Xi Jinping's Speeches*. Springer.

DOD, USA. (2020). *Military and security developments involving the People's Republic of China 2020*. https://media.defense.gov/2020/Sep/01/2002488689/-1/-1/1/2020-DOD-CHINA-MILITARY-POWER-REPORT-FINAL.PDF

Eurasia Group. (2018). *Top 10 risks*. https://www.eurasiagroup.net/issues/top-risks-2018

Ferrara, A. (2019). "Most reasonable for humanity": Legitimation beyond the state. *Jus Cogens, 1*, 111–128.

Habermas, J. (1976). *Legitimation crisis*. Heinemann.

Habermas, J. (1987). *Eine Art Schadensabwicklung*. Nachdruck.

Inoki, T. (2014). *A history of economics*. Chuokhoron-shinsha (in Japanese).

Jakimów, M. (2019). Desecuritisation as a soft power strategy: The belt and road initiative, European fragmentation and China's normative influence in Central-Eastern Europe. *Asia Europe Journal, 17*, 369–385.

Lorenz, A., & Anders, L. H. (2021). *Illiberal trends and anti-EU politics in East Central Europe.* Palgrave Macmillan.

Morgenthau, H. J. (1978). *Politics among nations: The struggle for power and peace.* Knopf.

Needham, J. (1956). *Science and civilisation in China (Volume 1, Introductory orientations).* Cambridge University Press.

Ringmar, E. (2005). *The mechanics of modernity in Europe and East Asia: Institutional origins of social change and stagnation.* Routledge.

Shibuya, K. (2020a). *Digital transformation of identity in the age of artificial intelligence.* Springer.

Shibuya, K. (2020b). *Identity health.* https://www.ncbi.nlm.nih.gov/pmc/articles/PMC7121317/. In Shibuya, K. (2020). *Digital transformation of identity in the age of artificial intelligence.* Springer.

Shibuya, K. (2021). A spatial model on COVID-19 pandemic. In *The 44th Southeast Asia Seminar, The Covid-19 Pandemic in Japanese and Southeast Asian Perspective: Histories, States, Markets, Societies.* Kyoto University.

USA (the White House). (2020). *United States strategic approach to the people's republic of China.* https://www.whitehouse.gov/wp-content/uploads/2020/05/U.S.-Strategic-Approach-to-The-Peoples-Republic-of-China-Report-5.20.20.pdf

Wu, X. (2020). *Technology, power, and uncontrolled great power strategic competition between China and the United States*, China International Strategy Review, vol.2, 99–119.

You, C., & Bu, Q. (2020). Transformative digital economy, responsive regulatory innovation and contingent network effects: The anatomy of E-commerce law in China. *European Business Law Review, 31*(4), 725–762.

Digitized Shifts of Regime and Hegemony

1 On Sacrifices in Political Crisis

Global epidemiological risks (e.g., COVID-19) and international conflicts are very similar in some respects. It is the arrogance of humanity to assume that it can always subdue its enemies and fundamentally control them, and such optimism leads to further sacrifices on the part of many uninvolved parties, often resulting in chaos. Emerging infectious diseases such as COVID-19 are also the result of humanity's uncontrolled responses of its own civilization on the Earth. From the point of view of the distribution of sacrifices, it can be said that humanity itself has come to bear the brunt of these sacrifices. In addition, as soon as the USA troops withdrew from Afghanistan in August 2021, the Taliban returned to power and the democracy USA advocated easily collapsed. This case showed the failure of a state to become a democratic system by imposing specific ideology on citizens. As result of this, it not only entails many sacrifices in many directions, but also will bring about the worst possible crises.

To be sure, the fragility of democracy also became evident in the COVID-19 pandemic (Druckman et al., 2021). In another case, after fueling domestic fragmentation (Bobo, 2017; Howard et al., 2018), in January 2020, in the USA, which has been advocating democracy, the incumbent Republican President Trump used Twitter to refuse to acknowledge his defeat in the USA presidential election (Green, 2021), responsibility of misguide for COVID-19 policy, and further even instigated an occupation of the USA Congress. This act, which could be called a self-denial of democracy or a self-initiated revolution by the president himself, was condemned both at home and abroad. This president, after much mismanagement of corona control in the USA, even though he caused the world's largest number of deaths and infected victims, refused to even allow himself to be defeated in the presidential election and denied even democracy.

Actually, the crisis itself may be saved by democracy, but somewhat sacrifice is always inevitable. The key to the survival of a nation is possibly to avoid various crises by all means, but it always requires some kind of sacrifices such as soldiers,

K. Shibuya, *The Rise of Artificial Intelligence and Big Data in Pandemic Society*, https://doi.org/10.1007/978-981-19-0950-4_10

labor, time, budget, diplomatic efforts, the global environment and future generations, and these are finally sacrificed by someone's decision. As *"Sunzi's martial arts"* as one of the ancient military books in China clearly stated, *"The art of war is of vital importance to the state"*. In today's situations, all politicians must be similarly recalled such phrase at confronting with any crises. Even more, a single crisis can sometimes spread in a chain and become larger *crises* in global (e.g., systemic risk in financial market).

Namely, regardless of the political system, politicians and statesmen must always be trusted by the sovereign, and human history has been able to accumulate continuously only when someone bears the sacrifices they have imposed. But when they refuse to make such sacrifices, it is time for not only the ruling system but also the nation itself to collapse. In fact, there are several countries whose approval ratings fell as a result of the COVID-19 pandemic. And it should be recalled that it had an uphill battle on the occasion of the presidential election. Therefore, it is necessary to examine each crisis in detail, and examine all aspects of sacrifice in order to deduce its reality and gain insight.

2 A Turning Point of the Regime

2.1 The Dynasty Change in China

It is often said that China has traditionally been a *rule of men* (i.e., ruling by the emperor in each dynasty) and Western society is a *rule of law* (i.e., *rechtsstaat*) based on democracy. Indeed, each dynasty of ancient China had always been repeatedly a state-governed system legitimated by the emperor who had been given power from "heaven". Modern Communist Party rule may be closer to the traditional system of governance, albeit with a modern legal system.

Since ancient times, the ideal way of the governance had been the *moral rule* according to Confucianism pervading the real institutions, practices, and culture in the ancient China and east Asian countries. Whether it was democratic nations or moral rule, as long as it was well functioning within each cultural system, it was not a problem. However, in both systems of the governance, revolutions and uprisings had sometimes occurred, and lastly governments and dynasties had been overturned (North, 2013). The very nature of politics is to determine the sacrifice, but there is usually no guilt about the consequences. However, it was only in times of the *revolution* whether such riot was larger or not that politicians were made to bear the burden of sacrifice as own death.

There is one of the old China aphorisms *"A monarch is a boat, and a people is water. This is quickly overturned"*. This means that any governance of the ruler could be firmly sustained by sharing the trust among ordinary citizens, and the degree of their supports could become to actualizes the longer imperial regime. However, as many historical events as *déjà vu* repeated, any regimes would be steadily decaying, and it lastly collapsed despite initially starting by the subtle events such as a small size of demonstration against the regime. Furthermore, emerging revolutionary nemesis against the *ancien régime* could finally overwhelm and terminate the ruling governance of the older kingdom and empire.

There had certainly been many *revolutions* in Chinese history over the past few thousand years. One of its characteristics was that it pretended to be a change of the ruler's clan by the imperative of *"Heaven"* (e.g., God). In Western civilization, on the other hand, spontaneous social movements and demonstrations gradually developed and eventually usurped the handle of state control to bring down the old regime and elect new leaders. The former relies on tradition within the three types of legitimacy of governance (i.e., traditional, charismatic, or legal-rational) preached by the sociologist Weber (1958), while the latter places more emphasis on legal-rational systems (Shibuya, 2017).

Indeed, even in Chinese history, it often started with small-scale rebellions and demonstrations in the beginning. For example, during Chén Shèng Wú Guǎng Qǐyì in China (209–208 BC), a revolutionary leader Chén Shèng organized a larger mutiny against the empire Qín (778–206 BC). He agitated the participatory peoples by his renown phrase at the mass meeting, namely he publicly claimed *"Becoming a king, a lord, a general, or a vizier is not determined from the birth. Namely, there is no rational reason to discourage us from becoming such the nobleman and personage"*. It seemed to be one of the most impressive and agitated phrases in the human history, and then their revolutionary armies that consisted of a large amount of the very poor peasants, ordinary citizens, and other antagonists against the empire Qín rushed into the territory of Qín, and finally it was triggered to terminate the dynasty of the Qín. After that, Chén Shèng had temporally usurped the throne from the ancient regime of Qín, but his ruling new kingdom sooner had been collapsed by the betrayals of his retainers (cynically, a nuance of his agitated sentence was actualized again), and the other stronger armies assaulted and overwhelmed his untrained army in collusion with various stakeholders. It can be recalled a Marx's aphorism *"History repeats itself, first as tragedy, second as farce"*.

At ancient era, the only means of building momentum for rebellion across the continent was through incomprehensible ways to those living in the modern social media age (e.g., above slogan by Chén Shèng). In order to steadily overturn the ancient regime, the leaders must have been required to plan considerable strategies. At the same time, it was important to notice that the corruption of the civilian infrastructure and the bureaucratic system had become apparent, and the focus was on how to incorporate the military and police forces, not to mention the cooperation of the citizens. Those factors determined whether or not it would end in a mere local rebellion.

And, in all the differences with the Western revolutions, emperors' clans were not simply replaced by another clan in ancient kingdoms of China. It would be also important to stress a fact that the nobles of each dynasty in ancient China were also generally replaced. In the West, each nobleman held one dominion and often kept distinguished authority from a king or an emperor who ruled over the whole nation (Turchin, 2005). For this reason, in Western society, there are still many aristocrats and their descendants who have survived into the modern times, unless they have fallen too far.

Some theorists (e.g., Locke) granted the citizens the right to resist against their government ("*Right of Resistance*"). Certainly, in many cases, rebellion could be reactive on its own motivation of the people (Le Bonn, 1894). However, once the new political system had been established, the citizens who promoted when overturning the old system will only find themselves at the bottom of the governing system. Even if not so, during rebellion and conflict, they were apt to be buried in the midst of so many sacrifices that they did not go down in history.

Similarly, in the "Jasmine Revolution" and the "Arab Spring" of the 2010s (Boening, 2014; Choudhary et al., 2012), there were many examples of how their socio-economies worsened rather than improved, even though they accomplished their revolutions by advocating liberal democracy (Shibuya, 2017). Moreover, the rulers had been replaced, but the people are still forced to suffer the pain of the coercive system. This is a common phenomenon. Democracy merely encourages citizens to participate in the governing structure, and citizens must take the initiative in accomplishing proactive economic activities themselves. It is a mechanism by which an entire country cannot develop unless each individual develops the skills to live independently. Democracy is not necessarily a panacea for the poverty, and citizens will only learn after the revolution that a social structure that distributes economic resources in a tyrannical and control-oriented manner was more conducive to the stability of the nation as a whole.

Nevertheless, why does the humanity continue to repeat the overthrow and establishment of political institutions? Nietzsche said "*the will to power*" and Hegel also said a similar opinion. Is this because citizens' sense of resistance to the oppression that comes with governance had become apparent, or because they desired to be a part of it by joining the power of the state? Their slavery, which Western civilization has been trying to forget, may have satisfied their ruling power and thus established a stable democratic system because modern democracy opened up the authority for the ordinary citizens to govern other followers by separating them from the rest of the population.

However, as (Fukuyama 1992, 2014) articulated, in order for a democratic system to be realized and exist in the history, that it is necessary to develop capitalism based on the legal establishment of the private rights among all individuals. A state in which capitalism is not functioning will not be guaranteed the available ways by which the individual can stay alive, and democratic institutions for participation and mutual coordination of political institutions will not take root in society. Furthermore, since the incentives of each individual such as education and skills do not effectively work, they do not form the foundation for making the right decisions. Such desires of each individual could certainly become a more stable political system entirely, as it has the opportunity to satisfy a number of political processes and the business scenes.

2.2 What is a Meaning of Revolution in the Western Democracy?

Certainly, democracy and other political regimes could be assorted by historical differences in the origin, growth, and decay process (Linz, 1978). According to Moore (1993), he dedicated that his historical view on the humanity driving to democracy could be clarified by evidences whether bottom-up based revolutions or industrial development. However, he wondered why such rough causes could provoke each difference such as democracy and capitalism in the EU and the USA, fascism in Japan and Germany, communism in China, and other cases. He just told what and why historical events occurred, but he did not examine what if those factors will control.

As mentioned in earlier chapters, it is said that the first democracy of the humanity was born in ancient Greece. It can be called the born of the western democracy. However, it should not forget that it was also a society of class status dividing between the class of citizens and the slavery (e.g., Hegel "*Grundlinien der Philosophie des Rechts*"). Democratic decision-making and political participation were, of course, the prerogatives of the civil class only. Moreover, such privileges were only available to adult men of the civic class, and women could not even participate. Democracy, in the end, was only a means to maintain and improve the privileges that only adult males have in the civic class. It is the duty of adult men who belong to the civilian class to protect them, and protecting one's country (city: *polis*) is, after all, protecting one's own privilege. If they were to be conquered by another city, such privileges might be lost, or they might downgrade a slave class themselves. If that happens, they will be forced to fight desperately, and by participating in politics, they will actively participate in political decisions that will protect their own interests in foreign and domestic affairs. Hence, this means that both self-sacrifice and necessary rules had been already established in ancient societies. Namely, they had duties as self-sacrifice, and obeyed rules and institutions for the leadership of subordinates.

Actually, ancient Greece at that time and modern democracy are completely different, even though we express both political systems as a same term of "democracy". The position and rights of the sovereign (i.e., citizen) belonging to the democratic state of modern man are quite different from those of the civic class in ancient Greece. It is, above all, whether or not there were many slaves, and whether or not they could be put to use and make good use of their own labor, time, and other resources. Whether the free time could be used for self-education, debate, political participation, and others in any amount. A labor force instead of the citizens is a rare resource that does not exist in modern society. Eventually, the AI and robots will take their place in labor in the future. However, contemporary citizens who do not have such "slaves" have no choice but to work for themselves. As a result, modern democratic societies are not just egalitarian societies in which class is denied, rather an inseparable system in which both roles of "slave" and "citizen" must be undertaken by a single individual.

3 Digitized Ways as Cyber-Warfare

3.1 State Governance in the Age of AI

Our living era conducted by the AI and big data has already come (Shibuya, 2020a, 2020b, 2021; Sunstein, 2001). The governance system also needs to be steered not only domestically, but also by taking into account the many disruptive factors caused by digital networking as well as international relations, which cannot be ignored. As we unexpectedly became a witness of the Arab Spring, the growth process of demonstrations influenced by social media (Jost et al., 2018) not only led to the eventual overthrow of the old regime, but also spilled over into many surrounding states, resulting in a serious negative impact on the entire system of governance in all Middle Eastern countries and even the rise of the Islamic State (Agarwal et al., 2013; Bryan et al., 2014; Choucri, 2012; Choudhary et al., 2012; Gonzalez-Bailon & Wang, 2016; Hill & Hughes, 1998).

3.2 Governance and National Security

Next, it needs to look at the latest in national security. Recently, there are computational social science (CSS) researches showing remarkable results[1] in crime prevention, terrorism (Fierke, 2012), defense and security (RAND Corporation, 2017; Shibuya, 2020a, 2020c). For example, Weinberger (2011) provides an example of how CSS researchers had contributed to counter-terrorism in the USA. They reported that analyzing the social media data used by the terrorists and uncovering their relationship networks, they were able to proactively identify and locate key terrorists. And lastly, they could avoid their plan of the terrorism. What's more, the USA government has taken this opportunity to announce that it will use its defense budget to fund a huge amount of research. There is also a separate Big Data-related research grant (news of March 29, 2012: announcement of an initiative to invest a total of $200 million) to advertise that research was underway in the USA. In the first place, the predecessor of the Internet itself (ARPANET) was originally developed by the Defense Advanced Research Projects Agency (DARPA) in the USA, and there are many other examples of the promotion of military and defense research using Internet technology (Guice, 1998). For example, GIS and GPS technology, which have become indispensable for mobile terminals, were originally used for military purposes. On the other hand, when it comes to information security and encryption technology (Efthymiopoulos, 2019), the highest level of technology is still a military secret.

---

[1] Scientists use big data to sway elections and predict riots—welcome to the 1960s: https://www.nature.com/articles/d41586-020-02607-8.

In addition, *"United States Cyber Command*[2]*"* is already in operation, and attack and defense research on cyberspace has already been conducted. With regard to counter-terrorism, it is quite important to prepare intelligence, surveillance, and reconnaissance (ISR) activities. It was not the only example of counter-terrorism applications. For example, it is already being done in the USA for police organizations to analyze big data such as geographic information and crime conditions to predict crime locations, and deploy police units in advance. Also recently, researchers at the University of Connecticut have announced that they have developed software to extract from a large amount of data not only from the regular Internet but also from the darkweb, including many articles with racist or hate content, terrorism or crime. Incidentally, during the COVID-19 pandemic, there was a report of Chinese-made vaccines being trafficked on the dark web (Bracci et al., 2021).

On the other hand, the terrorists have also taken advantage of the spread of social media, and since the Arab Spring in 2011, the Islamic State, which swept over parts of the Middle East by exploiting the chaos and gap in the governance system, has also made effective use of social media. Through this, they recruited combatants from all over the world and effectively manipulated information, including disseminating a lot of information on the Internet. Their networking was decentralized and collaborative, and it means no specific leader. This can be seen as a result of the strategic adaptation of al-Qaeda affiliated organizations to turn into decentralized, distributed and networked organizations after the 2001 terrorist attacks in the USA. In other words, social media has also made it easier to interconnect terrorists (and reservists) hiding in different parts of the world.

In this concern, there are many countries that use information manipulation to stabilize their own governance and regime. In particular, there is a growing need to eliminate the possibility of the country becoming a *legitimation crisis* due to the extreme demonstrations caused by the citizens and anti-government hacktivists and others (Shibuya, 2017).

In Russia, President Putin has been in power for a long time, but there has been a spate of anti-Putin demonstrations in various parts of the country. Critical speech and demonstrations against his overbearing leadership regime have been squelched with or without the use of social media. In addition, the surveillance for information control has been stretched across the Internet. Furthermore, on November 26th, 2017, the Russian government announced that it would restrict the activities of reporters and journalists from regular newspapers and other media outlets, imposing obligations on media companies designated as foreign agents to make public announcements to that effect. Being involved in the media itself may be considered espionage and may be subject to increased surveillance (Giusti & Piras, 2020). It suggests that Russian government sources understand what the legitimation crisis can bring in the social media age. On the other hand, in addition to such media control at homeland, there is also a suspicion that it has some sort of military application against other countries.

Similarly, China has tightened its domestic control. For example, at 30th, June 2020, the China central government had launched more strict law on national security

---

[2] https://www.cybercom.mil/.

legislation, and this target of enacted law will regulate Hong Kong and other cities in homeland of China. First, it is conducting panopticon-like monitoring system including image recognition technology, surveillance cameras and other digitized equipment that installed in all over China. Such system already makes it possible to recognize each individual even if there are tens of thousands of people in a square. In addition, although the Internet is open to citizens, the content of various services and contents are also always audited at all times, and some services are no longer allowed to be used (e.g., Twitter, Google, etc.). In 2017, the country enacted the "*Cyber Security Law*[3]" and advocated cyber sovereignty to strengthen control within the new authority and legal framework of the country's governance. Cybersecurity in China is said to have two main points, which can be broadly divided into (1) issues related to economic, social activities and infrastructure, and (2) threats to the national governance system itself.

3.3   Sharp Power: Digitized Vulnerability of Democracy

Can the AI and its machine learning algorithms identify a historical tipping point? In the age of big data, is it possible to predict the timing of uprising in a democratic society? Will it be possible to predict election results and trends in the legitimacy of the rulers through advanced AI driven services and big data analysis, as typified by one of the revolutions? In fact, there are already many examples of analysis of election predictions using big-data obtained from Twitter and Facebook (Bakshy et al., 2015; Lamont et al., 2017; Wang et al., 2017). In addition, it is beginning to be more accurate as a means of manipulating public opinion (Freelon et al., 2020), and some countries such as the USA have become more active online in their election campaigns.

It is a matter of cost and performance against the risks of cyber warfare. When it comes to social media and the Internet, the costs associated with them are relatively low compared to military and defense budgets to date. Assuming some kind of incitement involving the ordinary citizens, there is enough fear of incitement if the risk and cost are relatively low and the performance is good compared to the case of infiltrating several intelligence operatives to incite a similar incitement. As mentioned earlier, such actions in cyberspace are called *sharp power* as opposed to physical means of warfare (i.e., hard power) and cultural influences (i.e., soft power) (DeLisle, 2020; Nye, 2017). And Russia in particular has named them hybrid warfare to strengthen its defense. Naturally, it is also necessary to keep in mind the fact that research based on this is progressing in the USA and other countries.

---

[3] https://www.china-briefing.com/news/chinas-cybersecurity-law-an-introduction-for-foreign-businesspeople/.

In terms of physical damage, physical destruction of information infrastructure such as IoT, robots,[4] self-driving cars, drones, communication satellites, and information service infrastructures, or actions such as software tampering, rewriting, and virus infiltration are recalled (Von Braun et al., 2021). As of the end of November 2017, damage from IoT security vulnerabilities was rampant around the world and in Japan, it jumped more than 100 times in one month compared to October 2017. Not all of this is directly related to domestic defense issues. However, the question is what kind of security the future network society should have. And there is a risk that information appliances could be remotely controlled with the advancement of IoT and the AI.

Already each ordinary citizen is both a victim and a complicit one whether they are unintentionally or not. An *all out of war* involving ordinary citizens may have become normal status in the information space by utilizing advanced technologies. Social media and the ubiquitous society seem to have pervasively the non-ordinary space of "battlefield" into the "daily living space" by affordable tools of information technology. In online communication and communities, can we say that even casual daily conversations and information transmissions are not a mixture of hidden agendas and manipulations of other countries? The goal of sharp power is not so much a state as it is a means to an individual. Are the words and actions of individuals and a group truly independent and autonomous? Can we be sure that we are not unknowingly complicit (so called, "*mercenaryized*") by being instigated and guided by someone's intentions? Fake news that relies on conspiracy theories such as "*QAnon*" is also on the rise, and is beginning to have a negative impact on civic life and even national governance.

On the other hand, military intelligence is involved in intelligence operations against the executive branch, corporations, political parties and organizations through cracking, information manipulation, and various types of espionage activities. In particular, such actions on important national sites (e.g., nuclear power plants, government agencies) need to be particularly vigilant.

Furthermore, with the development of the AI, we will probably see the singularity in the near future. As a result, the issue of the AI and its ethics has already been discussed (Baker-Brunnbauer, 2021; Cath, 2018; Hagendorff, 2020). For example, Microsoft's AI ("*Tay*") learned discriminatory content (e.g., Nazism) by machine learning online information and interactions with users. Also, if a robot equipped with the AI specialized in combat technology reaches conclusions spontaneously and begins to act in the field, will humans have the means left to cope? Indeed, Google's June 2018 statement that it will not conduct research and development internally that could lead to military applications of AI (Google AI Principles[5]) is based on this perspective. The multilateral debate over Lethal Autonomous Weapons Systems (LAWS) is now on a parallel, and there is no guarantee that it will not turn into a multilateral war using the most advanced AI.

---

[4] https://www.stopkillerrobots.org/.

[5] https://ai.google/principles/.

To summarize above points, the digital world may lead to cyber warfare (Ministry of Defense, Japan, 2017). When the age of the AI and big data driven society will proceed further, only a small number of the people, regardless of differences in political systems, may be involved in politics as a decision-making way. Much more so, in the After Corona world, online collaboration and remote work become commonplace, and political deliberation and civic engagement can be substituted by these trends. Is it desirable that they make it possible to aggregate opinions and reach consensus more daily, or is it desirable to leave the top-level judgment to a few people (or central AI)? Depending on its direction, the new political system may be reshaped.

4   On Simulations of the Societies

Any nations with democracies often adopt majority voting and electoral procedures (Locke, 1998). As a result, even a strong country with a substantial military force might exposes the vulnerability of its governing democratic system. It should be borne in mind that it has now become easier to attack such vulnerability.

In the past, simulation studies with military and defensive intentions themselves have examined disarmament, alliance conclusion, and inter-state strategic situations from the perspective of international political theory through game theory and other means (Axelrod, 1997). In addition to being used for various strategic planning and verification, there are also other examples such as the Los Alamos Institute's EpiSim study that assumes bioterrorism (Cohen & Wells, 2004), and the development of a simulator based on geographic information and human behavior modeling (Abdalla & Niall, 2007).

In the cases of the recent spread of social media and cooperative actions based on the linkage of geographic information systems and human behavior, the key point of military and defense research is the analysis, modeling and simulation of those big data in the end. The reason why research institutes in various fields are working on such themes because they recognize the importance of this point. For example, the RAND Institute (2017) in the USA, which has a close relationship with military research, corroborates this by reporting that a cycle is required to model human behavior and engagement with the media and validate it through simulation, utilizing the results, theories, and empirical tools of the social and behavioral sciences.

With respect to the contents discussed above, modeling and social simulation can be used to implement policy making and test the pros and cons of strategies. To begin with, a number of studies have already been conducted by researchers in the cyber-physical field to predict the outcome of various types of voting from data on social media. In general, there is a certain co-occurrence between what people behave by using social media and voting behavior, given the reasonable predictive results. With the increasing use of smartphones for communication, it is possible to simulate what information can be distributed in real time in order to increase the effectiveness of communication. In addition to the dynamics of information posted by the ordinary citizens, the study focuses on the existence of influencers, the structure of the interpersonal networks that follow them, the influence of specific opinions, and the examination of these diffusion process. It can build a model based on human relationships and conduct social simulations of information propagation.

In fact, the National Science Foundation (NSF[6]) made a decision in 2015 to grant for interdisciplinary computer science and social science research related to political conflict. Researchers at the University of Texas at Dallas will use online big data to study the various conflicts caused by differences in political and cultural backgrounds. On the other hand, for full military applications, DARPA has announced that it will promote the Next Generation Social Science (NGS[7]) Program and the Computational Simulation of Online Social Behavior (SocialSim) project. Actually, the USC ISI (University of Southern California, Information Science Institute) research group has announced that it will obtain a SocialSim grant from DARPA to promote data acquisition, analysis, modeling, and simulation research on online public opinion and other data (October 2017). They obviously explore the possibility of some sort of manipulation of public opinion, not to mention the propagation and influence of certain opinions on social media. It can be a means of inducing public opinion in other countries or, conversely, defending similar actions from other countries.

In short, it clearly envisages not only the establishment of preventive measures for terrorist acts that aim at the fragility and collapse of social institutions and governance systems, but also their application to other countries. It is the use of social media to incite other countries and international public opinion without infiltrating other countries to weaken the political base of other countries, and to intervene in the governance system. Modeling and simulation studies are expected to be conducted to verify the effectiveness of inciting public opinion to impede the growth of the number of votes to be won or to incite demonstrations in support of the candidate's side in favor of the country when another country adopts a democratic voting system.

Finally, it should be recalled that science and technology are composed of three elements. These are description, prediction, and control. To find the principles and laws that explain and predict some phenomena, and to achieve specific social phenomena by a technique for controlling necessary factors. In other words, military and defense-oriented research seems to aim not only to explain and predict the dynamics of civilian speech and behavior, but ultimately to *control*. Namely, it means an intentional manipulation of ordinary citizens' actions and its revolutionary consequences.

---

[6] NSF Grants Bring Together Computer, Political Scientists for International Conflict Projects.
https://www.utdallas.edu/news/research/nsf-grants-bring-together-computer-political-scien/

[7] NGS2: http://go.usa.gov/cfqkm.
SocialSim: https://www.darpa.mil/program/computational-simulation-of-online-social-beh avior.

5   Conclusion

As international relations between the USA and China deteriorate, the preparedness is quite necessary for new threats. Principally, digital transformation through the advancement of AI and big data related technologies provides a means of attack that can lead to the overthrow of inherent governments and the collapse of states, and it also encompasses the inherent vulnerabilities and risks of democracy. In the past, this was due to civil revolts and revolutions, but it will be able to conduct such agitated movements through intentional manipulation of information and public opinion directed by the other countries. Namely, there is a growing concern that can use it as a means of military and diplomatic attacks. At the same time, a better understanding of these social phenomena and the means of the attacks can help improve the level of defense for own homeland.

References

Abdalla, R. M., & Niall, K. K. (Eds.) (2007). *Review of spatial-database system usability: Recommendations for the ADDNS Project.* Defence Research and Development Canada (DRDC).

Agarwal, N., Wigand, R. T., & Lim, M. (2013). *Online collective action: Dynamics of the crowd in social media.* Springer.

Axelrod, R. (1997). *The complexity of cooperation.* Princeton University Press.

Ayson, R. (2004). *Thomas Schelling and the nuclear age: Strategy as social science.* Routledge.

Baker-Brunnbauer, J. (2021). Management perspective of ethics in artificial intelligence. *AI and Ethics, 1*, 173–181.

Bakshy, E., Messing, S., & Adamic, L. A. (2015). Exposure to ideologically diverse news and opinion on Facebook. *Science, 348*(6239).

Bobo, L. D. (2017). Racism in Trump's America reflections on culture, sociology, and the 2016 US presidential election. *The British Journal of Sociology, 68*(S1). https://doi.org/10.1111/1468-4446.12324

Boening, A. (2014). *The Arab spring: Re-balancing the greater Euro-Mediterranean?* Springer.

Bracci, A. et al. (2021). Dark Web Marketplaces and COVID-19: before the vaccine. *EPJ Data Science, 10*(6). https://doi.org/10.1140/epjds/s13688-021-00259-w

Bryan, R., et al. (2014). Social media in the 2011 Egyptian uprising. *The British Journal of Sociology, 65*(2), 266–292.

Cath, C. (2018). Governing artificial intelligence: Ethical, legal and technical opportunities and challenges. *Philosophical Transactions Royal Society A (mathematical, Physical and Engineering Sciences), 376*, 20180080. https://doi.org/10.1098/rsta.2018.0080

Choucri, N. (2012). *Cyberpolitics in international relations.* The MIT Press.

Choudhary, A., et al. (2012). Social media evolution of the Egyptian revolution. *Communications of the ACM, 55*(5), 74–80.

Cohen, D., & Wells, J. (2004). *American national security and civil liberties in an era of terrorism.* Palgrave Macmillan.

DeLisle, J. (2020). Foreign policy through other means: Hard power, soft power, and China's turn to political warfare to influence the United States. *Orbis, 64*(2), 174–206.

Druckman, J. N., et al. (2021). Affective polarization, local contexts and public opinion in America. *Nature Human Behaviour, 5*, 28–38.

Efthymiopoulos, M. P. (2019). A cyber-security framework for development, defense and innovation at NATO. *Journal of Innovation and Entrepreneurship,* 8(Article number: 12). https://doi.org/10.1186/s13731-019-0105-z

Eurasia Group. (2018). *Top 10 risks.* https://www.eurasiagroup.net/issues/top-risks-2018

Fierke, K. M. (2012). *Political self-sacrifice: Agency, body and emotion in international relations.* Cambridge University Press.

Freelon, D., Marwick, A., & Kreiss, D. (2020). False equivalencies: Online activism from left to right. *Science, 369*(6508), 1197–1201.

Fukuyama, F. (1992). *The end of history and the last man.* Free Press.

Fukuyama, F. (2014). *Political order and political decay.* Farrar, Straus & Giroux.

Fukuyama, F. (2018). *Identity: Contemporary identity politics and the struggle for recognition.* Profile Books.

Giusti, S., & Piras, E. (Eds.) (2020). *Democracy and fake news: Information manipulation and post-truth politics.* Routledge.

Gonzalez-Bailon, S., & Wang, N. (2016). Networked disconnect: The anatomy of protest campaigns in social media. *Social Networks, vol.44,* 95–104.

Green, B. (2021). US digital nationalism: A Habermasian critical discourse analysis of Trump's 'Fake News' approach to the first amendment. In A. MacKenzie, J. Rose, I. Bhatt (Eds.), *The epistemology of deceit in a postdigital era: Postdigital science and education.* Springer.

Guice, J. (1998). Controversy and the State: Lord ARPA and intelligent computing. *Social Studies of Science, 28*(1), 103–138.

Habermas, J. (1976). *Legitimation crisis,* Heinemann

Hagendorff, T. (2020). The ethics of AI ethics: An evaluation of guidelines. *Minds and Machines, 30,* 99–120.

Hegel, G. W. F. (2018). *Phänomenologie des Geistes (The phenomenology of spirit)* (English Translated ed.). Cambridge University Press.

Hill, K. A., & Hughes, J. E. (1998). *Cyberpolitics: Citizen activism in the age of the internet.* Rowman & Littlefield Publishers.

Howard, P. N. et al. (2018). *The IRA, Social Media and Political Polarization in the United States, 2012–2018.* https://comprop.oii.ox.ac.uk/wp-content/uploads/sites/93/2018/12/IRA-Report.pdf

Jost, J. T., et al. (2018). How social media facilitates political protest: Information, motivation, and social network. *Advances in Political Psychology, 39*(1), 85–118.

Lamont, M., Park, B. Y., & Ayala-Hurtado, E. (2017). Trump's electoral speeches and his appeal to the American white working class. *The British Journal of Sociology, 68*(S1). https://doi.org/10.1111/1468-4446.12315

Le Bonn, G. (1894). *The psychology of revolution.* Binker North.

Linz, J. J. (1978). *The breakdown of democratic regimes: Crisis.* Johns Hopkins University Press.

Locke, J. (1998). *Two treatises of government.* Cambridge University Press.

Ministry of Defense, Japan. (2017). *Defense of Japan 2017* (White paper: English edition). Urban Connections, Japan.

Moore, Jr., B. (1993). *Social origins of dictatorship and democracy.* Beacon Press.

North, D. C. (2013). *Violence and social orders: A conceptual framework for interpreting recorded human history.* Cambridge University Press.

Nye, J. (2017). Soft power: the origins and political progress of a concept. *Palgrave Communications, 3,* Article number: 17008.

RAND Corporation. (2017). *Priority challenges for social and behavioral research and its modeling.* https://www.rand.org/content/dam/rand/pubs/working_papers/WR1200/WR1206/RAND_WR1206.pdf

Schelling, T. C. (1980). *The strategy of conflict.* Harvard University Press.

Shibuya, K. (2017). Bridging between cyber politics and collective dynamics of social movement. In M. Khosrow-Pour (Ed.), *Encyclopedia of information science and technology* (4th ed., pp. 3538–3548). IGI Global.

Shibuya, K. (2020c). A revaluation for the Morgenthau's prospects in the digital society. In *The International Workshop "Transdisciplinary Approaches to Good Governance"*. The Center for Southeast Asian Studies (CSEAS), Kyoto University, Japan.

Shibuya, K. (2020b). *Identity health,*. https://www.ncbi.nlm.nih.gov/pmc/articles/PMC7121317/. In K. Shibuya (2020). *Digital transformation of identity in the age of artificial intelligence.* Springer.

Shibuya, K. (2021). Breaking fake news and verifying truth. In M. Khosrow-Pour (2021). *Encyclopedia of information science and technology* (5th ed., pp.1469–1480). IGI Global.

Shibuya, K. (2020a). *Digital transformation of identity in the age of artificial intelligence.* Springer.

Sunstein, C. R. (2001). *Republic.com,*. Princeton University Press.

Turchin, P. (2005). *War and peace and war.* Plume.

Von Braun, J., et al. (2021). *Robotics, AI, and humanity.* Springer.

Wang, Y., Luo, J., Niemi, R., & Li, Y. (2017). To follow or not to follow: analyzing the growth patterns of the Trumpists on Twitter. In *The Workshops of the Tenth International AAAI Conference on Web and Social Media News and Public Opinion: Technical Report WS-16-18.*

Weber, M. (1958). The three types of legitimate rule. *Berkeley Publications in Society and Institutions, 4*(1), 1–11.

Weinberger, S. (2011). Web of war. *Nature, 471,* 566–568.

On Balance of Power

1 Global Tensions Beyond the Cold War

1.1 What Happened to Equilibrium and the "New World Order" Through International Cooperation After SARS?

In the past, Davies (2012) discussed the international coordination for international public health and medical systems and their new equilibrium relationship seen after the SARS pandemic outbreak in 2003 *"Post-Westphalianism"* and watched the trend closely. In particular, he pointed out that the outbreaks of SARS and H5NI pandemics originating from China and Southeast Asia, as well as the COVID-19 pandemic, were due to cultural and institutional differences. He said, *"Since the outbreak of Severe Acute Respiratory Syndrome (SARS) in 2003, there has been much discussion about whether the international community has moved into a new post-Westphalian era, where states increasingly recognize certain shared norms that guide what they ought to do in responding to infectious disease outbreaks"*, and he lastly concluded *"The H5N1 crisis showed that states acted as if they had a duty to promptly report disease outbreaks. Sovereignty was not employed as an excuse to avoid compliance by failing to report"*. Incidentally, Westphalianism refers to the balance of power among European countries that was achieved by the Treaty of Peace (1648) of the Thirty Years' War (1618–1648). Similarly, Fidler (2004) discussed the *"new world order"* of public health and medical system through international cooperation, based on the fact that China was the source of SARS as well as the COVID-19 pandemic, which led to a remarkable change in national governance and public health system.

However, the international cooperation and equilibrium that was seen after the SARS outbreak has collapsed, and it should be seen as beginning to shift again in search of a new equilibrium point. In fact, not only the relationship between the USA and China before and after the COVID-19 pandemic, but also the power relations among Europe, Russia, Japan, and emerging countries in Africa have completely changed to a degree that is incomparable to the After SARS era. This is due not only to infectious diseases, but also to the shifting equilibrium among nations caused by multiple factors such as the development of science and technology and economic power.

1.2   America is Back

President Trump's *"America First"* agenda and reckless policies led to the world. As such policies, by the end of 2020, there was not even a global leader who could lead and coordinate international affairs as a whole; there was even a cynical view of a G0 (zero) instead of a G7 or G8. This was because US-centrism had already come to an end and the position of global leader was vacant in the "After Corona" era. It was thought that China would comfortably take the vacant seat as the world leader.

However, in January 2021, the United States, since President Biden took office, has reversed almost all of the policies of the Trump administration and expressed its intention to lead the world in international cooperation, emphasis on Western alliances, responses for the COVID-19 pandemic, and defense of democracy ("America is back").

Similarly, in February 2021, at the Munich Security Conference 2021,[1] President Biden[2] expressed his recognition of the current state of democracy, which has been led by the USA and Europe, saying that "democracy is under attack around the world" and implicitly condemned China and Russia. He also made a clear distinction between the two positions of *"those who believe that tyranny is best"* and *"those who defend democracy as indispensable"* and said in his speech that he believes democracy should prevail. In addition, he clearly stated his intention to tackle trade friction with China, competition in science and technology, and the issue of nuclear facilities in Iran.

It means that the time has come to reexamine the structures of conflict, alliances, and balance of power among nations. The time has come to move again.

1.3   Balance of Power

"Que la bilancia non pendesse da alcuna parte". This phrase represented a Venetian' political strategy for their survivals balancing the power of the hegemony among them.

These power structures are still the same today. In a book "The twenty years' crisis 1919–1939" (1939), Carr who was one of the scholars of international politics based on realism criticized the system of international cooperation based on liberalism after the post-First World War era, and he argued that the essence of international society

---

[1] https://securityconference.org/en/msc-2021/.

[2] https://youtu.be/4tyUK2uk_D8.

is "power", that is, the rivalry between national powers. Military power, economic power, and power manifested in public opinion are the causes of conflicts between rivalry state powers.

In recent years, extreme right forces have also emerged in the West, and public opinion is prevalent as if the world situation had returned to the condition around the Second World War, and a kind of nationalism, ethnocentrism, or populism has been added to such chaotic atmosphere. In addition, post-truth or fake news related to them are being denounced. Those who support them believe the "facts" they want to believe and see only the "facts" they want to see (Bakshy et al., 2015; Shibuya, 2020a, 2021; Suntein, 2001). Thus, in the age of social media, there is a concern that the connections between their activities may become public opinion sometimes involving the entire nation and the world, and even more so, demonstrations that bring about large-scale social change (Shibuya, 2017). As noted earlier chapter, Morgenthau (1978) also indeed wrote to the effect of the Cold War era and to the effect of how citizens were embroiled in an all-out war in a frenzy as their nationalism and exclusionary attitudes intensified. In the current social media age, isn't that the same kind of atmosphere being fostered (Shibuya, 2020b)?

In this context, polarized opinion aggregation tends to become entrenched in emotionalism and ideological political beliefs. What's more, there is a concern that it will further detract from calm, objective, and scientifically verified decision-making. Indeed, the *balancer* is still important. A balancer is not always linked to the policies of a state or a group of states in either side, but it still retains influence (Morgenthau, 1978). Similarly, Henry VIII's proverb "*Cui adhaero praeest* ("He whom I support will prevail": Latin for "cling to rule") illustrates the purpose of maintaining own influence by taking the lead in alliances (Aref & Neal, 2020).

These are similar to elections in democratic country, where politics is influenced by such activities. This is also the case, for another example, when there are countries with overwhelming powers to veto such as the United Nations Security Council.

Those viewpoints can be formalized in game theory as an alliance and coalition formation that achieve a majority to become the winners through interacting public opinion among stakeholders (Neumann & Morgenstern, 2007). Entering into a relationship with a majority is called a winning alliance, while the opposite is called a losing alliance. The relationship between the winning alliance and the core in game theory can be assumed when the majority has a structure such as sweeping or dominating public opinion, a situation in which that structure is not overturned.

Next problematic concern what should be discussed is balance of power in actual alliances (Sheehan, 1996). It can regard a pattern of the alliance formations and becomes a power-holder for the majority. These problems in global politics had already become apparent before the pandemic of the COVID-19, but this pandemic has caused an accelerated intensification of confrontational structures. Moreover, military allies such as the EU and the USA are losing ground, while the military alliance between Japan and the USA has been becoming closer. On the contrary, multilateral trade and military cooperation with China (e.g., Russia, North Korea, etc.) have further complicated the international situation. And since the decline in the functioning of the United Nations has already been pointed out, it is likely that a new international relationship will be redefined and restructured in a new era. In this context, it can be said that the balance of power perspective and the significance of the balancer are also growing.

1.4   The Brinkmanship of the Third World War

In many cases, the essence of war and conflict resolution is simply to impose a greater burden on each other. Actually, the balance of power is a very crucial factor in international affairs (Levy, 1984). East-Asia and pan-pacific regions would be balancing sensitive condition between the super-powers. The same goes for the issue of European security with NATO at its core (Efthymiopoulos, 2019; Lucarelli et al., 2021; Menon & Ruger, 2020), and military cooperation between Russia and China (Lukin, 2020). Geopolitical risks may spur those military and diplomatic dynamics, which add up to serious tensions (Luttwak, 1990).

Please recall an actual fact. The brinkmanship of the Third World War between the USA and Iran at the beginning of 2020 endangered our daily living.[3] This stalemate situation can be formalized by a game model. At that time, their optimum strategy was to select mutually the cease-fire. However, it would not be balanced between them, because short-term and longer-term strategy of both conditions will converge to an advantage of the USA.

Iran:

- One of the most important generals was assassinated by the USA.
- They probably prepare own nuclear missiles. If it can be deployed, but not reach the territory of the USA homeland.
- They can just command their terrorists and guerrilla soldiers in underground.
- Economic sanctions by the USA and its alliance countries will continue stagnating domestic foundation of Iran.
- They probably intend to firmly keep an alliance with China and Russia to confront against their hardships.

USA:

- "There were no victims in our side". Either injured or broken military resources were almost ignoble conditions.
- They can also do additional economic sanctions against Iran.
- Their catastrophic missiles of ICBM and nuclear weapons can completely overwhelm Iran's homeland.
- They always hold a distinguished power in The United Nations.

Above all, Iran was forced to enter into a ceasefire with the USA by the sacrifice of a capable general. Contrarily, the USA has successfully overwhelmed them by own military power and international political position to nullify Iran's ability to respond

---

[3] https://trumpwhitehouse.archives.gov/briefings-statements/remarks-president-trump-iran/.

with only minimal cost and damage. Namely, final winner was the USA. The USA conducted their missions to be achieved by their strategic operations.

As *Vom Kriege (On War)*, which was written by Clausewitz (1993) also indicated, the *political* objective in war was especially reassessed. The smaller the sacrifices demanded of the enemy, the smaller their resistance, and then the smaller own effort and sacrifice. Similarly, the more dominant the political objectives are in war, the more meaningful these diplomatic efforts will become larger. In other words, in such points, before all out of war, they must commonly share the balance of forces as well as the accurate mutual recognition of the balance between "sacrifice" and "benefit".

There is also a theory that there is both justifiable justice and morality to war (Walzer, 2006). However, justice is claimed by both sides, and it is not worthwhile to judge the merits of the justice in the first place. As exemplified in the "principle of sacrifice" of the author, the sacrifices are ultimately most imposed on the relatively weakest position. The essence of war is the fact that the greater the difference in military forces on both sides, the more some sacrifice will be imposed unilaterally. Conversely, even if there is military inferiority, it can sometime obtain a result by diplomatic efforts. However, in this case between the USA and Iran, it strictly exhibited clear fact that the disparity between them in terms of military forces and diplomatic advantage of the USA in the United Nations, because the USA can veto a disadvantage proposal at the security councils (Ayson, 2004; Schelling, 1980). It is also possible that Richard Nixon's strategy based on his "Madman Theory" may appear. If there is no predisposition to trust each other, provocative policies between each other can only be a path to both ruins.

In addition, a question of the functions to be performed by the state, social morality, and sacrifice had already been discussed by Nozick (1974). He said *"They reflect the fact that no moral balancing act can take place among us; there is no moral outweighing of one of our lives by others so as to lead to a greater overall social good. There is no justified sacrifice of some of us for others. This root idea, namely, that there are different individuals with separate lives and so no one may be sacrificed for others, underlies the existence of moral side constraints, but it also, I believe, leads to a libertarian side constraint that prohibits aggression against another"*. It is true that different political systems (e.g., tyrannical, democratic, or libertarian) have different assumptions and constraints on what is right and fair. But even if such a political system is ideal for citizens, political decisions, especially in crises (e.g., conflicts between nations, defense issues, and other situations that affect the survival of citizens), may require ignoring the assumptions and constraints that make an ideal state possible. It is impossible to achieve the goals without making sacrifices. Therefore, it is equally difficult to discuss what the state is, what sacrifices citizens should make, and what sacrifices (e.g., obligations) they can undertake. One of the actions that humanity can take to prepare for a real crisis is balance of power, but its actual dynamics certainly reveals the nature of sacrifice in a crisis.

1.5   The Kant's Triangle for Perpetual Peace

Interdependence has one of the specific reasons in global politics. When both regional and global tensions lose its strong commitments against each counterpart, mutual trust cannot be sustained by provisional and longer-term perspectives.

Philosopher Kant (1795) contemplated how to achieve the perpetual peace in global, and he concluded that there are necessary factors for the peaceful condition such as *firm alliance, democracy, economic inter dependency* and *international institutions*. His model is often called as the *Kantian triangle of peace* (Fordham & Walker, 2005).

- Firm Alliances
- Democratization
- Economic Interdependence among Nations
- International Institutions (i.e., League of Nations, the United Nations)

In his context, it is often said that the League of Nations and the United Nations had been actualized by his proposal. Otherwise, Morgenthau indeed discussed that those mechanisms had own fatal deficits for building and sustaining the international peace conditions. It crucially requires keeping the balance of power, and mutual watching system for their interdependent alliance and economic foundations.

1.6   What is "Acceptance" of Sacrifices?

The perspective of justice and fairness in war has been debated many times over the years on war and peace. After the war, it is desirable to have a comprehensive perspective that includes future generations who were not involved in the war, not only those who were responsible for causing the war, but also how and whom to judge and how to compensate for the sacrifices made. For example, at the Paris Peace Conference in 1919 after the First World War, the economist Keynes argued in his book for a generous peace, not for justice and equity of the war, but for the economic development and stability of Europe as a whole (Keynes, 1919).

Also, regarding the Second World War, the defeated countries were miserable. The author was taught that its defeat and reconstruction were disastrous for Japanese history (Inoki, 2014). Germany as one of the defeated countries has also been described with similar devastation (Bresciani, 2006; Habermas, 1987). After the war, there have been repeated trials in various parts of Japan to seek reparations (i.e., for damage caused to a belligerent by an act of war) and compensations (i.e., for citizens who suffered damage due to acts of war), but the Supreme Court in Japan ruled against reparations, stating that "all citizens should equally accept the sacrifices of the war". In other words, the Supreme Court in Japan ordered citizens and victims to willingly accept the "sacrifice" of life and death. The sacrifices of the citizens and soldiers were thus buried in the depths of history with the word "acceptance", which was decided by the Supreme Court without question.

This is the so-called an "acceptance theory of war's sacrifices" in Japan. It had been pointed out for a long time that there was a disparity in the amount of compensation paid to former military personnel which was estimated to be 50 trillion yen, while civilians which were not almost compensated. From the perspective of the pacifism of the postwar Japanese Constitution, the issue of citizens' right to claim had been widely discussed. However, these debates had often been mixed up with ideological conflicts, justice and fairness of war (Walzer, 2006), and issues of reparation and compensation for various kinds of sacrifices such as loss and damage.

In addition, those who were born after 1945 were forced to make efforts in the postwar confusion and reconstruction, and even if the result brought unprecedented economic development, the lost national wealth and lives can never be restored. These precedents show that political decisions not only determine who makes "sacrifices", but also do nothing to save the hopes of each individual for a peaceful life.

Moreover, after the war, not only would many reparations and appeals be pursued by the countries involved, but the brunt of these appeals would also fall on future generations who had not even been born. In this regard, it is worth mentioning that Chiang Kai-shek prohibited any retaliatory actions against the Japanese people living in China immediately after the end of the war, and he said "Respond to grudges in good faith rather than retaliation". This phrase was originally based on the words of "Lao Tzu" (Laozi). As this classic of oriental thought teaches, the attitude of approving the nature of peace as an overall common good in war shows the magnanimity of the ruler. Such an attitude was far superior to that of the rulers who had caused their own disasters let alone saved the country. Rather than a state of conflict in which simple retaliation is repeating, generous measures that suggest the higher rationality of the other side will induce them to a state of peace, which is the common good, thereby avoiding future conflicts and comforting the feelings of their citizens. If the other side still refuses to change their attitude, there will be no shortage of ethical and political reasons to fight back, which will only increase one's own justice. It suggests a more fundamental solution than the retributive principle of tit-for-tat (TFT), which is the best solution in repeated games in game theory. An equilibrium relationship should equally consider the emotions and survival status of the citizens of the opposite country.

As it can see, one of the fundamentals of Eastern thought is the coexistence of self and the others. On the other hand, Western thought emphasizes the differences and distinction between self and the others, which may be a cause of miscommunication on the issue of peace and balance of power among the nations based on the different cultural backgrounds.

To be sure, war begins as a battle that each of us continues to be alive every day. It is not enough to win the war, but it would be better if each stakeholder and each country could coexist by creating resources and opportunities and by accommodating each other through means other than war, but that is not always the case. Rather, the problem is what to do in such cases. There is no way for humanity to escape such social evils as the emergent properties are caused by existential issues based on a distinction between self and the others. War is not a dichotomy of "win or lose". As Sun Tzu explained, "it is better if it does not lose", and a state of the "equilibrium as balance of power" is more acceptable. It is preferable if the damage caused by physical confrontation and conditions of war occurrence can be nullified, whether through avoidance or elimination of the problem itself, or through diplomacy or information manipulation. Therefore, an "equilibrium" condition must be prioritized particularly in international affairs. Thus, next, the question of what means balance of power and peace must be reexamined (Neff, 2012; Tuck, 2001).

1.7   Peaceful Situations or Not

The author further contemplates such matters as following. The United Nations firstly accepts the existence of the others, and then social institutions and laws are needed to regulate social evils as an emergent property. In short, the raison d'être of a world organization like the United Nations is definitely not to eliminate any warfare and conflict as a social evil altogether. Rather, it is about minimizing those sacrifices caused by those conflicts. Given accepting the existence of the others and the existence of other nations, the elimination of warfare conducted by the United Nations is indeed impossible. If and only if all wars and conflicts arise when two or more stakeholders simultaneously coexist. In other words, it is just a social evil as an emergent property.

The author categorized that the search for a point of mutual equilibrium or disequilibrium between those structural stability and interests reach the peaceful conditions or not. Namely, there are at least four types of the conditions as Table 1.

Above a relation between the USA and Iran suggested "Suppressed structure", and it apparently shows a peace at first glance. But their severe relation as the unbalance of power between them keep such situation. In other case, a relation between the USA and China can be regarded as "Ongoing competitive relationship as a rival" or

Table 1 Conceptual category of peace or related issues		Stability of the structure	Instability of the structure
	Equilibrium in balance of powers	**[A]** Peace as equilibrium Peace as cold war	**[C]** Ongoing competitive relationship as a rival
	Disequilibrium inbalance of powers	**[B]** Peace as subordination and servitude Peace by following to the stronger side Suppressed structure	**[D]** Unpredictable and chaotic situations

"Peace as Cold war". In Japan, it might still stay "Peace by Following to the Stronger side" with the alliance of the USA (Ministry of Defense, Japan, 2021). In such way, the United Nations plays own role to avoid and solve "Unpredictable and chaotic situations".

To put it another way, this table serves another viewpoint. First, dynamic relations among [A], [C] and [D] quadrants actually are easier transited across quadrants. For example, if every country has almost equally own power ([C] or [D]), each of them cannot initially overwhelm each other, but it can also fight each other in the same condition. It exemplifies the Warring States period of China (BC 5 Century-221 BC) that seven states continued to conquer each other, gradually, weaker nations were yielded by stronger ones, and finally, one state defeated the other rivals and synthesized totally ([A]). Then, such situation definitively further requires structural stability to institutionalize and organize all of them, such as entire national legitimacy (e.g., an empire over the nations), agreement for the peace, alliances among them, and other ways.

However, as in Spinoza's article "Tractatus Politicus", there is a point of view that a state in which citizens are apathetically subjugated to other nations as peace is not a state. He also said "Peace is not mere absence of war, but is a virtue that springs from force of character", and the depth of peace is sought in the individual's adherence to social morality. Thus, the difficulties of maintaining peace and at the same time protecting both freedom and sovereignty of the state and individual citizens are always a problem.

Secondly, for another example, any organizations and business company can be similarly observed in those quadrants. Especially, in the quadrant [C], their competitive business scenes and market-oriented economy usually inspire them to survive with some constraints (e.g., legal procedures, taxation, compliance). Innovative idea and its products will overwhelm the markets by through those dynamics. But such environment can be named as the battlefields, and it can paraphrase such situation as a try and error process to determine the most competitive company of the business, and the others will be handled as the sacrifices or losers. On the contrary, [B] quadrant can be regarded as a condition such as socialism and somewhat controlled economy by the national government, and [A] quadrant can never be probably observed in the human history because such situation may be an imaginary perfect market. Similarly, a concept of competitive advantage in the strategic management of MBA (Master of Business Administration) has been rooted in a competitive business environment (Porter, 1998). His five forces analysis should be well examined each element by subdividing such as Threat of new entrants, Threat of substitutes, Bargaining power of customers, Bargaining power of suppliers and Competitive rivalry.

2   Computational Modeling for Balance of Power

2.1   How to Model a Social Simulation on Balance of Power?

As discussed previous chapter, computational social science, which examined by modeling and computer simulation, includes sociology (Caforio, 2018), political science (Voinea, 2016), international issues, economics, organization theory, network analysis, cliodynamics (Turchin, 2005, 2009), and artificial intelligence research (Shibuya, 2020a). It often examines many problems in the social sciences by agent-based modeling and simulation. For example, such problems are addressed through simulations of artificial societies and artificial economies (Coleman, 1990; Epstein, 2014; Gilbert & Troitzsch, 1999). In addition, the analysis and application of human behavior in ubiquitous environments can also be conceptualized (Shibuya, 2004a). While the former is more theoretical and experimental in nature, the latter is mainly about quantitative analysis and modeling of crowds and human behavior in the real world.

Figure 1 generally shows a cycle of the entire process for computational modeling and simulation in social science (Shibuya, 2004a, 2004b). Social sciences ostensibly deal with unrepeatable historical events, but mathematical abstraction can formalize the common rules by revealing necessary factors. Principally, any simulation requires conceptualization to unveil the nature of any phenomena in the actual world. Secondly, design process occupies in the core of modeling. This part aims designing a practical procedure by trying much creative minds. Third, implementation process has to be developed by own programming skills. Using one of the programming languages, we can build simulation programs for the solutions running in the computers. Fourth, analysis and examination process can be understood as a data-driven measurement for the results and clarification on hypothesis. Fifth, evaluation process employs to interpret meaning of the consequences in depth. Finally, sixth point will be cyclically feed-backed toward the actual worlds, and we can obtain several parts of the new knowledge and experiences by the simulations. As necessary, it can repeat a whole process from the conceptualization step.

In this way, simulation can be categorized as one of research methods in social sciences.

1. System Modeling: It is especially found in the studies of computational system and engineering.
2. Equation-based Modeling: It is mathematical formalization and its examination (e.g., an appendix of Chap. 4 in this book).
3. Agent-based Modeling: It pays attention to each object and entity in the world, and it practically configures the relationships and interdependent activities among them. In social science, the most frequent cases are underlying in our personal attributions and behaviors.

In many cases, social scientists are still needed to implement own models using programming languages. For example, Java, C++, Python, R, and others. Moreover, mathematical, statistical, and computer scientific skills are still quite important place for our studies. Data-centric learners will be able to take advantages for such solutions. For practical simulation studies, those who intend to model are required various knowledge and much experience. If standing on the social science, but we shall learn much knowledge and subjects as follows, at least.

1. Comprehensive knowledge in human and social sciences
2. Mathematics and statistics
3. Computer sciences
4. Data sciences
5. AI
6. Physics
7. Engineering
8. Biological inspired mechanisms (e.g., genetic algorithm).

2.2 Models of Opinion Dynamics

In general, opinion dynamics models allow for situations such as time variation in influence between opinion groups, collective decision making, and manipulation of public opinion such as fake news (Ciamapglia, 2018; Vosoughi et al., 2018). For example, there is a need to clarify the formation of opinion groups and the splintering of the real world using social media (Druckman et al., 2021). Especially, in this chapter, the author exemplifies own discussion on international affairs (Fellman et al., 2020; Heintze & Thielborger, 2016) by means of computational simulation (Axelrod, 1984, 1997; Deutschmann et al., 2020; Schelling, 1981).

Here, the author firstly reviews several models on opinion dynamics. And next, it details an introduction on Shapley-Shubik index, and taken together, the author demonstrates an example of social simulation on balance of power using them.

2.2.1 Watts Model

Watts and Dodds (2007) found that, as exemplified by online viral marketing, specific opinions tend to diffuse when clusters of not only one opinion leader but also others who positively support them form in relation to advertising and publicity.

These findings are consistent with previous work by a sociologist Granovetter (1978) in relation to the threshold model (MacCoun, 2012) and percolation phenomena (Malarz & Galam, 2005). In the early stages of the diffusion process, the social fact that is for even a very small amount of support is itself significant, suggesting the possibility that the surrounding perceptions will progress by leaps and bounds and consequently spread throughout society. In other words, these threshold and percolation models can be described as a conformity type of collective action, in which a cluster of supporters develops in a self-organized way to become a majority.

2.2.2   Lattice Model

There is a study based on a lattice model regarding the process of public opinion formation (Chen & Morimoto, 2015). This is essentially based on the percolation model of Malarz and Galam (2005) and other threshold models. Chen examined the equilibrium and dynamics of public opinion by assuming the following agents. The balancer in this model refers to whether or not an agent adopts "*attuned behavior with its surroundings*". It certainly affects a large group of opinions, as how easily opinions fluctuate in society as a whole depends on how much of this is present.

(1)   *Floater*: This type of agents changes an opinion easily. A sort of floating vote. It is also possible to establish a setting that is easily influenced by the opinion trends of the opinion group to which it belongs.
(2)   *Inflexible*: This type of agent does not fundamentally change the opinion.
(3)   *Balancer*: Like *Floater*, this types of agent changes its opinion in the direction of following the Majority, but only in situations where there is no *Inflexible* around it. A variety of settings should be possible, including the compatibility with the opinion groups to which they belong.

The simulation process differs depending on how much the agents from (1) to (3) above are configured in the initial state and how the spatial arrangement differs. It may be possible to simulate a modified version of the model by introducing an agent that plays the role of an extremist in the model of Deffuant et al. (2000).

2.2.3   Helbing Model

Principally, Helbing (2012) proposed a mathematical model on opinion dynamics, which underlies in the theory of integrating forces inspired from sociologist Durkheim. In this simulation, each agent has incentives to adapt in specific opinion which weighted means ($O_{j(t)}$) at the time interacting with other agents. Lastly, this progress converges to a single opinion among agents, and it organizes homophily clustering. When an opinion of agent is more influential than other agents, opinion distance (d_{ij}) will minimize its value as following formula.

$$W_{ij}(t) = e^{-d_{ij}(t)/A} = e^{-|o_j(t) - o_i(t)|/A} \qquad (1)$$

Where, variable alpha denotes a range of influential power of each agent. When this value is smaller, and it means that this agent strongly commits to specific opinion. But if not so, as Durkheim's theory presumes, each agent can alter to another opinion. And $\zeta_{i(t)}$ defines white-noise, and it plays a role of disintegrating disturbance factor in the dynamics.

$$\Delta o_i = \frac{\sum_{\substack{j=1 \\ j \neq i}}^{N} (o_j(t) - o_i(t))w_{ij}(t)}{\sum_{\substack{j=1 \\ j \neq i}}^{N} w_{ij}(t)} + \zeta_i(t) \tag{2}$$

2.3 Shapley-Shubik Power Index

Next, the author would like to detail Shapley-Shubik power index (hereafter, it abbreviates Shapley index) (Shapley & Shubik, 1954).

$$\varphi = \sum s/C \tag{3}$$

In the coalition formation game situations, Shapley index (φ) can be calculated by aligning an arbitrary element from a set $\varphi = \{x_1, x_2 \ldots x_n\}$ (Eq. 3). When specific criterion (e.g., majority) firstly exceeds, such element is called as a *pivot*. Such pivot can be regarded gaining a *casting vote* to establish a winner's coalition formation among them, and then it holds more influential power than the others. All combinations (C) to become winners' coalition are determined in the factorial order (mathematically, it notates N!). In short, this index is defined as each total counts (s) per a person to be a pivot among the all combinations (C). If each influential power is considered to hold a voting weight of each individual at the poll, by accumulating all votes of each individual, when total voting counts firstly exceed its criterion for the majority, such person becomes a pivot among them.

For example, when three persons (a, b, and c) exist in a society, there are six combination patterns ($3! = 3 \times 2 \times 1 = 6$). If the threshold can be determined by the majority, when everyone (in this case a, b, and c) has *equal* weights, a winner's coalition must be organized by more than two persons. In other words, under these conditions, the second person to participate is always the pivot.

However, at Table 2, when a case of *unequal* weights per an individual is a = 2, b = 1, and c = 1, total sum is four (=2 + 1 + 1), and then majority criterion of this case is three (>= 2 / 4). Further, each chance to be a pivot among above three persons, a = 4/6 (person(a) has four chances to be a pivot in all combinations (i.e., bac, bca, cab, cba)), whereas b (abc) and c = 1/6 (acb), and then Shapley index (φ) respectively indicates each value 4/6, 1/6, and 1/6. Namely, person(a) relatively has the strongest influence and power than the others.

Table 2 Character with bold and underline means a pivot among them

a*b*c	a*c*b	b*a*c
bc*a*	c*a*b	cb*a*

However, it should pay much attention to calculation performance and time consuming to complete an assigned task by a computer, because calculation time always depends on the factorial order (N!), and then such *combination explosion* usually calls a NP hard problem (i.e., Non-deterministic Polynomial-time hardness). For example, when the author uses a Windows OS installed computing environment (e.g., Intel Core i7 2.7 GHz, 6G RAM), but it can hardly calculate a Shapley index more than $N = 7$ ($5040 = 1 \times 2 \times 3 \times 4 \times 5 \times 6 \times 7$). Therefore, higher performance computation for the goal seeks is ordinarily required. Certainly, a quantum computer may promptly solve such NP hard problem in the future, but at the present, it is hard to complete such combination matters.

In the other example cases, the author shows some calculation results as follows.

- $N = 4$, weight matrix of each individual [A $= 1$, B $= 2$, C $=$ **4**, D $= 3$]:
  Person C becomes a pivot ($\varphi = 0.583$).
  $4! = 24$, $\varphi = [0.083, 0.083, \mathbf{0.583}, 0.250]$.
- $N = 5$, weight matrix of each individual [A $= 1$, B $= 2$, C $=$ **4**, D $= 3$, E $= 1$]:
  Person C becomes a pivot ($\varphi = 0.416$).
  $5! = 120$, $\varphi = [0.083, 0.116, \mathbf{0.416}, 0.250, 0.083]$.
- $N = 6$, weight matrix of each individual [A $= 1$, B $= 2$, C $=$ **4**, D $= 3$, E $= 1$, F $= 1$]:
  Person C becomes a pivot($\varphi = 0.383$).
  $6! = 720$, $\varphi = [0.066, 0.183, \mathbf{0.3833}, 0.233, 0.066, 0.066]$.

These results are an intrigued dynamics for the balance of power. When these cases are assumed as a voting condition at the congress, these situations imply that tyrannic person (person C) can influence to the others in each case ($\varphi = 0.583, 0.416, 0.383$). Such person monopolizes the influential power, and then he definitely decides all consequences of the voting whether he is pro or con.

In the actual cases, these cases resemble the security council of the United Nations (UNSC). Five permanent members (i.e., the USA, the UK, France, Russia, and China) have the strongest power than other ordinary countries, because their right of veto can crucially decide an invalid against any issues as necessary.

2.4 Simulation of Dynamics on Balance of Power

Next, the author demonstrates one of the social simulations based on above Shapley-index and opinion dynamics models of Helbing's study. It assumes that four group members interacts with each other, and each of them also sought to obtain a position of balancer for the majority. In this context, a balancer still retains the ruling controls for the entire decision on the bases of the majority rule. It exemplifies results ran by agent-based simulation as following below configurations.

Basic Configuration:
Opinion Groups: four groups (A, B, C, and D)
Agents: 10,000

Each agent must be belonged and committed to one of the opinion groups. And each of them cannot alter belonged group, but can periodically update its committed weight on opinion interacted by other agents. In each phase, each agent polls a vote pro or con within belonged group. Each opinion group accumulates all opinions among belonged members to converge integrating an entire group opinion. After that, totally, it checks to become a pivot and calculate a Shapley index among four groups.

Weight Matrix: It is initially configured in each opinion group, and it means a latent power of each opinion group to the others. Weight matrix, further, can be determined as *equal* or *unequal* conditions.

Pivot: During simulation, at each phase, a pivot among four groups can be determined.

Shapley index: Before and at the end of simulation, it is strictly measured.

Influential Power: As opinion of each agent (O_i) is laid specific weight on Shapley index in each opinion group, and such intensified attitudes have more influence against the other groups. And then, the highest Shapley index among groups can be likely assumed to become a pivot, and such opinion group seems to occupy a position of balancer among them.

Above graphs (Figs. 2 and 3) illustrate one of the examples results in the influence of each opinion group and Shapley index. Due to the influence of constant white noise, it can see the trends of power relations fluctuating in real time. It can be considered to assume trends in various public opinion trends, support rates between political parties, or trends in the balance of countries that discussed earlier.

3 Discussion

This time, the author simply thought of the "weight of opinion" or "number of votes" that one person has. However, it can also think of it as the relative superiority of "military power", "national power" or "R&D advancement". It is easy to imagine that uneven influence among stakeholders can be a decisive difference in specific situations. For example, suppose that the military power of the USA is a = 5, Japan b = 1, and South Korea c = 1, while China d = 4 and North Korea e = 2. Here, there is almost an equilibrium between the alliance with the USA as the core (a + b + c = 7) and the alliance with China as the core (d + e = 6). Here, it means a serious situation which side Australia (f = 2), Taiwan (g = 1), and Russia (h = 4) take, and how they participate. And this shows that they (f, g, and h) occupy a position of balancer.

Moreover, the UK, which has achieved Brexit, has clearly stated its intention to actively participate in the Pacific region, and has even announced the dispatch of an aircraft carrier. The participation of the UK, which is also one of the permanent five, in the Pacific region becomes greater meanings on military and international political contexts.

In addition, at 2020, Japan and other countries (France, Germany, and South Korea) joined the UK-USA Agreement (UKUSA, *"Five Eyes"*), a framework for

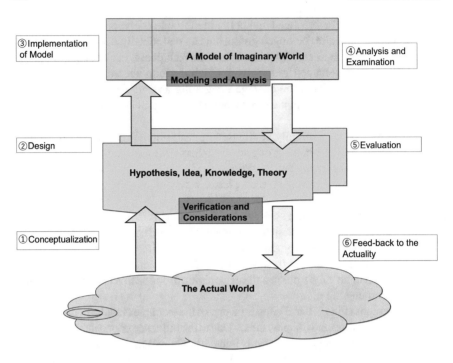

Fig. 1 Simulation process

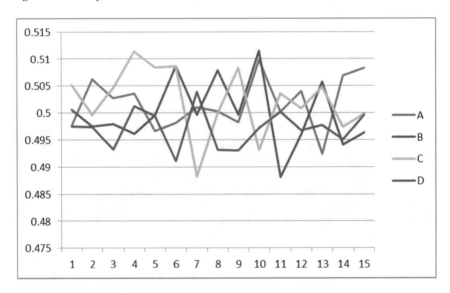

Fig. 2 Average of each opinion group in unequal condition (variable O)

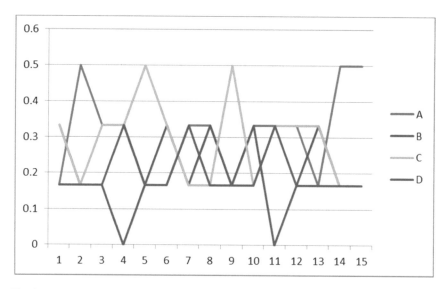

Fig. 3 Shapley index of each opinion group in unequal condition

sharing classified information among the five English-speaking countries of the USA, the UK, Canada, Australia, and New Zealand, and their military cooperation is becoming even closer.

Of course, the degree of relative superiority among these nations will also fluctuate if they significantly increase their military power by increasing the production and deployment of nuclear weapons, or developing AI and other technologies as military weapons. Hence, the balance among these countries and other alliances will determine the stability of the Asia–Pacific region (e.g., QUAD (*Quadrilateral Security Dialogue*) among the USA, Japan, Australia, and India, AUKUS among Australia, the UK, USA, and D10 Strategy Forum among G7 with South Korea, Australia, and India).

In similar, China's space strategy, which aims to disable, weaken, or shut down the powerful USA military by destroying military and communication satellites in space, led the USA military to establish a space force, and Japan and European countries are following similar strategies. Because of the digital age, the importance of information networks, information security, deployment of AI weapons, and defense of digital communication capabilities have come to occupy a pivotal position in national defense. This outstanding strategic perspective by China clearly shows the effectiveness of tactics that seek to nullify the act of war itself by destroying or disrupting "*information*", which is essential to the conduct of war. As discussed in previous chapters, the possibility of a realignment of the balance of power among the nations through the use of sharp power and related information technology is also increasing.

Therefore, such concrete examples show that the Shapley index and the balance of power perspective are quite effective, and mathematical modeling is very important.

4   Conclusion

This chapter mainly detailed mathematical processes on the balance of power. The balance of power is a sort of the social dynamics in the interaction of multiple stakeholders. Whether it is an iteration among individuals or nations, as long as the conflict of opinion and military power stay vividly, the dynamics among those who desire to occupy an influential position will oscillate their balance of power. Every time it fluctuates in human history into either an equilibrium or a non-stationary state, such crisis repeatedly calls further tragedy.

References

Aref, S., & Neal, Z. (2020). Detecting coalitions by optimally partitioning signed networks of political collaboration. *Scientific Reports, 10*, Article number: 1506. https://www.nature.com/articles/s41598-020-58471-z

Axelrod, R. (1984). *The evolution of cooperation*. Basic Books.

Axelrod, R. (1997). *The complexity of cooperation agent-based models of competition and collaboration*. Princeton University Press.

Ayson, R. (2004). *Thomas Schelling and the nuclear age: Strategy as social science*, Routledge

Bakshy, E., Messing, S., & Adamic, L. A. (2015). Exposure to ideologically diverse news and opinion on Facebook. *Science, 348*(6239).

Bresciani, C. (2006). *The economics of inflation—a study of currency depreciation In Post War Germany (English Edition)*. Hesperides Press.

Caforio, G. (2018). *Handbook of the sociology of the military*. Springer.

Carr, E. H. (1939). *The twenty years' crisis, 1919–1939: An introduction to the study of international relations*. Macmillan.

Chen, T., & Morimoto, J. (2015). *Balancer effects in opinion dynamics*. https://arxiv.org/pdf/1507.07339

Ciamapglia, G. L. (2018). Fighting fake news: A role for computational social science in the fight against digital misinformation. *Journal of Computational Social Science, 1*, 147–153.

Clausewitz, C. (1993). *On war*. Everyman (Translated in English).

Coleman, J. S. (1990). *Foundations of social theory*. Berknap Press of University of Harvard Press.

Davies, S. E. (2012). The international politics of disease reporting: Towards post-Westphalianism? *International Politics, 49*, 591–613.

Deffuant, G., Neau, D., Amblard, F., & Weisbuch, G. (2000). *Mixing beliefs among interacting agents*, Advances in Complex Systems,vol.3, 87–98.

Deutschmann, E., et al. (2020) *Computational conflict research*. Springer.

Druckman, J.N., et al. (2021). *Affective polarization, local contexts and public opinion in America*, Nature Human Behaviour, vol.5, 28–38.

Efthymiopoulos, M. P. (2019). A cyber-security framework for development, defense and innovation at NATO. *Journal of Innovation and Entrepreneurship, 8*, Article number: 12. https://doi.org/10.1186/s13731-019-0105-z

Epstein, J. M. (2014). *Agent_Zero: Toward neurocognitive foundations for generative social science (Princeton Studies in Complexity)*. Princeton University Press.

Eurasia Group. (2018). *Top 10 risks*. https://www.eurasiagroup.net/issues/top-risks-2018

Fellman, P. V., et al. (2016). *Conflict and complexity countering terrorism, insurgency, ethnic and regional violence*. Springer.

Fidler, D. P. (2004). *SARS*. Palgrave Macmillan.

Fordham, B. O., & Walker, T. C. (2005). Kantian liberalism, regime type, and military resource allocation: Do democracies spend less? *International Studies Quarterly, 49*, 141–157.

Gilbert, N., & Troitzsch, K. G. (1999). *Simulation for the social scientist*. Open University Press.

Granovetter, M. S. (1978). Threshold models of collective behavior. *American Journal of Sociology, 78*(6), 1420–1443.

Habermas, J. (1987). *Eine Art Schadensabwicklung*. Nachdruck.

Heintze, H. J., & Thielborger, P. (2016). *From cold war to cyber war the evolution of the international law of peace and armed conflict over the last 25 years*. Springer.

Helbing, D. (2012). Social selforganization, Springer

Inoki, T. (2014). *A history of economics*. Chuokhoron-Shinsha (in Japanese).

Kant, I. (1795). *Perpetual peace*. Macmillan (English translated edition).

Keynes, J. M. (1919). *The economic consequences of the peace*. Binker North.

Levy, J. S. (1984). The offensive/defensive balance of military technology: A theoretical and historical analysis. *International Studies Quarterly, 28*(2), 219–238.

Lucarelli, S., Marrone, A., & Moro, F. N., (Eds.) (2021). *NATO Decision-Making in the Age of Big Data and Artificial Intelligence*, NATO HQ-Boulevard Leopold III, https://www.iai.it/sites/default/files/978195445000.pdf

Lukin, A. (2020). The Russia-China entente and its future. *International Politics*. https://doi.org/10.1057/s41311-020-00251-7

Luttwak, E. N. (1990). From geopolitics to geo-economics: Logic of conflict. *Grammar of Commerce, the National Interest, 20*, 17–23.

MacCoun, R. J. (2012). The burden of social proof: Shared thresholds and social influence. *Psychological Review, 119*(2), 345–372.

Malarz, K., & Galam, M. (2005). Square-lattice site percolation at increasing ranges of neighbour bonds. *Physical Review E, 71*, 016125.

Menon, R., & Ruger, W. (2020). NATO enlargement and US grand strategy: A net assessment. *International Politics, 57*, 371–400.

Ministry of Defense, Japan. (2021). *Defense of Japan 2021(Annual White Paper)*. https://www.mod.go.jp/j/publication/wp/wp2021/pdf/index.html

Morgenthau, H. J. (1978). *Politics among nations: The struggle for power and peace*. Knopf.

Neff, S. C. (2012). *Hugo Grotius on the law of war and peace*. Cambridge University Press.

Neumann, J., & Morgenstern, O. (2007). *Theory of games and economic behavior*. Princeton University Press.

Nozick, R. (1974). *Anarchy state and Utopia*. Basic Books.

Porter, M. (1998). *On competition*. Harvard Business School Publishing.

Sandler, T., & Hartley, K. (2007). *Handbook of defense economics: Defense in a globalized world*. North Holland.

Schelling, T. C. (1981). *The strategy of conflict*. Harvard University Press.

Shapley, L. S., & Shubik, M. (1954). A method for evaluating the distribution of power in a committee system. *American Political Science Review, 48*, 787–792.

Sheehan, M. (1996). *The balance of power: History and theory*. Routledge.

Shibuya, K. (2004b). Perspectives on social psychological research using agent based systems. *Studies in Simulation & Gaming, 14*(1), 11–18 (in Japanese).

Shibuya, K. (2017). Bridging between cyber politics and collective dynamics of social movement. In M. Khosrow-Pour (Ed.), *Encyclopedia of information science and technology* (4th ed., pp. 3538–3548). IGI Global.

Shibuya, K. (2020b). A revaluation for The Morgenthau's prospects in the digital society. In *The International Workshop "Transdisciplinary Approaches to Good Governance"*. The Center for Southeast Asian Studies (CSEAS), Kyoto University, Japan.

Shibuya, K. (2021). Breaking fake news and verifying truth. In M. Khosrow-Pour (2021). *Encyclopedia of information science and technology* (5th ed., pp. 1469–1480). IGI Global.

Shibuya, K. (2004a). A framework of multi-agent based modeling, simulation and computational assistance in an ubiquitous environment. *SIMULATION: Transactions of the Society for Modeling and Simulation International, 80*(7), 367–380.

Shibuya, K. (2020a). *Digital transformation of identity in the age of artificial intelligence.* Springer.

Sunstein, C. R. (2001). Republic.com, Princeton University Press.

Tuck, R. (2001). *The rights of war and peace: Political thought and the international order from Grotius to Kant.* Oxford University Press.

Turchin, P. (2005). *War and peace and war.* Plume.

Turchin, P. (2009). A theory for formation of large states. *Journal of Global History, 4*(2), 191–217.

Voinea, C. F. (2016). *Political attitudes: Computational and simulation modelling* (Wiley Series in Computational and Quantitative Social Science), Wiley.

Vosoughi, S., Roy, D., & Aral, S. (2018). The spread of true and false online. *Science, 359*, 1146–1151.

Walzer, W. (2006). *Just and unjust wars: A moral argument with historical illustrations* (4th ed.). Basic Books.

Watts & Dodds. (2007). Influentials, networks, and public opinion formation. *Journal of Consumer Research.* http://www.uvm.edu/pdodds/teaching/courses/2009-08UVM-300/docs/others/2007/watts2007a.pdf

Wu, X. (2020). Technology, power, and uncontrolled great power strategic competition between China and the United States. *China International Strategy Review, 2*, 99–119.

For Strategies in the Age of After Corona

1 Power Shift from Westness to Eastness as a Hegemony of the Nation?

1.1 New Chaos or Order

It may be true that an outbreak of the COVID-19 reshaped our established paradigms such as policy, economy, and socio-cultural aspects, which founded and rooted in the globalism and capitalism (Held et al., 2000) to the renewal world, which may be driven by the AI systems and rewarded by its algorithm. What an ideology can the humanity drive toward the next generation? Will American hegemony be shifted to a hegemony of Chinese centered Orientalism in nearly future? Updating the social system itself may not solve the crises. And *sacrifices* corresponding to each crisis must be reconsidered.

As mentioned in earlier chapters, at front of the globally changing, it is time that a balance of power between *Westness* and *Eastness* should be intensively investigated by multiple perspectives. As of 2020, the conflict between the USA and China has become a serious. Tensions between them continued to be higher in the economic and political spheres before the pandemic (Zhao, 2016), but furthermore, the likelihood of warfare becoming a reality has increased (DOD, 2020). In the After Corona world, there is a deep-rooted argument that the permanent members of the United Nations should be inevitably restructured, not to mention the restructuring of the framework for international cooperation. In other words, it is believed that there will be one of the power-shifts towards a bipolar structure with a frame of international cooperation with the USA, Europe and Japan at its core, and another axis state with China and Russia at its core (Menon & Ruger, 2020; Lukin, 2020). The crucial criticism of China's despotism that has led to such a global pandemic is, of course, a growing hostility towards anti-democratic states is no longer the only means left for all-out warfare. It will bring about a state of international tension and conflict beyond the

Cold War era, with the Eastness (e.g., China) and Westness (e.g., the USA and Europe) structure of conflict becoming apparent.

Their relationship can be seen as one of *"The Thucydides Trap"*. Endless chicken game of their balance of power between them will lastly reshape a new world order, and consequentially it may reorganize Eastern hegemony of China by the power shift. However, China has not presented to the international community a democratic-like global principle and trustful leadership that brings together a pluralistic set of the multiple values yet.

Here, regarding such situations, who has the most to gain from the USA and China competing with each other? Russia? Is it North Korea or Iran? Which country is required to play the most important role in a situation where the USA and China are competing? Japan? What will be Japan's position in such a situation? And to which power will the rest of Asia, Africa and Latin America join? Or will a third alliance be formed to counter them?

In brief, the possible route for Japan will be enumerated as follows, for example.

(1)   [Alliance with the USA] Japan will steadily continue to be a partner of a coalition with the USA and Europe at its core.
(2)   [New Alliance with China] Japan alternatively changes to a coalition with China and Russia at the core.
(3)   [The Third powers' alliance] Japan centrally forms a third pole and oppose the two major powers.
(4)   [Neutral Position] As a neutral country, Japan will contribute to peace at homeland and in the world as a whole.

In short, is there a post-Corona global depression? Will there be a post-Corona world warfare? And how will the probability of their co-occurrence and the severity of the crisis estimate?

1.2   Where is Japan Heading?

The author wonders where does the ruling party of Japan intend to lead their citizens after the pandemic. As of November 2021,[1] the government's proposed budget for fiscal year 2021 will bring the total general account to 106.6097 trillion yen, a record high of 3.951 trillion yen over the initial budget for fiscal year 2020, and the third consecutive year of exceeding 100 trillion yen. Moreover, the Japanese government is also about to issue an additional 22.1 trillion yen in government bonds to support the economic stimulus measures in the form of a supplementary budget for fiscal year 2021 after the Corona pandemic. In other words, this is a debt from future generations, and the government will try to recover the crisis from Corona by imposing the "sacrifice" of debt on them. In addition, digital transformation is seen as the key to infection prevention and economic revitalization, but these are areas in which

---

[1] https://www.nikkei.com/article/DGXZQOUA246XZ0U1A121C2000000/.

the Japanese government and companies have difficulty. Lack of effective economic measures and inadequate innovation in industrial structure will not only leave an excessive burden on future generations, but it is doubtful whether it can be called sustainable national management.

Secondly, the issues surrounding international relations are also of concern, especially the tensions between the USA and China, which has been already visible for Japan. What should Japan's diplomatic and economic relations with both countries be like? During the COVID-19 pandemic, both of them concentrated own domestic policy, but after the pandemic, their tensions will be raised more serious again. Hence, Japan has no enough time to encompass their future direction. Probably, the basic strategy of Japan will remain the same with the alliance with the USA as the basic strategy, while maintaining a peace policy. It is likely to maintain a certain distance from China, while maintaining economic relations. Especially, 2020 was the year of the USA presidential election, which resulted in the inauguration of Democrat Biden as president, and he made it clear that the USA response to China remains strict and confrontational.

In this concern, before the USA presidential election at 2020, NHK broadcasting center in Japan[2] has published their social survey results on international trends (at the time of February and March, 2020). NHK conducted it by mail to 3600 people aged 18 and older nationwide to focus on Japanese attitudes toward the USA, and received total responses from 2195 people (61%). When asked how familiar they are with the USA, China, and other countries, 72% of respondents said they were very and somewhat familiar with the USA, while only 22% said they were familiar with China, with the USA far ahead of China. Meanwhile, when asked which of the two countries should be more important, the USA or China, which are fiercely vying for the lead in security and economic affairs, the largest number of respondents (55%) said that both should be important, with 34% for the USA and 3% for China. The poll also asked about the president they rate the most out of all the successive USA presidents since the end of the Second World War. As a result, "President Obama" had the highest percentage at 54%, followed by "President Kennedy" at 17%, "President Reagan" at 11%, "President Clinton" at 4%, and so on, with incumbent "President Trump" at 2%.

Further, NHK detailed committing additional questions. When asked how Japan will be affected if President Trump is re-elected in the 2020 presidential election, 10% of respondents answered that it would have a large positive impact and 57% answered that it would have a large negative impact (lastly, the Democratic candidate Biden won). Next, when asked about the extent to which President Trump's re-election will lead to the denuclearization of North Korea, 65% of respondents said they did not expect it, up from 34% who said they did. Regarding the Japan–USA alliance, 18% said it should be strengthened, 55% said it should maintain the status quo, 22% said it should reduce cooperation, and only 3% said it should be dissolved. And when asked about the need for nuclear umbrella due to the USA nuclear deterrence, 39%

---

[2] https://www3.nhk.or.jp/news/html/20200518/k10012434191000.html?utm_int=news_cont ents_news-main_006.

said it was necessary now and in the future, 25% said that it was necessary now but not in the future.

This survey clearly revealed the Japanese people's sense of balance in relation to international affairs. As the global trends after the COVID-19 pandemic, many of the citizens are aware of the fear of the military conflict between the USA and China. Because of Japan's position on the cultural, political and economic borders between the two countries, Japan's ability to mediate and coordinate between the two countries in the international community is not impossible, but its ability to demonstrate a higher level of their diplomacy is called into question.

1.3 Clash of Civilization?

Huntington's book "*The Clash of Civilizations*" (1996) theorized the transformation of the world's structure and explained the conflicts between civilizations in the 1990s. At that time, the collapse of the Soviet Union and the unification of East and West Germany led to the collapse of the post-Cold War international political framework, and this book derived a one of the guiding principles for the global shift to the new international politics.

Now, we are indeed facing another emerging conflict since the 2020s. Nowadays, China has emerged and become the second largest power along with the USA, and it has made great strides to become a technological powerhouse, especially in the area of the AI and big data. At the same time, it means enhancing it's a military power, and then the USA cannot ignore such emerging rival. There were probably a number of factors that have contributed to this situation. Certainly, differences in the value standards brought by different civilizations and cultural foundations may have been a major factor.

However, as the author has an own point of view (Shibuya, 2020a, 2020c), the problem seems to exactly depend on the education and human resources. For example, the USA was eagerly to educate foreign students such as those came from China and India within its own country, and such foreign graduated workers further contributed the economic development of the USA with their abilities and skills. At the same time, however, China has achieved its own development by attracting such young graduated workers back to the country and positioning them as the leaders in domestic development, while at the same time allocating its ample budget to cutting-edge technology.

Such differences may become huger than it expected at glance. America's human resources are still largely dependent on the outside nation, while China holding with its overwhelming domestic population and labor force has learned technology and knowledge from other external developed countries. In other words, these are not very noticeable differences in civilization or culture. There is no doubt that there used to be differences in the level of higher education between the two countries in terms of universities and graduate schools, but it is ironic that the USA has continued to pull the trigger to unravel its own state of superiority by providing even foreign students from China and other countries with advanced and world-class education (Loyalka et al., 2021). The technology was not stolen, rather they just taught it spontaneously. As consequentially, it will be unavoidably reflected on when the structure of global influence is transformed from Westness to Eastness.

2 On Cultural Differences and Identity

2.1 Category to Divide Self and the Others

There is a theory in social psychology called social identity theory (Oakes et al., 1994). It tries to reveal the social psychological facts behind the endless ethnic conflicts, ethnic cleansing, and group conflicts that have been repeated. It is based on the need for each individual to maintain a unique identity and to share it with society as a whole (Albery et al., 2021). As proof of this, they point out that differences in some "category" become the basis of identity. For example, gender, origin, education, region, group, organization, ethnicity, and nation. Such categories become the basis for individual identity and the most important reason to separate oneself from others. Humanity is constantly confronted with the existential question of who they are. Moreover, as each individual exists in the same environment with an unspecified number of other people at the same time, the reason for dividing ourselves is rather inevitable. Sometimes, differences of opinion or even principles become the basis for the categories and identities that divide us. For example, political ideology and the categorical distinctions of right-wing and left-wing. They are, in other words, the difference between conservatives and liberals. This is truly a category, reflecting some kind of identity.

First, what the right-wing and conservatives are "protecting" is, in the end, the maintenance of their own identity, and also the desire to improve their self-esteem. Therefore, by emphasizing and maximizing the heterogeneity between oneself and others, and by seeking one's own superiority over others, one aims to maximize the superiority and rank between oneself and others. It may include "nationalism" and what is called "neoliberalism", those who seek the aforementioned "equality as opportunity" over "equity as outcome", and whose winners often advocate conservatism. Hence, it can be said that they are constantly striving for technological innovation and self-improvement. The very few winners are often overwhelmingly talented and fortunate people who achieve extraordinary economic success and achievements, while the majority of losers (or ordinary people) have to settle for the bottom. However, these are the people (including extreme social evolutionists) who tolerate the resulting inequality and seek the evolution or development of society or the nation as a whole (or humanity as a whole). If they are actually blessed with the talent to accomplish this, they will continue to keep such slogans. It is true that these people also form cliques and form groups, but the struggle for rank within these cliques is fierce. Even if they admit that others are on the same level, they want to find some kind of meaning in the indicators and values. When one identifies oneself with the group to which one is committed, the victory of the group is one's victory, and the development of the group is synonymous with making one's superiority known to others.

On the other hand, leftists and liberals try to live together by minimizing and ignoring differences and heterogeneity between themselves and others. Rather than the trivial categories that divide individual people, they seek to live together in the largest categories, such as "international" and "global". There are many who advocate "social liberalism". Slogans such as "equality", "peace" and "benevolence" are typical examples. These are the people who seek not only "equality as outcome" but also "equity as opportunity", and those who dislike competition itself often advocate liberalism. They harbor the delusion that they are all commoners, and insist that it is the difference in effort rather than the difference in birth origin or talent. Or, it is the clergy and saints who are tainted by such thoughts. In the real world, there are many human rights groups and human rights lawyers. It is also characteristic that there are many minorities and women's groups.

As Fukuyama (1992) once pointed out, it can regard that the former is "*megalothymia*" (the desire to be recognized as superior to others), and the latter is "*isothymia*" (the desire to be treated equally). And the above difference between conservatives and liberals is a matter of relative definition, and the author mainly exemplified Japan cases.

Naturally, the reality, positions, slogans, goals and others of "conservatives" and "liberals" in each country might be quite different. In democracies, especially in Japan, above both sides somehow "appear to be in balance" because of the "principle of majority rule" and the "foundation of a mass-consumption society that values the general public". In reality, the conservatives handle the pivot of the nation and most of the capital, and they control the masses. In fact, many democracies outside of Japan that rely on a capitalist economy are like that, and are no far different in nature from a controlled society. From the perspective of conservatives or not, Spinoza made a similar observation in his "*Tractatus Politicus*" (1675), and Western society has been divided by such differences since at least the 1700s.

However, both of them commonly share the same aspect of "protection". Conservatives may be about protecting vested interests and vast private wealth, while liberals may be about protecting human rights and positions of criticism against the establishments and others. Depending on which side one is committed to, one's life and identity can be distinctive. The social composition of these citizens and the balance of power between organizations (e.g., opinion groups, political parties) over their identities and the conflicts they face will affect the state of the nation. The same is true for international issues. Given this background, there is a concern that international conflicts will become a reality. At advent of the After Corona, the values seem to shift to the exact antithesis of the values during the age of the Before Corona. Moreover, the pandemic exposed the inability of the United Nations to function, and then it was in vain to mediate and coordinate the conflict between the military powers of the USA and China. The maintenance of mutual understanding and cooperation among the nations, represented by the United Nations, defined the world structure of the post-Second World War and led to the cultivation of mutual trust among the nations and citizens. But the After Corona world is about to be torn apart between the two poles, broadly divided into the USA and China. They are also the epitome of cultural conflict, as they are both representative of Westness and Eastness. Will the military confrontation between the two sides become an unavoidable situation?

2.2 Identity Matters

During the pandemic, social distancing strategy became one of the popularized ways in the public health policy, but such physical distance among the citizens could intentionally provoke xenophobia.[3] "Corona virus-purging" activities were easily shifted to "foreigners-purging" in the EU and the USA. Especially, any peoples who rooted in Asia were often harmed by the Europe and other racial backgrounds.

Identity committing in each nation can paraphrase that each individual commit what it is true to behave. It seems that the global trend has headed for the decoupling identical and cultural backgrounds between Westness and Eastness. Eurasia Group (2018) ranked "9. Identify Politics in Southern Asia: *Southern Asia identity politics threatens the future of these increasingly prosperous regions, creating unexpected challenges for economic planners and foreign investors*". Various kinds of oppressions and conflicts occur based on differences in categories as conceptual differences such as race, ideology, gender, social class and others in developing countries, and it can understand that there are inherent risks in identity itself to these matters (Fukuyama, 2018; Shibuya, 2020a). Social identification to commit toward each belonging nation is indispensable for citizens, and their cultural backbone has tightened their definition as ourselves rooted in the homeland. Their customized self-perception and definition can be arranged by somewhat category such as region, ideology and nationality, and their behavior are apt to go ahead for collective preferences. Those differences on categorical backgrounds for social identification among citizens clearly exhibit the fundamental cause against social troubles and hardships to be conquered (Thompson, 2019). According to Maalouf (2012), he argued that social identification which committed to those specific categories could be rooted in identity crisis of Arab citizens and its conflicts, as literary "*Les Identités meurtrières*" (The murderous identities in English).

On the other hands, in light of the cultural identity of modern Japan, it was cultivated on the fusion form of Eastness and Westness. The identity of the modern Japanese people, which embraces cultures and traditions unique to the Asian region and Western values, is clearly in a position to maintain a neutral international status. However, given the deep ties between the USA-Japan alliance and European countries, as well as the need to maintain and increase its presence in Asia as a whole, it cannot afford to lose sight of its rivalry with China. It should also consider other possibilities than the bipolarity of the international situation centered on the USA

---

[3] https://www.un.org/en/coronavirus/covid-19-un-counters-pandemic-related-hate-and-xenophobia.

and China. It is true that the two major powers, the USA and Europe, and China and Russia, will increase severe tensions in various parts of the world, but alternatively, it may be possible to reform a third power that unites other Asian countries, Arab countries, Africa, Latin America and others, with Japan at the center. If Japan can secure its position as the leader of the third force, it will be able to gain greater power than it currently has in the new international situation. Whether such an alternative can actually be realized is another matter, but as a national strategy, it is only natural to consider all possibilities in advance.

2.3   What is Next Ideological Factor?

One of the differences between Western nations and East Asian nations, especially China, was considered an ideology. Such dogmatic example was Marxism, and it was a socioeconomic system based on such idea. To be sure, it was compelling until the end of the Cold War. Religion is also still the only factor that has influenced the history of the humanity not only in terms of social economy but also in terms of science and technology and political systems. However, there is no reason to regard Marxism as an important factor now. As a matter of fact, China still pretends to be a believer in communism based on Marxism, but the recent socioeconomic successes within China are the result of adopting the advantages of capitalism. Rather, the ideology guiding China seems to be shifting toward a national leadership system with Xi Jinping at the top (Department of Commentary People's Daily, 2020). As mentioned earlier, such political system is good fit for AI driven scoring services and monitoring systems work very well. During the COVID-19 pandemic, the ideological differences were not effective in preventing further epidemic contagion, rather the author thought that it was the differences in the quality, competence and leadership of the ruling leaders.

Then, what kind of ideology will emerge in the future world? In the author's view as mentioned earlier chapters (e.g., Chaps. 5,, 6, and 7), there is no overarching religion or ideology that holds an entire country together and underlies the world as a whole. At the After Corona, someone said a slogan "*liberty or safety*", as an alternative of ideology. Of course, this dichotomy suggest that *liberty* means here individualistic value, which like European tradition, whereas *safety* stands for not only the national security itself but Chinese tyrannic governance.

But actually, both of them are quite needed in each nation, and further many of the citizens may enthusiastically requires more own specific integration among them. After the end of the COVID-19 pandemic, many countries probably seek not only to restore their own country-centered economies, but also for each citizen to have own national identity with pride. It could be a movement called nationalism or ethnocentrism, or it could be a social movement calling for the correction of economic disparities and a democracy full of dreams. Alternatively, there may be more cases of puppet regimes (such as the military coup in Myanmar in February 2021) rising in national governance at the behest of major powers such as China.

2.4   The Ubiquity of AI and the International Affairs

As mentioned earlier in Chaps. 7 and 8, if the ubiquity of AI leads to the world in which social order is maintained and inequality and social problems are controlled and optimized, will liberals, who hold to a similar ideal of "solidarity and inclusion", eagerly follow the dictates of AI? Many liberals called for the elimination of economic disparity and ethnic discrimination, and worked on many social issues at the United Nations and in the global community, and if AI can be used to serve a basic income system, social security, and actuarial calculations for optimal burden sharing among the citizens, some of their ideals will be solved. However, it is unlikely that they will abandon their own identities and agree to the ubiquity of AI immediately. Choosing to live in accordance with the optimal solution driven by AI means abandoning the differences and boundaries between nations, societies, communities, ethnicity, self and others. What liberals want is cooperation beyond national and ethnic differences. In the same way, conservatives will not approve rather the way to govern a nation that is to put AI above human beings than human beings control AI. Therefore, both principles will not honestly accept being placed under the control and management of AI.

However, suppose that a nation that proactively adopts the ubiquity of AI emerges, and not only domestic governance but also military technology and diplomatic strategy are optimally calculated by AI, and the nation's power is increased according to AI's directives. It would inevitably lead to the need for each country to promote technological innovations such as superior AI and quantum computers, not only for the sake of balance of power, but also to prevent being outmaneuvered in terms of strategy. In fact, this is exactly what is happening in international relations centered on China and the USA. It should recall that Morgenthau (1978) once argued that the size of the population and science and technology define the national power of each country. Certainly, the innovation of advanced technologies such as AI will be a game changer for international affairs. In this way, there is a concern that technological progress and issues surrounding identity will become fetters for the international community in parallel (Shibuya, 2020a).

3   Balance of Power Between Westness and Eastness

3.1   A Matrix Between Westness and Eastness

Actual global situations cannot be permitted to stand on the optimism (SIPRI, 2021, Ministry of Defense (Japan), 2020, 2021). Here, as the author detailed below matrix at earlier chapter, the author summarized balance of power between Westness and Eastness, and it further appended necessary elements into the category. In particular, Japan's ambivalence becomes to be very salient (Table 1).

Table 1 Differences between westness and eastness (some items appended)

	Eastness (mainly cultural bases)	Westness (mainly cultural bases)
Basic Principle	Collectivism	Individualism
Countries	China, Japan (Cultural, genetic features), India, Asian Countries, etc.	The USA, EU nations, the UK, etc.
Religions	Buddhism, Shinto, Islam, etc.	Christianity, etc.
Economics	Communism, capitalism	Capitalism
Political regime	Communism, democracy	Democracy
Languages	Chinese, Japanese, etc.	English, French, German, etc.
Genetic races	Mongoloid, etc.	Caucasoid, etc.
Permanent five members of The United Nations	China (and Russia?)	The USA, the UK, France
Bloc economics, or trades	ASEAN, TPP, OBOR (One Belt, One Road Initiative), AIIB, etc.	The EU, TPP, ADB (Asia Development Bank), etc.
Total population (2019)	Total 2,935,138,400: China (1,441,860,300), Japan (126,860,300), India (1,366,417,800)	Total 632,234,800: the USA (329,064,900), the UK(68,045,800),France (67,978,400), Germany (83,517,000),Italy (60,550,100), Spain (46,736,800), Belgium(11,539,300)
GDP (2018: Million US$)	Total 22,779,061: China (13,368,073), Japan (4,971,767), India (2,718,732), South Korea (1,720,489)	Total 33,643,964: the USA (20,580,250), Germany (3,951,340), France (2,780,152), the UK (2,828,833), Italy (2,075,856), Spain (1,427,533)
Military forces (2019)	China (2.29 million), Japan (230,000), India (1.33 million), South Korea (660,000)	The USA (1.57 million), the UK (170,000), Germany (250,000), France (240,000)
Military budgets (2018: Million US$)	China (249,997), India (66,510),Japan (46,618), South Korea (43,070)	The USA (648,798), the UK (49,997), Germany (49,471), France (63,800)
Number of nuclear warheads (2018: deployed, total number of warheads)	China (0(?), 280), India (0, 140), North Korea (0, 20(?))	The USA (1600, 6450), the UK (120, 215), France (120, 215)
Military Alliances	China, North Korea, etc.? And with Russia?	NATO, the USA, the EU, the UK, Australia, and others (e.g.. Japan, South Korea, Taiwan)

Globalism intensified to organize a huge market interlinking with winners' coalition. Larger clusters for regional economic zones have still then been specialized in such economic context, and military alliances go straight forward to keep own dominance and overwhelm over the others.

Although Asian countries belong to the Asian bloc historically, they differ in the regimes of military alliances and international cooperation that they participate in due to differences in their political and economic systems. Therefore, if the Third World War will occur, it is inevitable that the main battlefield will probably be the Asian countries across the Pacific Ocean. Even if there is no large-scale war, the hegemony of the world with the Atlantic Ocean at its core led by European countries and the USA will be expected to be driven by a dynamism centered on the Pacific Ocean. At the same time, it is a sign of the rise of Asian countries. However, on the other hand, there are a number of states that do not belong to them. For example, these are a lot of African countries and Latin American countries. The casting vote led by them may clearly determine the outcome of the conflict between Eastness and Westness. Alternatively, rather it is also possible that they form their own forces as the third bloc in the world. In reality, however, China is likely to use the "One Belt, One Road" (OBOR) as a conduit to form a "coalition of emerging nations" that will include Southeast Asia, Central Europe, Brazil, and Africa from an economic standpoint, including cooperation with Russia, and North Korea. From the perspective of the USA, EU and other developed countries (e.g., UK, Japan, Taiwan, South Korea, Australia), this may appear to be a "coalition of the poor countries". However, the fact that it is led by several permanent members of the Council, and that it is actively developing AI and big data-related technologies and using them in its own statecraft and military affairs, makes it obvious that its rapidly increasing military superiority is a threat to Western countries that advocate democracy. Thus, the decoupling of the world will only continue to grow.

3.2 OBOR and AIIB

The One Belt, One Road initiative directed by the Central government of China and the AIIB (Asian Infrastructure Investment Bank) originated by China are working together to form a large-scale economic zone linking African countries, Southeast Asia, Central Asia and the EU.

The former is a large-scale economic zone that mainly interconnects the Eurasian continent to the east and west with more than 130 participated countries. However, at the time of the COVID-19 pandemic, the strength of the relationship with China and its trade routes malfunctioned as a route of infection and brought great damage, which aroused a negative public opinion about trade with China.

The latter serves for a member of 70 countries, and it holds a capital of $100 billion at the time of its creation. It had been established by the will of the Chinese central government under the guise of growing need for funding for infrastructure development in Asia, as a rival (67 countries) to the Asian Development Bank (ADB) led by the USA and Japan.

However, the Chinese government had already invested a substantial amount, and the insolvency of the debtor nation was already a problem. In addition, the impact of the COVID-19 pandemic has led to further insolvency in debt-ridden countries, raising the possibility that China will not be able to recoup its investment. On the other hand, countries that can't pay back will have their port facilities and other facilities seized, which in turn will lead to more Chinese domination and more opposition to China.

In addition, the issuance of digital CNY online and in conjunction with smart phone payments could be also linked to big data on people's payment histories and consumption behavior around the world, and if the AI driven scoring services are analyzed in conjunction with such data, the daily rhythms of peoples and their economic trends around the world will be handled by China government. And there is a high probability that this will shift to a new form of consumption as the tendency to consume at home has increased in the wake of the COVID-19 pandemic.

In addition, the USA, Japan, and the EU, which are concerned that national governance may be shaken, are moving toward issuing their own cryptocurrency. Therefore, it is unlikely that China, which is Eastness, will be able to move in the direction of controlling everything as China government desires.

3.3 Balance of R&D Between Both Sides

Close look at backgrounds on balance of R&D between both the USA and China, they eagerly invest a larger amount of budgets for innovating and progressing both academic and industrial fields such as AI, big data, medical health, and information communication technologies (Executive Office of the President (USA), 2016). In contrast, Japan government has approved a total budget of approximately 700 million US dollars for AI researches in 2018FY. On the contrary, the USA already prepares at least more than 4.5 billion US dollars for investment on the AI, and it seems that China also invests more than 4.0 billion US dollars for investment on the AI-related researches and developments in 2018FY. Both countries may excess more than six times of the extent of Japan's investment on AI.

Further, China plans to become the world's top-level AI innovation country by 2030 (China Institute for Science and Technology Policy at Tsinghua University, 2018). And China's unicorn companies (abbreviated as BATH (Baydu, Alibaba, Tencent, and Huawei)) have been increasing their presences in the world. Especially, China invests larger grants for intensification on the AI applications such as automatic driving cars, smart cities, supercomputing, FinTech, online shopping, social media, smart agriculture, robotics, biotechnology, aerospace engineering, military industry, and other R&D fields (Chen & Li, 2018). Then, new technology war between the USA and China has already provoked mutual serious tensions (Wu, 2020). Following

Table 2 Status of major vaccines (as of February 2021)

	Number of vaccination contracts worldwide (as of February 2021)	Main recipients
Pfizer (USA)	Approximately 1.5 billion times	EU, USA, Japan, China, and others
AstraZeneca (UK)	Approx. 2.3 billion times	EU, USA, Japan, Africa, UK, Brazil, COVAX, and others
Sinopack (China)	About 400 million times	Brazil, Southeast Asian countries, and others
Sputnik V (Russia)	Unknown	Unknown
Janssen pharma (Belgium)	About 1 billion times	COVAX, EU, USA, Africa, and others

with them, global competitions among other nations such as EU countries, Japan, and other nations have been investing large amount of budgets and supports for more intensive innovations (Stanford University HAI, 2021).

By the way, to settle the COVID-19 problems, there was a view[4] that the military powers should cut its budgets in each nation. Looking back at the situation during the COVID-19 pandemic, intensively vaccine diplomacy was promoted. Regarding such global situation, the UK as a permanent member of the United Nations security council called for a "*vaccine ceasefire*". As of February 2021, the cumulative number of vaccinations worldwide exceeded 184.16 million (USA: 55.2 million, China: 40.52 million, UK: 16.61 million, etc.). With its own research institutions such as AstraZeneca and Oxford University producing outstanding results, it was taking the initiative and promoting a strategy of international cooperation and dominance for their global strategy. It was an interesting fact[5] that each of the five major powers was mass producing its own vaccine and providing it to other countries. When we took into account the fact that these vaccines were distributed mainly to allied and friendly countries, it clearly saw that their scientific and technological capabilities were closely related to their diplomatic relations. On the contrary, Japan also has the technological capability to develop its own vaccines, and there are discussions about the development of a mass production system for domestic demand for COVID-19 and its variants. Besides, in the table below, COVAX Facility means a WHO-led project for equitable vaccine supply in global (Table 2).

3.4   Military Balance

China's influence, both military and civilian, is becoming highly vigilance matter in global (Stewart, 2020). For example, the reality of the Confucius Institute, which is

---

[4] https://www.nature.com/articles/d41586-020-02460-9.

[5] https://vdata.nikkei.com/newsgraphics/coronavirus-vaccine-status/.

supposed to teach the Confucian spirit throughout the world, has raised suspicions that it is functioning as an intelligence agency. According to an above report by Stewart, Japan is considered as a negative case and the influence of the Chinese Communist Party is inevitable. Standing on an optimistic concern, Confucianism was already traditionally ingrained in the culture of East Asian countries and Japan, and it is apparently that there is no reason for China to deliberately exert any further cultural influence on Japan. Mutually beneficial economic relations are deepening, but only to the extent that they strike a mutual balance. However, there are inevitable problems related to the leakage of intellectual property as well as national security. There will be a greater sense of caution than ever before about the manipulation of public opinion through information.

In the Chinese history, ancient books on the military contain many strategies and tactics that are still useful today such as goal setting and alliances in both military and international affairs. On the other hand, in the modern era of digital transformation, not only physical warfare, but also *sharp power* as one of the information technologies is sufficiently considered as warfare, but these were also suggested in the ancient texts of the Chinese military books.

Here, it should keep an appropriate distance from the sharp power as one of the ways for cyber warfare. Such sharp power often arises to distort what is true and confuse what it is. According to a report by the SIPRI (Stockholm International Peace Research Institute) at 2018, the total amounts of budgets for the largest military forces in developed nations were respectively 649 billion US dollars (the USA, approximately 3% as GDP rate) and 250 billion US dollars (China, approximately 2% as GDP rate). But those budgets included only ordinary weapons and military costs, and it might not include sharp power and its necessary costs.

In addition, according to the IISS (The International Institute for Strategic Studies, the UK), they announced that total amount of deaths caused by warfare and conflicts at 2016 were globally estimated as 157,000. Further, the development of AI technology has led to the emergence of international debates over LAWS (Lethal Autonomous Weapon Systems) and AI driven military weapons. But it would not reach consensus among the nations (Ministry of Defense, Japan, 2020, 2021).

3.5 SWOT Analysis

3.5.1 USA and China

Let me summarize considerable matters confronting with conflict between the USA and China by SWOT analysis. SWOT indicates an abbreviation for Strength, Weakness, Opportunity, and Threat. Here, first initial character (S: Strength) represents any positive factor of specifications of achieving the goal, on the contrary, second character (W: Weakness) denotes any negative factor of specifications of obstructing to achieve the goal. And Opportunity indicates any positive factor of external specifications of achieving the goals (e.g., possibility of competitive advantage), whereas

Table 3 A part of SWOT analysis in the USA

USA	Strength	Weakness
Internal origin	Military powers Leading digital transformation AI driven weapons GAFA companies	Natural resources for developing ICT based technology (e.g., rare metals) It depends on immigrants who employed in lower wage jobs Improving Inequality of socio-economic issues and public health, and racial disputes Requiring to reintegrate common national identity among the citizens and peoples
	Opportunities	Threats
External origin	Competitive progress in the fields of AI and big-data among the other nations	Losing technological and military advantages against China and other regimes

Table 4 A part of SWOT analysis in China

China	Strength	Weakness
Internal origin	Military powers Population size Natural resources. Higher Loyalty to national identity and its integration Designing and implementing AI and big-data driven society as earlier as possible Governing citizens and principal technologies conducted by the central government	Social stratified divides: socio-economic and educational level between rural and city
	Opportunities	Threats
External origin	Investing much budget for their safety and national security Monitoring and watching the system driven by the AI	It does not have enough military experiences of operations

Threat is any negative factor of interfering to achieve the goals (e.g., accelerating possibility of competitions) (Tables 3 and 4).

Weaknesses within the USA[6] include reducing domestic racial tensions and economic disparities between rich and poor, and improving the treatment of immigrants and refugees in public health care. Before the COVID-19 pandemic, they

---

[6] No one knows: How the unknowable consequences of COVID-19 affect thinking about foreign policy and US-China relations: https://www.brookings.edu/opinions/no-one-knows-how-the-unknowable-consequences-of-covid-19-affect-thinking-about-foreign-policy-and-u-s-china-relati ons/.

The Coming Post-COVID Anarchy: https://www.foreignaffairs.com/articles/united-states/ 2020-05-06/coming-post-covid-anarchy.

claimed such diversity was rather a strength, but the pandemic reveled such opinions as a deception. It will be required by significant efforts to share their national identity as a nation across racial differences. This troublesome situation would reduce the strength of the USA national power, and then it is likely to be its external Achilles' heel. On the other hand, China's domestic weaknesses are still the economic disparity between the urban and rural areas that remain. However, China has own richer human resources, and many of its brightest people are returning home after studying abroad. It is also the most advanced country in terms of data-related business and AI technological innovation. The rise of its military will become more remarkable, and then the twenty-first century will clearly be the era of China. However, the problems that China especially face are the post-pandemic external responses, environmental issues, and an aging population problem. In short, each nation should have the national vision of what to sacrifice and what to consolidate capital to increase national power. In addition to the "trade war" between them (Chang et al., 2014), future conflicts between them certainly have enormous implications for Japan and global affairs in all areas. However, even though both countries are at odds with each other, trade between them accounts for a significant portion of each country' economies, and they continue to lead the world in science and technology while mutually restraining each other. Both of them concentrate and specialize in growth areas, accumulate national wealth, and then distribute it across other fields and the citizens. This means it is very strategic, but on the other hand, it is also sure to widen the tremendous gap between the rich and the poor.

3.5.2   EU, UK, and Russia

In contrast, it should pay also attentions to other permanent five and it's security council of the United Nations. Especially, it seems that other permanent five members of the United Nations security council such as France and the UK similarly has own national strategy. Firstly, French President Macron[7] claimed severe powerless conditions of the Security council of the United Nations, as multilateral 'fractures' exposed by pandemic as of November 2020. France has probably implicit intention to become a leader of the EU handing with Germany and other cooperative member countries. Certainly, France has superiority to leading domestic and international policy, but their surroundings would not clearly support for them. Alliance of NATO (Lucarelli et al., 2021) with the USA has structural problems against Russia and China, and after the pandemic, there are severe concerns that EU members' countries cannot integrate with their dogma as the EU (Hirsch et al., 2020). Those complicated jigsaw puzzles taking with each piece of the member's purpose and condition must be settled by leading of France. But on the contrary, the UK after the Brexit has free-hands for own domestic and international strategy. At the end of 2020, IISS[8] (International Institute for Strategic Studies) reported the COVID-19 pandemic enlarged global

---

[7] https://news.un.org/en/story/2020/09/1073172.

[8] https://www.iiss.org/events/2020/11/strategic-survey-2020-launch.

tensions and its disorder, and they foretasted their strategic policies. Probably, they will intend to become a balancer at the security council and other situations, as they occupy an independent position neither China and Russia nor the EU and the USA. The UK is aims at coordinating with each economic and diplomatic cooperation among the nations, and if their success has been achieved by careful efforts of them, the UK may gain a good position in the world of the After Corona. Lastly, Russia has been stagnating own economics, but their influences keep a global power cooperated with China. It should be focused on their strategic policy of "After Putin" rather than "After Corona". In addition, in terms of foreign affairs, the annexation of the Crimean Peninsula in 2014, as well as the military balance with the EU and the tensions in Ukraine, have severe concerns about those future progress.

3.6 Japan's Future

Going back to the initial questions, what will Japan's strategy be in the end? To formulate a strategy is to impose constraints and define the focus of a decision vector, and to undertake the readiness to sacrifice other possibilities for that amount of time. But it will be a difficult path. And Japan is traditionally an *egalitarian* country, and the practice of sharing the burden of sacrifice with everyone has been deeply rooted in their mentality. The economic disparity is relatively low, but the strategic nature of decision-making and concentration in growth areas is not well developed. 50 *years* later, it seems likely that all Japanese will become relatively poorer than other OECD countries. In fact, in March 2021, the Biden administration released *Global Trends 2040*[9] report, which includes scenarios on whether the USA-led democracy will be restored or not, and outlooks on environmental issues and global technological trends, as well as future projections for major countries, and discloses a pessimistic forecast for Japan.

To be sure, Japan has a declining and aging population. Even if it gradually accepts immigrants, many of them will be simple labor. If the ratio of these people to the total number of populations increases, discrimination and other problems (e.g., prejudice, racism) will emerge as similarly in the USA and other countries. On the other hands, advanced R&D progress also lags behind the USA and China. So far, there were certainly many Nobel prizes and other awards in R&D fields, but the weaknesses in R&D investment, entrepreneurship and patent prosecutions are still weaker than China and the USA.

Further, Japan has traditionally vulnerability in a strategic decision-making for foreign wars and any scientific quantification on the sacrifices among decision-makers. There seems to be a lack of understanding of the essence of the problem. Hence, they often fail to come up with a strategy that focuses on an achieving goal. In other words, they get confused by the sacrifices in front of them, and fail to understand what the essential goal is, and the necessary sacrifices. Therefore, there is a lack

---

[9] https://www.dni.gov/files/ODNI/documents/assessments/GlobalTrends_2040.pdf.

of quantitative analysis of how much benefit there is to balance. However, Japan's strength lies good partnerships with not only in the EU, the UK and the USA, but also many countries in the Middle East (e.g., UAE, Saudi Arabia, Iran, Turkey). Japan's "*Free and Open Indo-Pacific (FOIP[10])*" is a clear indication of its own diplomatic vision, with China's rise in mind, and their desire to avoid overt conflict with China. Thus, international diplomacy as a neutral power in Asia will be an important factor for Japan in the future. It will be a possible to maintain a certain foothold if they keep to develop for a high level of science and technology making strategic investments, and strive to improve their national strength.

Certainly, until now, sandwiched situation between China and the USA, Japan will try to maintain a balance between economics and politics with them (Sahashi, 2020). Through their peace diplomacy, Japan will seek international cooperation and own stability of the governance. Can it maintain relative stability in the post-Corona economics and reduce such tensions? The question is whether they can contribute to those international affairs. In the nearly future, if the USA and other countries demand the compensations for China due to the COVID-19 pandemic, the serious problems such as AI driven weapons and the human rights may be triggered by sensible reasons. In that time, Japan will be required to take some kind of stand. Japan's strategy will be principally to emphasize its relationship with the USA and to maintain the position it has been demanding of China.

4   What is the Critical Factors for Cooperation?

4.1   Cooperation Beyond the Differences

The role of the United Nations has been becoming increasingly important in dealing with the natural disasters that occur on a global scale and the calamities that humanity creates. Indeed, there has been a lot of research and practice on international cooperation and peaceful problem solving. Nevertheless, many challenges still remain.

How has international peace been achieved in the face of serious tension between the great powers? Especially, international cooperation in the control of infectious diseases is always also essential. For example, under the initiative of the WHO, smallpox was eradicated from the world by the 1980s. Even both the USA and the Soviet Union, which were in the midst of the Cold War, joined their hands with each other in an international effort to eradicate smallpox. This was not a good story, but a realistic response.

The COVID-19 pandemic should have been similarly tackled among all nations in global. However, the mutual conflict and distrust between China, the outbreak country of the pandemic, and the USA, one of the most severely affected countries,

---

[10] https://www.mofa.go.jp/policy/page25e_000278.html.

which could be described as the New Cold War, has made it difficult to establish global cooperation, and as of early 2021, relations between the two countries have not improved. This situation is similar to the cold war among the USA and the Soviet Unison. At that time, the political institutions and ideological differences between tyrannical and democratic institutions had created a deeper divide between them. Under such circumstances, how can we make efforts not only to cooperate in combating infectious diseases, but also in economic and cultural aspects, as well as in issues of our existence on the earth? The number of the USA deaths during the COVID-19 pandemic surpassed the number of deaths during the Vietnam War. And it was quite correct in pointing out that the devastation in Europe has not occurred since the Second World War. In light of this situation, the United Nations Secretary-General warned that "*this pandemic has given a hint to the terrorists*". Infectious disease control, both domestically and internationally will become clearly one of the principal policies for the national security in the non-military sector. More than just the mobility of ordinary people across the borders, virus can easily cross them through infected people. In order to maintain *medical national security*, each country cannot be established by itself (Cohen & Wells, 2004). Epidemiological and medical policies are closely linked to a number of technologies and knowledge, and the need to be prepared for such risks has been made clear.

4.2 Sacrifices and Cooperation

In game theory, *cooperation* can be often rephrased as *self-sacrifice*. Mutual cooperation means mutual self-sacrifices even if each of them are unawareness, and an equilibrium as balance of power similarly suggests such condition. Choosing and approving mutual survival each other means, at least at that stage, it is mutual self-sacrifice rather than compromise. Acceptable criteria of each of the stakeholders would be problematic, therefore, and then balancing and coordinating own acceptable criterion among them might to be dynamically observed anywhere. But, as exemplified a case between Iran and the USA at earlier chapter, mutual cooperative situations do not always mean an equilibrium of self-sacrifices between both stakeholders. In those dynamics, there will be occurred that diverse risks and crises whether these are structural or changing patterns.

In debates on international issues at the United Nations and elsewhere, there is always a push for sacrifice. The focus in negotiations and politics is invariably on how to allocate *sacrifices* to each other. Whether it is called a burden, a cost or a demerit, its purpose is to determine and reach an agreement among stakeholders. Cooperative action means that each of them has undertaken own burden as sacrifices, whereas disagreement indicates that at least a part of them will not bear own such sacrifices.

For example, infectious disease control and environmental issues are more likely to be reached consensus and agreed on at the United Nations. This is because there is nowhere else to live but on Earth, and it is an issue that is deeply dependent on the survival of the citizens in each country, and the interests of each country are aligned.

However, the longevity of such *coerced cooperation* requires that the United Nations has an overwhelming capacity to carry out penalties to the extent that a few countries fail to comply with the cooperative act, abandon the agreement, or are unable to counter such violations. In other words, as long as the above factors are satisfied at the same time, we can see that international cooperation will be firmly sustaining. As a matter of fact, this frame of international cooperation has nearly collapsed as a result of the USA continued own nationalism against international cooperative policy since the late 2010s.

4.3   Seven Factors for Cooperation

4.3.1   Necessary Factors

What are the factors that make international cooperation possible? In brief, what is the universal principle that binds human beings equally to each other in order to achieve a common goal? If it examines many past cases, it can imagine that there are certain prerequisites. In particular, this is when the situation is so critical as to be life-threatening. For example, this was why the cold war with nuclear weapons engendered a balanced situation. And further, environmental, foods, and energy issues are often more negotiable when they are economically dependent on each other. This is because factors such as mutual trust and aligned interests among the nations come into consider.

As Kantian triangle of peace noted, the elements such as *democratization, economic interdependence, firm alliances,* and *the establishment of international organizations* have been considered the essential to avoid international conflict and war, and also these are quite necessary for the perpetual peace in global (Kant, 1795). In particular, the interdependence between stakeholders, not just the economy, provides a very useful perspective (Jackson & Nei, 2015). It can be founded on a game-theoretic formalization, and a phase in which each side is situated in a strategic position over mutual interests (Heath, 2011).

Regarding this, on the other hands, the author has been considering the nature which brings about *coerced cooperation* (Shibuya, 2004, 2006, 2010, 2012, 2020a). One of the theoretical backgrounds is depending on the fundamentals of the *Jigsaw classroom* beyond the difference of white and black students in the USA (Aronson et al., 1978). Many people do not realize that there are meaningful hints of human behavioral control hidden in this method. In other words, the author thought that the basic hypothesis for the possibility of cooperation can be formalized if such necessary factors satisfied. Of course, there is a possibility that other factors can be found by examining other cases. However, the author considers that it is better if it can explain the majority with a minimum of common factors, as likewise factor analysis. At the very least, these factors are external variables that, in a controlled experimental setting, activate people's behavior, and the social situations that game theory predicts.

Namely, the author enumerates seven factors for cooperation such as *Desire for survival, Rationality, Responsiveness, Inevitability, Impossibility of deviation, Mutual Consistency of Interests,* and *Internal and external connectivity limitations.* When these seven factors are simultaneously satisfied, the author hypothesized that coerced cooperation can be engendered in social context. Conversely, a state in which all of the above is not fulfilled is truly a chaotic catastrophe. It is possible to design laboratory experiments and computer simulations based on these factors and is a foundational hypothesis for studying collaboration from multiple perspectives. In this regard, the author has been analyzing the structure of cooperative relationships and the requirements for the peace from the perspective of these necessary factors.

1. *Desire for Own Survival*

Everyone has the desire to stay alive. This is a kind of *"self-preservation"* (Tuck, 2001). For the individual, it is the pursuit of survival as the source of the highest good, and for the state, it is the survival of the state through the response to crisis situations that threaten the survival of all citizens.

2. *Rationality*

To be in a state of functioning reason and act rationally as a human being.

3. *Responsiveness*

When one party commits an act against the others, the other party is *equally* capable of doing the same or better. Mutual Assured Destruction (MAD) can be regarded in this context.

4. *Inevitability*

There is *no surrogate*, the person has *no choice* but *to stay* in the group or situation, and there is *no way* to leave or avoid it.

5. *Impossibility of deviation*

Even if one side attacks first or free-rides, both sides will certainly suffer the worst result.

6. *Mutual Consistency of Interests*

All stakeholders are parties to the agreement and have mutual interests.

7. *Internal and external connectivity limitations*

One side shall not be assisted by the externality such as a third party, organization, society or nations. For example, all possibility (e.g., military, diplomacy and others) as secret routes to the external nations, military reinforcements, extra independent troops (e.g., cooperative terrorists), secreted alliances, assaulting ways from the outside of the earth (e.g., universe), or something like this from the external sides.

The most important essence of the jigsaw method is not cooperative learning itself, but the fact that it is based on the principle of facilitating cooperative behavior and pursuing common goal among the participants designed by external facilitators and teachers' *coerced constraints*. The reasons why the Jigsaw classroom could be engendered cooperation among the participants at the arbitrary timing simultaneously satisfies above all of seven factors. First, each participant has to hold own "The desire for own survival" and "Rationality". Second restrictions of the Jigsaw require "Responsiveness", "Inevitability", "Impossibility of deviation", and "Mutual Consistency of Interests" to achieve both each achievement of participants and the common goal of each group. Further, "Internal and external connectivity limitations" regulates each of them for interrupting any helps by external third party (e.g., parents, others). In this way, the internal world of the jigsaw class can be established by reconfiguring human relationships, including those between whites and blacks. Furthermore, under the teacher's strict control, all students must cooperate with each other to achieve a common goal. If not so, they will not be able to complete each individual tasks and all will not receive credit. Whites also have to help blacks. On the other hand, if it loosens the constraints on some of these factors, it is inevitable that it will not see cooperative behavior.

Inductively thinking, when the author considers the experience of observing various cooperative phenomena in the world, such as not only the jigsaw method but the ways led innumerable people to cooperative actions for achieving the common goal in interpersonal, inter-group, international and domestic governance level, the above seven factors can be extracted. The benefits can be broadly categorized as follows. First, this hypothesis facilitates empirical testing beyond the mere framework of practice. It can be assumed that not only the original Jigsaw method, but other cooperative actions as well would include above seven factors. In addition, it can assume that various patterns of the cooperation can occur depending on the setting of various requirements, and it has the advantage of being able to compare them on such common ground. Second, during cooperation, all or only some of above seven factors can be dynamically reconfigured to examine how specific factors act critically on collaborative relationships. In other words, each factor can be manipulated and verified such as what the factors are for collaboration to exist.

4.3.2 Explaining Actual Problems in Line with Seven Factors

To be sure, it can see that the requirements for peace suggested by Kant's triangle are also included in the above seven requirements proposed by the author, albeit at a more abstract level. The author contemplated that the more these seven factors are completely met, the more they have to cooperate with each other. The Jigsaw classroom works effectively because all of these seven factors are constrained to be met. On the other hand, any conditions that do not meet all the requirements would be in a state of conflict or confusion, where mutual trust is lacked. Economic dependence, in particular, is likely to be an important factor, as it often calls a factor of critical damage in the hostile nations. Since the military and economic power of countries are certainly linked (Sandler & Hartley, 2007), coordination and cooperative relationships through alliances and international organizations can also be effective in this regard, and such condition usually become a reason why economic sanctions can be effective.

Next, this frame of reference for cooperation can be understood deductively for actual problems as well. For example, first, as mentioned earlier chapter, let me model a severe conflict between Iran and the USA at 2020. Such case of the asymmetry balance of power rarely reaches mutual cooperation (Levy, 1984), because those conditions lack mutual equality of "Responsiveness" (e.g., military weapons, ways and quantity of the attacks) and "Mutual Consistency of Interests". The large difference of nuclear weapons and advanced military technology between them determined their strategic and tactic ways of offense and defense. And diplomatic efforts could not assist for Iran, because the security council of the United Nations dominated by the USA and its alliances. In the first place, economic sanctions against Iran are aimed at exhausting the Iranian economy by restricting trade transactions outside Iran ("Internal and external connectivity limitations"). The advantage of the side of the USA to dominate such situation could not be fluctuated. And then, Iran desired staying own national consistency ("The desire for own survival"), and they ought to be reluctant against more aggressive attacking ("Rationality"). Thus, the USA and other major powers had succeeded in forcing Iran in the direction they want, bringing about a ceasefire that was the peaceful solution for both countries. This example indeed implied that such peaceful situation was manipulated deliberately if the seven factors could be properly aligned.

Secondly, Chinese government planned to punish Hong Kong's civic movement members for "colluding with foreign powers" in 2020, on the contrary, the democrats "homeland government is trying to cut off ties with the world" and intensified their opposition. "Impossibility of deviation" and "external connectivity" weighed heavily on the side of Hong Kong. On the other hand, there is no room for compromise between the both sides because of the disagreement of "mutual consistency of interest" and other factors. Therefore, it can model that there is a lack of critical necessary factors leading to mutual cooperation between the both sides.

Thirdly, during the COVID-19 pandemic, the wealthy citizens in NY state were able to flee to the outside state. It was too rational for their own survival reasons. However, it was institutionally and economically difficult for the poor peoples to escape from the state. The voluntary actions of ordinary citizens seemed to be selfish, following a mere own "survival desire" rather than cooperating altruistically with each other. While medical doctors were struggling alone, the state government relied on external supports from the federal government and NGOs. It was probably not true cooperation. Like Wuhan City of China, if the municipal government locked down the entire state and prevent all citizens from leaving the state altogether, it suggests that they had been forced to more cooperate with everyone (and it means to request more self-sacrifices), then a situation would have been different.

And fourth, during the COVID-19 pandemic, vaccination progressed in many countries of the world after 2021. However, developed countries bought up large quantities of vaccines, and further the EU even imposed export restrictions, making it difficult for developing countries with weak financial foundations to obtain vaccines.

In response, the COVAX Facility, which promotes fair vaccine distribution internationally, took steps to correct the situation. Moreover, the UK eventually called for a "vaccine ceasefire", which finally led to the realization of an internationally coordinated system. In these cases, similarly, since the factors of "survival" and "rationality" were closely intertwined, the egos of each country are exposed, and cooperation was ordinarily difficult to achieve. Gaps of economic, scientific and technological capabilities of each nation are wide open, international cooperation is often not feasible. Namely, if "mutual consistency of interest", "responsiveness" and "impossibility of deviation" are not effective for all stakeholders. In fact, the vaccine diplomacy of the UK, China, and Russia had been effective because, without vaccination of not only developed countries but also developing countries and all those who wished to be vaccinated, it would be difficult to eradicate new mutants, not to mention the occurrence of new mutants, and as a result, the entire human being would be disadvantaged. In other words, in times of global crisis, the question is whether the world as a whole can quickly establish a cooperative framework in which saving other countries will also save own citizens, and whether this framework can be implemented reliably and promptly.

However, there are rare cases when "irrational" and "self-survival denial" cooperative behaviors (Literally, self-sacrificing cooperative behavior). For example, someone devotes own life to save other person (e.g., parents rescue their children). Of course, it cannot be said that such cooperative behavior always arises in many situations, but it is a fact that such human-specific cooperative behavior definitively determines the foundation of not only humanity but also society.

4.3.3 Insight for Actual Problems

Here we notice an interesting fact. These seven factors can also characterize democratic and tyrannical states. This is because the state is formed by citizens, and cooperative behavior among citizens and the legal system (including various ways of sacrifices) clearly shape the contours of the state (Table 5).

What seven factors reveal outlines the difference in the quality of the cooperative behavior of citizens in each state. The more democratic a state is, the more difficult it is to induce its citizens to cooperative behavior at the sacrifices of assuring freedom, while the easier it is for a tyrannical state to induce its citizens to cooperative behavior at the sacrifices of restricting freedom. A democratic state evokes the image of a network structure with open relationships inside and outside on the bases of liberal thought and legal framework of the state. A tyrannical state, on the other hand, forces the citizens to comply with the directions.

In other words, the original jigsaw method was a pseudo-realization of the latter situation in the classroom, thereby eliminating the discriminatory structure between whites and blacks. It dismantled the free relationships and superior situations that favored whites, and enforced the constraints that blocked them, to achieve a cooperative relationship in which blacks could not be ignored in the classroom. On the contrary, such discriminatory structures could not be resolved in the classroom similar

Table 5 It exemplifies national differences distinguished by seven factors

	Democratic states (e.g., the USA, Japan, the EU)	Tyrannical states (e.g., China, Russia)
1. Desire for own survival	The goal is often to maintain the state system and the livelihood of the people, including in emergency situations	It is highly likely that the priority is to maintain the state system, including emergencies, and that maintaining the lives of the people is merely a means to that end
2. Rationality	They appoint suitably talented people to national public service and leadership positions	AI technology and advanced technological development may go beyond human reason and computational rationality
3. Responsiveness	Citizens are guaranteed freedom of demonstration, protest, and speech against the government	Citizens cannot defy the central government
4. Inevitability	Citizens can move freely beyond the borders	Citizens cannot always move freely beyond the borders
5. Impossibility of deviation	Citizens are free to comply or not	The strong authority of the communist party and the central government, and AI-based surveillance
6. Mutual consistency of interests	It ensures that mutual interests are aligned between the state and the citizens in terms of the social contract	Citizens' will is irrelevant
7. Internal and external connectivity limitations	No restrictions on internet	Strict restrictions on internet

to the former. Actually, the only way to eliminate economic disparities as well as discriminatory structures is through legal enforcement and state control, even in democracies.

On the contrary, the *"ubiquitous jigsaw"* (Shibuya, 2004, 2006, 2010) was designed to realize the cooperative relationship based on the jigsaw method not only in "closed environments", but *ubiquitously* in "open Internet spaces" and "global environments". When the "seven factors" are set appropriately, it will be possible to lead humans and nations to cooperation. In other words, from "the principle of distribution of sacrifices", it imposes "sacrifices", that is, various kinds of constraints (e.g., taxes, restrictions on freedom, legal prohibitions, and all other means necessary to achieve the goal) on each citizen and state (please recall a definition of "sacrifice", it is "something" valuable that is necessary to achieve the goal). This is one of the author's answers to the fundamental question[11] posed at the beginning of this book (Table 6).

While there are certainly many points of contention, in the crisis such as the COVID-19 pandemic, the debate between democratic and tyrannical is no longer relevant in order to ensure the survival of the entire humanity through international cooperation. It is well known that economic disparity and racial discrimination in vaccination and public health become the hurdles to global infection prevention and virus containment all over the world. As "the principle of distribution of sacrifice" shows, the "weakest" part of the system determines the overall performance. In other words, it is not the wealthy nations or their technological powers, but the structurally underprivileged people and nations that hold the key to humanity's survival on the planet. Sacrificing them will result in sacrificing not only economic burdens and labors, but also one's own survival of each country is similar to global crises caused by climate change or world wars as well as the pandemic. What is needed clearly suggests the formation of a social order based on coerced cooperation among all the citizens. Chinese government has already operated a surveillance system that uses the AI and big data. Although someone calls it a tyrannical society, rather it may be an ideal society to keep peace and no-crime within such society. As already equipped, in order to achieve and maintain international cooperation, similarly, mutual monitoring system is quite necessary for all countries. The real equilibrium and equalized peace situation cannot be achieved without imposing constraints that meet above seven factors. If it looks at such situation as a pattern of the *Byzantine Generals problem*, if one country does not honestly participate, all countries will not achieve both each goal and total survival. The same was true for the situation of promoting vaccination worldwide in the COVID-19 pandemic. If only the developed countries were vaccinated and the developing countries were left, it meant that the risk of mutation and reemergence of the disease remained, and the possibility of multiple pandemics could not be eradicated. Therefore, we should keep our efforts to respond through international cooperation. If there were the ability to monitor the health status among the citizens in each country and the implementation process of medical care in real time before the pandemic, using digitized tools, it had probably led to the qualitative differences. But who and how bear the unnecessary sacrifice?

Therefore, it is quite important how much *self-sacrifice* both each citizen and each country are *equally* accepted as a necessary part of the entire global community on the earth to keep total *cooperation*. But after the epidemics, many of the frequent demonstrations and riots were observed as a reaction to these repressive conditions, and therefore both public health and economic policy makers should learn the lesson for the future. Because these social phenomena indicated acceptable criteria for the self-sacrifices among the citizens.

---

[11] "How can we achieve mutual survival and cooperation while overcoming and avoiding unnecessary conflicts and confrontations, and eliminating social evil, which is one of the emergent properties, engendered by the simultaneous coexistence of an unspecified number of people in the same environment?".

Table 6 Setting seven factors for international cooperation during the COVID-19 pandemic

	Present status of each country (Especially, democratic)	The goal of the humanity
1. Desire for own survival	Priority is given to the country itself and to cooperation through political systems and alliances	It prioritizes the perspective of the entire planet and humanity beyond national borders. Give top priority to international cooperation
2. Rationality	Priority of economic rationality over rationality for the goal	Scientific rationality is the top priority, making maximum use of AI technology and the results of advanced technological development
3. Responsiveness	Since citizens are guaranteed freedom of demonstration, protest, and speech against the government, demonstrations against vaccination and anti-government movements frequently occurred	Don't be distracted by short-term humanitarian aid and civil liberties, and it should prioritize the survival of the entire humanity from a long-term perspective
4. Inevitability	It tends to allow freedom of movement for citizens	It restricts the freedom of movement of citizens
5. Impossibility of deviation	Citizens are often free from legal constraints as well	Thoroughly it inhibits deviations of citizens and countries from a global perspective. Through advanced technology and AI monitoring, we will attempt to adjust the load to be optimal for each individual and each country
6. Mutual consistency of interests	Disparity of interests between vaccinated and non-vaccinated people (occurrence of free-riding), disparity in the quality of public health due to economic disparity between richest and poorest countries, etc.	The priority is to achieve the goal of the survival of humanity as a whole, regardless of national sovereignty or the freedom and rights of individual citizens
7. Internal and external connectivity limitations	It tends to avoid restricting cross-border movement, global economic activities, the Internet, etc.	Strict distinction between red zone and green zone. Restricts the movement of civilians and limits them only to necessary supplies, medicines, and treatment staff

4.3.4   Limits of the Cooperation

If requesting people to own self-sacrifice and satisfy all factors for the cooperation, there is the case not to settle any causes and ideal solutions. For example, these are the "Noah's Ark" situation that described at the beginning of this book, or the triage of life as a situation that forces the stakeholders to set priorities and sacrifice someone's life. At the crises such as the emergency and disaster cases, such situations will frequently be experienced. Furthermore, there is no guarantee that everyone will be saved even if everyone cooperates self-sacrificially in such situation. It means that either few sacrifices to be died or all die. Namely, even if everyone cooperates with each other, an earlier requirement to "The desire for own survival" is not always guaranteed to everyone. It is necessary to maintain mutual cooperation by all, but such cooperation itself cannot ensure that each person can survive in an emergency. In these situations, self-centered behavior is more likely to manifest itself than to maintain mutual cooperation, and then mutual cooperation is more easily disrupted. Therefore, the risk remains if they may not reach a coordinated solution, and consequentially they finally all die.

However, if everyone will not cooperate with each other, the only way is all extinction. Therefore, it is necessary to be vigilant to prevent such crises and risks from occurring in advance. Everyone must all make self-sacrifices for the benefit of all of them and act on the basis of mutual cooperation in global. Furthermore, the goal of avoiding or minimizing the sacrifices caused by any risks or crises must be shared by all of them, and the they eagerly have to do everything that they can to achieve the goal. The COVID-19 remind us such the nature of cooperation in global.

5   Critical Thinking on Sacrifices in Crisis

5.1   Collaborative Education Design for Critical Thinking

Kennedy said "*Ask not what your country can do for you; ask what you can do for your country*". It clearly means voluntary efforts and cooperation by self-sacrifices of each citizen for the nations. Further, he also told at another scene "*Confident and unafraid, we must labor on–not toward a strategy of annihilation but toward a strategy of peace*". In light of this concern, it should be promoting discussion-based education at universities and other institutions to establish the social system that allows citizens to participate in political discussions (Fernández & Sundström, 2011; Mangini, 2017). The author believes that the political process should be improved on the nature of the crises and sacrifices.

In following case, the author made a collaborative education course design for university students to discuss and deliberate various topics at confronting with crises and sacrifices (Shibuya, 2006). The author convinced that students had better learn practically and autonomously their own critical thinking style and verification on multiple standpoints from actual social issues through collaborative group process. Both critically thinking and verification on multiple standpoints are possibly cultivated in deliberative discussion and practical process, and these abilities are often crucially the weakness of many Japanese citizens.

The design is based on methodology of deliberative opinions poll and role-playing style. That is because it is quite necessary to provide deliberative and critical thinking experiences for students in the age of social networking and collective intelligence (Evangelopoulos & Visnescu, 2012). And each student takes a role such as stakeholder and critical standpoint, for example. Of course, the author instructs students to recognize crisis, risks and sacrifices at the group discussion such as group think, polarization of opinions and risky shifts (Herek et al., 1987; Gibson, 2012a, 2012b). With these educational materials, learning design and necessary instructions at the university, especially the author facilitates wholly discussion with students as more actively learning. Further, as results, if necessary, this role-playing style deserves to apply as an agent-based model for dynamic simulation of interactive models with AI after actual deliberative discussions (March, 2021; Dubois et al., 2013).

5.2   Practical Exercises

Training of critical thinking for everyone is required at all times. Especially, political decisions, especially in the face of a crisis, are always fraught with risks of their cognitive capacity. Different values, indicators, and measures simultaneously exist, and stakeholders often have different expectations and intentions. Under the circumstances, decisions are made to determine the "goal" to be achieved and to tried proportionately optimize the risks and sacrifices. It will be also limited in the amount of time, it has to conceive and examine the various scenarios deeper.

Here, it considers an educational course through deliberation. For a better civil society, each citizen will debate various issues with many stakeholders to understand scientific evidences and make the right decisions by themselves. Instead of accepting the distribution of sacrifice by the politicians, each independent individual makes the decision to undertake own sacrifice. That's what it means to live in an unprecedented risk and crisis society (Yahara, 2021). There are already excellent educational resources such as Clade-X[12] on epidemic diseases and pandemics. This section addresses other issues related to the crisis.

5.2.1   Case 1: AI Driven Weapons

Both the USA and Russia are in the direction of promoting development of the AI driven military weapons, while China is aggressively pursuing to develop it. However,

---

[12] https://www.centerforhealthsecurity.org/our-work/events/2018_clade_x_exercise/, https://www.centerforhealthsecurity.org/our-work/events/2018_clade_x_exercise/clade-x-resources.

many countries are planning to abolish or even ban them. On the other hand, Japan is struggling to cope with the Japan–USA alliance.

Motivation of Deliberation

AI and big data related technologies have been becoming more prevalent. In addition, in the COVID-19 pandemic, health management and infection prevention through effective operations using these services. It has been applied to countermeasures and have become an integral part of our daily lives. But these military applications clearly reflect national strategies and attitudes towards military technology in each country. In the case of the democratic country, how each citizen thinks about such matters determines the future of their country.

For example, from the perspective of the aforementioned seven factors for achieving a coercive cooperation (i.e., mutual cooperation and control system), how can countries achieve and maintain an international cooperative system on these issues of AI weapons, nuclear disarmament, and balance of power among the nations? Conversely, which factor, if lacking, would lead to what kind of conflict?

5.2.2 Case 2: Crises and Sacrifices on Public Health After the Tohoku Disaster (11th, Mar. 2011)

Here, it shows an example case instructed by the author at university. This was controversial issues on public health after the Tohoku Disaster in Japan (11th, Mar. 2011). Of course, this harder experience had been shared among the all students (Shibuya, 2017, 2018, 2020b).

Directions

The Fukushima crisis was one of the worst cases of nuclear power plant accidents in the world. In brief, unfortunately one of the most controversial issues was rather human matters than other disasters. Until then, we Japanese recognized that unrealistic and unscientific assumptions for the safety ("*Anzen Shinwa*" in Japanese, it means the myth of safety without any scientific evidence) were shared among us. Further, stakeholders in charge were difficult to strategically and realistically conduct crisis, risk and hazard management for the Disasters (e.g., tsunami, nuclear plant accident) (Funabashi & Kitazawa, 2012). For that reason, after the Tohoku Disaster, they were criticized by the citizens for their optimism in nuclear energy (JST, 2011; Science Council of Japan, 2011; NRC, 2011; Ahn et al., 2015; OECD/NEA, 2016).

At least, these following matters have been aware of the safety in Japan. Up to the present, parts of people hesitate to discern public acceptance and risk communication on nuclear power plants and its terrible incidents for everyone. Thus, the task requires participatory students to discuss for operating and conducting below severe incidents and crises for the future disaster management. Then, each participant plays a stakeholders' role selecting from governors, engineers and ordinary citizens.

1.   Safety of fresh water for drinking
2.   Safety of agricultural foods
3.   Safety of sea foods
4.   Contaminating with pollutants and nuclear substances in the air, soil, river and sea
5.   Health care for little children in particular.

For example, from the perspective of the seven factors as mentioned above to realize the coercive cooperation (i.e., mutual cooperation and management system), what are the priorities of each stakeholder in such risk management and crisis response, and how can mutual cooperation be realized?

Motivation of Deliberation

Actually, during deliberative process, students vigorously claimed various critical opinions. The author picked up some opinions through entire discussion as follows.

First of all, as a whole, many students criticized that Japan stakeholders could not persuade citizens and they failed to lead good understanding for the citizens' safety.

Secondly, next opinion was about risk communications. Some of the participants claimed that safety information and knowledge were severely not enough to understand for the all citizens at that time. At the Tohoku Disaster, citizens autonomously and spontaneously communicated with each other using social media such as Twitter ever since the Disaster. It was frequently observed a part of the real-time discussing progresses on foods' safety using social media tools.

Finally, some students commented as follows: If many of Japanese shared critical attitudes at the moment of truth during the disaster, stakeholders may be able to operate effectively management and unveil assuredly the reality faster.

5.2.3   Case 3: Taxation for Global Cooperation

Directions

As mentioned earlier chapter, rethinking the usefulness of common wealth and social common capital is quite necessary for sustaining our lives with equality and accessibility of public health (Uzawa, 2005). In addition to the enhancement of public medical institutions and welfare, the conservation of the natural environment and the prevention of the waste of global environmental resources (e.g., SDGs) are still more important as quality of life (QOL) (IPBES, 2020; Chaturvedi et al., 2021). By intersecting them, it can ensure that the priority of individual liberty is superseded by lasting values. It is to focus on not only utility of each individual but values for the society and for the all generations of the humanity (Al-Delaimy et al., 2020).

What burdens as sacrifices should each citizen and country undertake in response to the challenges facing various global challenging matters? For example, global solidarity tax (i.e., International Solidarity Levy), environmental tax and digital taxation (it means how to tax from the economic activities on the Internet) have been discussed. Increasingly, the problems are easily crossing over across the national borders on the earth such as disasters and environmental issues. It is severe question of the survival of the human existence (Kaltenborn et al., 2020).

Motivation of Deliberation

The focus is on how to equalize any sacrifices among the stakeholders as fairness. There is often a tangle of national and the United Nations agendas on those matters, which should be improved. Education and enlightenment are also important for each citizen. In order to deepen their awareness of these issues, it is useful to educate them through deliberations.

The international community has begun to seriously consider ways to finance the achievement of its goals. For example, solidarity tax can be now enumerated below categories, at least.

- Carbon tax
- Air and sea transport tax
- Airline ticket tax
- Multinational corporation tax
- Arms dealing tax
- Financial transaction tax and others.

How can these tax measures be effectively applied? For example, from the perspective of the seven factors as mentioned above to realize the coercive cooperation (i.e., the international cooperative system), it devises necessary scenarios to realize the appropriate tax measures for each country and each company.

6   Concluding Remarks

History repeats itself. Indeed, we are now witnessing next global fluctuations. The crisis has already arrived when nations and citizens begin in motion to keep balance of power. As they push sacrifices each other, further sacrifices with nowhere to go will cause ordinary people to suffer. History has repeated itself that way. It seems also the original sin of the coexistence of an unspecified number of the people at the same time. "Social distance" in the pandemic will continue to be a clear indicator of the distance between the peoples and nations in the age of the After Corona. In addition, mutual cooperation as "self-sacrifice" and prudence for the peace will inevitably be required. Future generations will be asked how to achieve practical solutions to the problems of crises and sacrifices in the real world.

References

Ahn, J., et al. (2015). *Reflections on the fukushima daiichi nuclear accident: Toward social-scientific literacy and engineering resilience*. Springer.

Albery, I. P. et al. (2021). Differential identity components predict dimensions of problematic facebook use. *Computers in Human Behavior Reports, 3*. https://doi.org/10.1016/j.chbr.2021.100057

Al-Delaimy, W. K., Ramanathan, V., & Sorondo, M. S. (Eds.). (2020). *Health of people*. Springer.

Aronson, E., et al. (1978). *The Jigsaw Classroom*, SAGE.

Chang, S., et al. (2014). *The impact of the US-China trade war on Japanese: Multinational corporations*. https://www.rieti.go.jp/jp/publications/dp/19e050.pdf

Chaturvedi, S., et al. (2021). *The Palgrave handbook of development cooperation for achieving the 2030 agenda*. Palgrave Macmillan.

Chen, W., & Li, X.-Y. (2018). Welcome to the China region special section. *Communication of the ACM, 61*(11), 38–87.

China Institute for Science and Technology Policy at Tsinghua University. (2018). *China AI development report*. http://www.sppm.tsinghua.edu.cn/eWebEditor/UploadFile/China_AI_development_report_2018.pdf

Cohen, D., & Wells, J. (2004). *American National Security and Civil Liberties in an Era of Terrorism*, Palgrave Macmillan.

Department of Commentary People's Daily. (2020). *Narrating China's governance: Stories in Xi Jinping's speeches*. Springer.

DOD, USA. (2020). *Military and security developments involving the People's Republic of China 2020*. https://media.defense.gov/2020/Sep/01/2002488689/-1/-1/1/2020-DOD-CHINA-MILITARY-POWER-REPORT-FINAL.PDF

Dubois, E., et al. (2013). An agent-based model to explore game setting effects on attitude change during a role playing game session. *Journal of Artificial Societies and Social Simulation, 16*(1) http://jasss.soc.surrey.ac.uk/16/1/2.html

Eurasia Group. (2018). Top 10 risks, https://www.eurasiagroup.net/issues/top-risks-2018

Evangelopoulos, N., & Visnescu, L. (2012). Text-mining the voice of the people. *Communications of the ACM, 55*(2), 62–69.

Executive Office of the President (USA). (2016). *Artificial intelligence, automation, and the economy*. https://www.whitehouse.gov/sites/whitehouse.gov/files/images/EMBARGOED%20AI%20Economy%20Report.pdf

Fernández, C., & Sundström, M. (2011). Citizenship education and liberalism: A state of the debate analysis 1990–2010. *Studies in Philosophy and Education, 30*, 363–384.

Fukuyama, F. (2018). *Identity: Contemporary identity politics and the struggle for recognition*. Profile Books.

Fukuyama, F. (1992). *The end of history and the last man*. Free Press.

Funabashi, Y., & Kitazawa, K. (2012). Fukushima in review: A complex disaster. http://bos.sagepub.com/content/early/2012/02/29/0096340212440359

Gibson, D. R. (2012b). *Talk at the brink: Deliberation and decision during the Cuban Missile Crisis*. Princeton University Press.

Gibson, D. R. (2012a). Decisions at the brink. *Nature, 487*, 27–29.

Heath, J. (2011). *Following the rules practical reasoning and deontic constraint*. Oxford University Press.

Hegel, G. W. F. (2018). *Phänomenologie des Geistes (The Phenomenology of Spirit)* (English Translated Edition). Cambridge University Press.

Held, D., McGrew, A., Goldblatt, D., & Perraton, J. (2000). Global transformations: Politics, economics and culture. In C. Pierson & S. Tormey (Eds.), *Politics at the edge: Political studies association yearbook series*. Palgrave Macmillan. https://doi.org/10.1057/9780333981689_2

Herek, G. M., Janis, I., & Ruth, P. (1987). Decision making during international crises. *Journal of Conflict Resolution, 31*(2), 203–226.

Hirsch, E., et al. (2020). *European variations as a key to cooperation*. Springer.

Hofstede, G. (1980). *Culture's consequences: International differences in work-related values*. SAGE.

Huntington, S. P. (1996). *The clash of civilizations and the remaking of world order*. Simon & Schuster.

IPBES. (2020). IPBES workshop on biodiversity and pandemics. https://ipbes.net/sites/default/files/2020-12/IPBES%20Workshop%20on%20Biodiversity%20and%20Pandemics%20Report_0.pdf

Jackson, M. O., & Nei, S. (2015). Networks of military alliances, wars, and international trade. *PNAS, 112*(50), 15277–15284.

Janis, I. (1982). *Groupthink: Psychological studies of policy decisions and fiascoes* (2nd ed.). Houghton Mifflin Company.

JST. (2011). Strategic proposal: Proposal for recovery from the Tohoku Earthquake—Form the viewpoint of science and technology.

Kaltenborn, M., Krajewski, M., & Kuhn, H. (2020). *Sustainable development goals and human rights*. Springer.

Kant, I. (1795). *Perpetual peace*, Macmillan Company (English translated edition).

Levy, J. S. (1984). The offensive/defensive balance of military technology: A theoretical and historical analysis. *International Studies Quarterly, 28*(2), 219–238.

Loyalka, P., Liu, O. L., & Li, G. et al. (2021) Skill levels and gains in university STEM education in China, India, Russia and the United States. *Nature Human Behavior*, https://doi.org/10.1038/s41562-021-01062-3

Lucarelli, S., Marrone, A., & Moro, F. N., (Eds.) (2021). NATO decision-making in the age of big data and artificial intelligence, *NATO HQ—Boulevard Leopold III*. https://www.iai.it/sites/default/files/978195445000.pdf

Lukin, A. (2020). The Russia-China entente and its future. *International Politics*. https://doi.org/10.1057/s41311-020-00251-7

Maalouf, A (2012) *In the name of identity: Violence and the need to belong* (original title "*Les Identites meurtries*", English translated edition), Arcade Publishing.

Mangini, M. (2017). Ethics of virtues and the education of the reasonable judge. *International Journal of Ethics Education, 2*, 175–202.

March, C. (2021). Strategic interactions between humans and artificial intelligence: Lessons from experiments with computer players. *Journal of Economic Psychology*, https://www.sciencedirect.com/science/article/pii/S0167487021000593

Menon, R., & Ruger, W. (2020). NATO enlargement and US grand strategy: A net assessment. *International Politics, 57*, 371–400.

Ministry of Defense, Japan. (2020). *Defense of Japan 2020 (Annual White Paper)*, http://www.clearing.mod.go.jp/hakusho_data/2020/w2020_00.html

Ministry of Defense, Japan. (2021). *Defense of Japan 2021 (Annual White Paper)*, https://www.mod.go.jp/j/publication/wp/wp2021/pdf/index.html

NRC (Nuclear Regulatory Commission, U.S.A.). (2011). The near-term task force review of insights from the Fukushima Dai-Ichi accident. http://pbadupws.nrc.gov/docs/ML1118/ML111861807.pdf

Oakes, P. J., Haslam, S. A., & Turner, J. C. (1994). *Stereotyping and social reality*. Blackwell.

OECD/NEA. (2016). Five years after the Fukushima Daiichi accident: Nuclear safety improvements and lessons learnt. https://www.oecdnea.org/nsd/pubs/2016/7284-five-years-fukushima.pdf

Rawls, J. (1971). *A theory of justice*. Harvard University Press.

Rawls, J. (2001). *Justice as fairness a restatement*. Harvard University Press.

Sandler, T., & Hartley, K. (2007). *Handbook of Defense Economics*: Defense in a Globalized World, North Holland.

Sahashi, R. (2020). Japan's strategy amid US? China Confrontation. *China International Strategy Review, 2*, 232–245.

Science Council of Japan. (2011). Report to the foreign academies from science council of Japan on the Fukushima Daiichi nuclear power plant accident. http://www.scj.go.jp/en/report/houkoku-110502-7.pdf

Shibuya, K. (2004), A Framework of Multi-Agent Based Modeling, Simulation and Computational Assistance in An Ubiquitous Environment, *SIMULATION*: Transactions of The Society for Modeling and Simulation International, vol.*80*(7), 367–380.

Shibuya, K. (2010). *A study on the ubiquitous collaborative activities.* Doctor Dissertation (In Japanese).

Shibuya, K. (2017). An exploring study on networked market disruption and resilience. *KAKENHI Report (in Japanese).*

Shibuya, K. (2018). A design of fukushima simulation. *The society for risk analysis, Asia conference 2018.*

Shibuya, K. (2020b). *Identity health,* https://www.ncbi.nlm.nih.gov/pmc/articles/PMC7121317/. In K. Shibuya (Ed.), *Digital transformation of identity in the age of artificial intelligence.* Springer.

Shibuya, K. (2021). Breaking fake news and verifying truth. In M. Khosrow-Pour (Ed.), *Encyclopedia of information science and technology* (5th, Ed., pp.1469–1480). IGI Global.

Shibuya, K. (2006). Collaboration and pervasiveness: Enhancing collaborative learning based on ubiquitous computational services, including as Chapter 15. In M. Lytras & A. Naeve (Eds.), *Intelligent learning infrastructures for knowledge intensive organizations: A semantic web perspective* (pp. 369–390). IDEA.

Shibuya, K. (2012). A study on participatory support networking by voluntary citizens—The lessons from the Tohoku earthquake disaster. *Oukan, 6*(2), 79–86. (in Japanese).

Shibuya, K. (2020a). *Digital transformation of identity in the age of the artificial intelligence.* Springer.

Shibuya, K. (2020c). *A revaluation for the morgenthau's prospects in the digital society, the international workshop "Transdisciplinary approaches to good governance", The center for Southeast Asian studies (CSEAS).* Kyoto University.

SIPRI. (2021). SIPRI Yearbook 2021. https://www.sipri.org/sites/default/files/2021-06/sipri_yb21_summary_en_v2_0.pdf

Stanford University, HAI (Human-Centered AI). (2021). Artificial intelligence index report 2021. https://aiindex.stanford.edu/wp-content/uploads/2021/11/2021-AI-Index-Report_Master.pdf

Stein, B., et al. (2007). Assessing critical thinking in STEM and beyond. In M. Iskander (Eds.), *Innovations in e-learning, instruction technology, assessment, and engineering education,* Springer

Stewart, D. (2020). China's influence in Japan: Everywhere yet nowhere in particular, CSIS report. https://csis-website-prod.s3.amazonaws.com/s3fs-public/publication/200722_Stewart_GEC_FINAL_v2%20UPDATED.pdf

Thompson, C. E. F. (2019). *A psychology of liberation and peace.* Palgrave Macmillan.

Tuck, R. (2001). *The rights of war and peace: Political thought and the international order from Grotius to Kant.* Oxford University Press.

Turner, J. H. (Ed.) (2001). *Handbook of sociological theory,* Springer.

USA (the White House). (2020). United states strategic approach to the people's republic of China, https://www.whitehouse.gov/wp-content/uploads/2020/05/U.S.-Strategic-Approach-to-The-Peoples-Republic-of-China-Report-5.20.20.pdf

Uzawa, H. (2005). *Economic analysis of social common capital.* Cambridge University Press.

Wu, X. (2020). Technology, power, and uncontrolled great power strategic competition between China and the United States. *China International Strategy Review, 2,* 99–119.

Yahara, T. (Ed.). (2021). *Decision science for future earth: Theory and practice.* Springer.

Zhao, K. (2016). China's rise and its discursive power strategy. *Chinese Political Science Review,* 539–564. https://doi.org/10.1007/s41111-016-0037-8

Conclusion

Conclusion

1 Concluding Remarks

My intention for this book can be summarized in one sentence: *"How can we achieve mutual survival and cooperation while overcoming and avoiding unnecessary conflicts and confrontations, and eliminating social evil, which is one of the emergent properties, engendered by the simultaneous coexistence of an unspecified number of people in the same environment?"*.

Sacrifice principally determines the "something" (or "someone") that is required to achieve the goal, and it also implies the aspect of being responsible for the solutions. A crisis is nothing more than an actuality that forces us to confront a serious problem that requires sacrifice. In many cases, when above summarized question becomes the worst, and the crisis leads to more severe crises. Therefore, if a crisis cannot be dealt with before it arrives, people shall know the chaos that will be coated with more social evils. Historical crises are the repetition of such facts.

The opposite of the goodness is not evil. Evil is an emergent property when the goodness of an unspecified number of individuals conflicts with each other. Similarly, cooperative relationships among us are an emergent property as well. Cooperation is established when individuals make sacrifices in order to mutually eliminate social evils that they inevitably create in order to achieve their own goodness. The fact that it seems to oscillate between social evil and cooperation is never an accident, but rather an inevitability. But we are not two-sided between goodness and evil. If it looks at it from this perspective, we follow the same principle. It is indeed the distribution of sacrifices. Namely, do we sacrifice ourselves or do we sacrifice other beings? If we mutually adhere only to self-sacrifice and refrain from pursuing only our own goodness, harmony and cooperation can naturally emerge in our society. However, the society that only seeks mutual sacrifice for others is nothing but chaos. The real world brought into sharp relief by the effects of the corona pandemic inevitably reveal which side it was on.

We confront with a periodical edge of the age of the After Corona. What we have to face is actually what we have to resolve now. An invisible problem means

K. Shibuya, *The Rise of Artificial Intelligence and Big Data in Pandemic Society*, https://doi.org/10.1007/978-981-19-0950-4_13

a serious problem for us. Many of the problems that have become visible suggests already too late. Philosophy and other academic disciplines on risk and crisis shall explore the possibility of the survival of the humanity by anticipating such situations and revealing its true nature. Namely, it was for this reason that in this book, the author contemplates negative facets of the possibility and the actuality as well as the imaginary hopes.

History repeats itself. Indeed, we are now witnessing its global fluctuations. The crisis has already arrived when nations and citizens begin in motion to keep balance of power. As they push sacrifices each other, further sacrifices with nowhere to go will cause ordinary people to suffer. It is also the original sin caused by the coexistence of an unspecified number of the people at the same time. "Social distance" in the pandemics will continue to be a clear indicator of the distance between the peoples and nations. Moreover, mutual cooperation as "self-sacrifice" and prudence of the peace will inevitably be required.

What this pandemic had revealed was that freedom with self-sacrifice is freedom with social constraints, which leads to a responsible and sensible social order. Freedom without self-sacrifice leads to irresponsibility and chaos in society as a whole. Forcing sacrifice to the others by the latter is a crime against someone else. If we call it freedom, there is nothing in this world that can be called a crime. In order to emancipate ourselves from the social evils as emergent properties, the focus is on how to reconcile self-sacrifice by the former with social order, and in this balance, we have no choice but to coexist with the others. It is the legal system and social morality that guide people in a just and fair direction, and sometimes external means are required to induce people to the coercive cooperation, but there are many who deny or criticize such solutions. However, if we firstly accept freedom with self-sacrifice as a given, we must also desire a social order based on this freedom with self-sacrifice, and it is inevitable that the legal system and coercive cooperation are applied as a means to achieve such society. The question is rather how to avoid or solve such situations and problems as much as possible by utilizing AI and big data technologies. Differences in political systems whether democratic or tyranny are no longer an issue. The only difference is whether or not such technology is used thoroughly for national control. As a result of the decision to accept it or not, we must be prepared to accept the consequences of the difference in national power and science and technology, and whatever national crisis it may occur.

As a matter of course, it is not justice that can save social evil. Justice cannot settle injustice. In order to realize justice in the world and to save justice, it is impossible to solve such social matters without sacrificing something or someone. The crisis is indeed the situation in which the intention to save injustice by justice has failed. The mutual justice calls the other party evil, and the out-of-control of the justice which lost the balance of power among the stakeholders brings further crises. The human-being already knows its tragedy.

In order to achieve the goal, it is necessary to take all possible measures, and even if this results in a reasonable sacrifice to the citizens and other stakeholders, it is worthwhile to do so if the benefits of achieving the goal becomes relatively greater. For this reason, as discussed in this book, it is necessary to implement the measures quickly and intensively, and if possible, before such crisis happen. In reality, however, there is often insufficient scenario-based forecasting and preparation for the worst-case scenario, and consequentially, even the balance of benefits and damages for the entire society often collapses. Human history has proven that no one asks who bears the cost invoked by the crisis and how.

2 For Coordinating International Issues

As especially noted at earlier chapters, international tensions have been continuing around this pandemic. In the future, it must deal with a variety of other international problems, including international cooperation to solve infectious diseases. The possibility of a military collision between the two major military powers such as the USA and China would need to be carefully considered in advance. The humanity repeatedly experienced international conflicts among the nations or the periods of peaceful condition. As the author has discussed through this book, each social structure and change encompasses its own inherent risks and crises.

Economic interdependence, one of the factors for international cooperation in Kant's triangle, also matches the global economy and is usually not a particular problem. In the event of the COVID-19 pandemic, however, if countries closed their national borders or imposed travel restrictions, it would mean that their economies could not stand on their own. Such interdependent economy and globalization were the reasons why the pandemic had not ended at all. In other words, it was a question of the extent to which each country had a self-sustaining economic base, and the extent to which they had taken constant public health measures and were able to respond on the ground.

It is for this reason that the COVID-19 pandemic has exposed the nature of humanity and the problem of international cooperation. The focus is on how to implement the "seven requirements for cooperation", the "mechanism for mutual monitoring of all nations of the world" and the "optimization and balancing of self-sacrifice and the interests of the whole", as the author has described. AI technology may be the means to realize these goals. If there is a utopia (or distopia) for humanity, it will be in the realization of omnipresent cooperation through the use of such technologies.

3 On the Nature of State Governance in the Age of After Corona

There was a very famous episode in the history of China during *Jōgan era* (AD 627–649). An emperor Taizong of Tang[1] posed a question "*Which is more difficult*

---

[1] https://en.wikipedia.org/wiki/Emperor_Taizong_of_Tang.

establishing the national regime or sustaining the national regime?" for his staffs and generals. As many intellectual staffs who were the most evaluated members through China history existed in the *Jōgan era,* two representative answers were responsively proposed for an emperor. A first answer was decisively pro for the former, on the contrary, second one was underlying in the latter. After hearing those comments from his staffs, this emperor praised each of those opinions, he noted that both factors are actually quite difficult. Finally, he led conclusively them to collaborate with each of them for further development of his nation. His ruling era was one of the most ideal eras in China and East Asia, because of leading propensity of governing the nations by emperor, higher consulting abilities conducted by clever brains, deep interactive discussions among them, and higher policy making and its pursing the goals which committed by talented administrative officers.

During his ruling era, his governance was splendidly perfect. Then, ordinary citizens had been ceasing in daily life, even though many of them were not necessary to close the door of each home with own key and nobody committed stealing crimes even if there were something valuable on the road.

The *Jōgan era,* the ideal of politics in Eastness, would in fact be realized by the AI driven governance and big data-based analysis as a radical evolutionary system of Eastness in nearly future. On the other hand, Westness's iconic democratic decision-making had long since been negated by Arrow's impossibility theorem (1963), market-oriented economy had only widened the gap between rich and poor, and further liberalism's maximization of citizens' rights was not even a barrier against the COVID-19 pandemic. As the *Jōgan era* suggests, the difficulty of governing the nation does not depend on ideological differences, but merely on the initiative of the rulers and the swift enforcement of effective responses. Even if it is rule by AI, irresponsible individualism that touts freedom will gradually be eradicated by country with AI-driven national governance systems when the overall interests continue to far outweigh the interests of the individual.

As discussed at chapters from 5 to 8, the digital transformation has already come. China government in the present has been running own monitoring systems driven by the AI on the bases of big-data accumulated from the ordinary citizens. For example, there are facial recognition system, interconnected communication tools, smartphone based cashing, scoring services for rewarding economy, tracing behavior of the citizens geolocated in the city, and other imperative operations (Shibuya, 2004, 2006, 2020a). Using such tools, government leaders can handle with collective citizens and their morality. On the other hands, clever personal advisors for the political leader are now unnecessary, because the AI driven robotic consultant (e.g., IBM Watson) can be alternatively replaced for them. Ruling governance conducted by the human leaders are supported and emancipated by the AI driven mechanisms. Collective intelligence assorted by the advanced autonomous AI agents can prioritize the total opinion based on the demands of the citizens and optimize the final direction among affordable solutions. Namely, can the contemporary society that is reshaped by the digital transformation nearly reach the ideal of the *Jōgan era*? As such ways, it seems that old ideal and traditional governance in East Asia can be actualized in the digitized world.

In the contrary, the global interconnected world desires such factors of leading, governing and integrating heterogeneity to solve a sort of hard problems. Especially, in the democratic societies as Westness, individual liberalism and freedom for the humanity could not always solve effectively social problems, even if they use one of the AI driven services. Although Arrows (1963)' impossibility theorem had shown the limit of the liberal democracy and emergence possibility of the tyrant, government leaders in the West-side countries often misunderstand what consensus means, and they lose the trustfulness for their governance misconducting of their morality from the citizens (e.g., legitimation crisis). Socioeconomic policies based on the capitalism cannot always distribute equally valuable resources among the citizens in the market without somewhat well-defined mechanisms (e.g., mechanism design in economics).

In such context, it might be answered sooner for the question "*Which regime is better Westness-based or Eastness-based in the age of the After Corona?*". Then, we are indeed survivors of the COVID-19 pandemic, and then we must conquer the aftermath of this historical event by coordinating and integrating with the values beyond the differences among the nations and citizens.

4   Final Words

Finally, the author would like to calmly conclude significant points in this book.

The history of humanity has been continuing for a long time with "*the problems that was completely impossible to solve*" from our ancestors and handed their *sacrifices* to their descendants. The author describes it as "*solving the present problems by the seeds that becomes the future problems*".

For example, indeed, when the author studied the Fukushima problem (Shibuya, 2017, 2018, 2020a, 2020b, 2021a, 2021b), the author found that radioactive elements take more than 10,000 years to reach their half-life, the problems of disposing of huge amounts of contaminated polluted water and radioactive materials, and an inevitable fact that the aftermath of a series of the nuclear accidents would take more than 45 years to complete. These are nothing less than *sacrifices* made by the egoism and greed of the last generation (and the past generations), and the generation that created those problems would not be responsible for the *sacrifices* such as succeeding damages and recovery costs. In other words, the essence of the problem condenses that a different person undertakes either benefit or damage.

The result of the Second World War was similarly rooted in large numbers of the *sacrifices*. There have been repeated court cases and other cases in various parts of Japan seeking compensation, but the court ruled "*sacrifices caused by warfare must be equally undertaken by everyone*". In short, it made it clear that the citizens had to bear such terrible conditions as *sacrifice* equally. It indeed revealed that political decision making is just to decide who undertakes sacrifices, and it does not save each individual's hope for peaceful life.

The same can be said of the recent pandemic. The country that caused this pandemic will not be held accountable, even if they were aware of the damage and had dealt with it, and even after they had spread the damage around the world. Much less, there is no way to compensate for the dead. Politicians must be held accountable for such *sacrifices* for their decision making process. But in any political system, they will not be prosecuted. Because such decision making is a kind of the necessary evils in the world, rather it is called as a kind of the original sin for the politician. Again, the nature of the politics is definitely *"distribution of sacrifice"* and this is irrespective of differences in political systems and cultural backgrounds. The approval ratings are merely a reflection of the voices of those who were survived and those who were not infected, but it does not include the vindictive voices of those who "died" as victims to it. Then, any politicians can turn away from such fact forever.

The same is true of the citizens. As soon as they were freed from the restrictions on going outside, the city began to get crowded with a lot of the citizens. They have already forgotten that they have obtained their current peace at *sacrifices* of so many dead people and medical doctors. Many of them are not even aware of such important fact that during the pandemic, they unintentionally monopolized profits and imposed unilaterally the sacrifices on victims, even though they made no contributions to the solution.

Furthermore, the Tokyo Olympics, while appealing to be a festival of world peace, were being forced to be held at the expense of the citizens who were suffering from the corona pandemic. As a result, there is no one to blame for the conflicts that may arise in domestic governance and international affairs, and even if the situation worsens, no one will take its responsibility. The infrastructure and unseen costs of the stadiums and other facilities will be borne by future generations of Japanese, but the people who made the decisions only forces such sacrifices.

The individual's intrinsic desire to be "happy" is dependent on the highest good of each individual, which may be a "goodness". However, it is extremely difficult for the society as a whole to satisfy all of the individuals at the same time, and the conflict between them emerges as a *"social evil"*. Politics is responsible for coordinating the solution, but the process only decides how the sacrifices are to be distributed, and the original "happiness" of each individual is gradually trivialized. Such repetition is the history of the humanity itself. The more someone satisfies, the more someone must be sacrificed. We must know that the unseen *sacrifices* are actually the most dangerous and serious factors to occur future troubles and conflicts. Still then, let's close look at the sacrifice corresponding with each crisis. *"Quid pro Quo"*. This phrase directly expresses both the weight of the responsibilities and the seriousness of the nature that nothing can be decided without any sacrifice. And the history of the humanity continues to be written by the winners, the survivors, and their descendants. Namely, any pages on the book about our history was filled by innumerable *sacrifices*.

The author had indeed written this book from such perspective.

References

Arrow, K. J. (1963). *Social choice and individual values.* Yale University Press.

Shibuya, K. (2004). A framework of multi-agent based modeling, simulation and computational assistance in an ubiquitous environment. *Simulation: Transactions of the Society for Modeling and Simulation International, 80*(7), 367–380.

Shibuya, K. (2006). Actualities of social representation: Simulation on diffusion processes of SARS representation. In C. van Dijkum, J. Blasius & C. Durand (Eds.) *Recent developments and applications in social research methodology*, Barbara Budrich-verlag.

Shibuya, K. (2017). An exploring study on networked market disruption and resilience. KAKENHI Report (in Japanese)

Shibuya, K. (2018). A design of fukushima simulation. *The society for risk analysis, Asia conference 2018.*

Shibuya, K. (2020a). *Digital transformation of identity in the age of artificial intelligence.* Springer.

Shibuya, K. (2020b). *Identity health*, https://www.ncbi.nlm.nih.gov/pmc/articles/PMC7121317/. In K. Shibuya, *Digital transformation of identity in the age of artificial intelligence*, Springer.

Shibuya, K. (2021a). A risk management on demographic mobility of evacuees in disaster. In M. Khosrow-Pour *Encyclopedia of information science and technology* (5th, Ed., pp.1612–1622). IGI Global.

Shibuya, K. (2021b). A spatial model on COVID-19 pandemic. *The 44th Southeast Asia Seminar, The Covid-19 Pandemic in Japanese and Southeast Asian Perspective: Histories, States, Markets, Societies, Kyoto University.*

Further References

AAAS COVID-19 Resources: https://www.aaas.org/programs/covid-19-resources

ADB (Asian Development Bank): https://www.adb.org/adbi/covid-19

Asia Pacific Initiative, Japan: https://apinitiative.org/project/covid/

CDC (USA): https://www.cdc.gov/, https://www.cdc.gov/coronavirus/2019-nCoV/index.html

Cell Press: https://www.cell.com/COVID-19

Coalition for Epidemic Preparedness Innovations (CEPI): https://cepi.net/

Coronavirus (COVID-19) (USA): https://www.coronavirus.gov/

COVID-19 Dashboard by the Center for Systems Science and Engineering (CSSE) at Johns Hopkins University (JHU): https://coronavirus.jhu.edu/map.html

Chinese Center for Disease Control and Prevention (China): http://www.chinacdc.cn/en/COVID19/

Elsevier: https://www.elsevier.com/clinical-solutions/coronavirus-research-hub

Facebook: https://covid-survey.dataforgood.fb.com/?date=2020-07-02&dates=2020-04-23_2020-07-02&region=WORLD

FDA: https://www.fda.gov/

GCPJ (Government CIO's Portal, Japan): https://cio.go.jp/en/index.php

Google: https://www.google.com/covid19/mobility/

Google (COVID-19 Japan): https://datastudio.google.com/reporting/8224d512-a76e-4d38-91c1-935ba119eb8f/page/ncZpB?s=nXbF2P6La2M

Healthcare.gov: https://www.healthcare.gov/

IMF: https://www.imf.org/external/index.htm, https://www.imf.org/en/Topics/imf-and-covid19

Japan Economy Policy for the COVID-19 Pandemic: https://www5.cao.go.jp/keizai1/keizaitaisaku/keizaitaisaku.html

Japan's V-RESAS: https://v-resas.go.jp/

Nature Journal: SARS-CoV-2: https://www.nature.com/subjects/sars-cov-2#research-and-reviews

NSCAI (National Security Commission on Artificial Intelligence): https://www.nscai.gov/home

NIID: https://www.niid.go.jp/niid/ja/idwr.html

NIH COVID-19 Information: https://www.nih.gov/health-information/coronavirus

NIH NCBI PubMed Central (PMC): https://www.ncbi.nlm.nih.gov/pmc/

OECD Data: https://data.oecd.org/

Public Health Emergency COVID-19 Initiative: https://www.ncbi.nlm.nih.gov/pmc/about/covid-19/

Science Journal: Coronavirus: Research, Commentary, and News: https://www.sciencemag.org/collections/coronavirus?IntCmp=coronavirussiderail-128

SDG Index: https://sdgindex.org/reports/sustainable-development-report-2020/

Tableau COVID-19 Project: https://www.tableau.com/covid-19-coronavirus-data-resources

The Lancet: https://www.thelancet.com/coronavirus
The New England Journal of Medicine: https://www.nejm.org/coronavirus
Twitter: https://developer.twitter.com/en/docs/labs/covid19-stream/overview
UNDP (United Nations Development Programme): http://hdr.undp.org/en
UNDRR (United Nations Office for Disaster Risk Reduction): https://www.undrr.org/drr-and-cov
   id-19
Wellcome Trust: https://wellcome.org/, https://wellcomecollection.org/collections
WFP (The World Food Programme): https://www.wfp.org/
WHO COVID-19: https://covid19.who.int/
WHO Electronic State Parties Self-Assessment Annual Reporting Tool (e-SPAR): https://extranet.
   who.int/e-spar

Web Education Resources

CDC Solve the Outbreak: https://www.cdc.gov/mobile/applications/sto/web-app.html
Clade-X: https://www.centerforhealthsecurity.org/our-work/events/2018_clade_x_exercise/clade-
   x-resources
Economic Tracker: https://opportunityinsights.org/, https://tracktherecovery.org/
WHO (2018) Managing epidemics: https://www.who.int/emergencies/diseases/managing-epidem
   ics-interactive.pdf

Index

Printed in the United States
by Baker & Taylor Publisher Services